Interface Oral Health Science 2009

T. Sasano, O. Suzuki
Editors

**P. Stashenko, K. Sasaki, N. Takahashi,
T. Kawai, M.A.Taubman, H.C. Margolis**
Associate Editors

Interface Oral Health Science 2009

Proceedings of the 3rd International Symposium for Interface Oral Health Science,
Held in Sendai, Japan, Between January 15 and 16, 2009
and the 1st Tohoku-Forsyth Symposium, Held in Boston, MA, USA,
Between March 10 and 11, 2009

Springer

Editors:

Takashi Sasano, D.D.S., Ph.D.
Dean and Professor
Tohoku University Graduate School
of Dentistry
4-1 Seiryo-machi, Aoba-ku,
Sendai 980-8575, Japan

Osamu Suzuki, Ph.D., M.Eng.
Professor
Tohoku University Graduate School
of Dentistry
4-1 Seiryo-machi, Aoba-ku,
Sendai 980-8575, Japan

Associate Editors:

Philip Stashenko, D.M.D., Ph.D.
President and Chief Executive Officer
The Forsyth Institute
140 The Fenway, Boston,
MA 02115, USA

Keiichi Sasaki, D.D.S., Ph.D.
Director of Tohoku University
Dental Hospital and Professor
Tohoku University Graduate School
of Dentistry
4-1 Seiryo-machi, Aoba-ku,
Sendai 980-8575, Japan

Nobuhiro Takahashi, D.D.S., Ph.D.
Vice-Dean and Professor
Tohoku University Graduate School
of Dentistry
4-1 Seiryo-machi, Aoba-ku,
Sendai 980-8575, Japan

Toshihisa Kawai, D.D.S., Ph.D.
Senior Member of the Staff
Department of Immunology
The Forsyth Institute
140 The Fenway, Boston,
MA 02115, USA

Martin A. Taubman, D.D.S., Ph.D.
Senior Member of the Staff and Head
Department of Immunology
The Forsyth Institute
140 The Fenway, Boston,
MA 02115, USA

Henry C. Margolis, Ph.D.
Senior Member of the Staff and Head
Department of Biomineralization
The Forsyth Institute
140 The Fenway, Boston,
MA 02115, USA

ISBN 978-4-431-54704-4 ISBN 978-4-431-99644-6 (eBook)
DOI 10.1007/978-4-431-99644-6
Springer Tokyo Berlin Heidelberg New York

Softcover re-print of the Hardcover 1st edition 2010

Printed on acid-free paper

Springer is part of Springer Science+Business Media (www.springer.com)

Preface

Since 2002, the Tohoku University Graduate School of Dentistry has proposed "Interface Oral Health Science" as a major theme for next-generation dental research. That theme is based on the following new concept: healthy oral function is maintained by biological and biomechanical harmony among three systems: (1) oral tissues (host); (2) parasitic microorganisms of the oral cavity (parasites); and (3) biomaterials. The concept implies that oral diseases such as dental caries, periodontal disease, and temporomandibular disorders should be interpreted as "interface disorders" that result from disruption of the intact interface among these systems. The uniqueness of this concept rests on the fact that it not only encompasses the field of dentistry and dental medicine, but also expands the common ground shared with other fields, including medicine, agriculture, material science, engineering, and pharmacology. We aim to promote advances in dental research and to activate collaboration with related fields by putting interface oral health science into practice. On this basis, we have already organized the 1st and 2nd International Symposiums for Interface Oral Health Science, which included inspiring special lectures, symposiums, poster presentations, and other discussions. The contents of the two symposiums were published as monographs entitled Interface Oral Health Science in 2005 and 2007.

The 3rd International Symposium was held in January 2009 as part of this project. With prominent researchers invited from Japan and other countries, the symposium included a keynote lecture by Professor Joji Ando of the Graduate School of Medicine of The University of Tokyo. In addition, there were three symposiums: "Novel Bioengineering," "Mechanobiology," and "Biomaterial Surface." In the poster session, more than 100 poster presentations (the largest number ever) were listed from a wide variety of fields related to interface oral health science, including "Social Interface" as a new section. In addition, the Poster Award for Young Researchers was newly announced, and the winners made a presentation at the Tohoku–Forsyth Symposium (the second part of the Sendai Symposium), held in March in Boston, U.S.A.

This book, containing the presentations at the symposium, is being published in 2010 as a serial entitled Interface Oral Health Science. We hope that our project, including the symposium and the book, will accelerate the progress of dental science and point the way for dental research for future generations.

In closing, I would like to extend our best wishes for the health and success of those who participated in this symposium and who presented such outstanding papers.

Takashi Sasano
President, 3rd International Symposium for Interface Oral Health Science
Dean, Graduate School of Dentistry, Tohoku University
Sendai, Japan
January 2009

Commentary to The 1st Tohoku-Forsyth Symposium

On March 10th and 11th, 2009, the Tohoku–Forsyth Symposium was held at The Forsyth Institute in Boston. The Forsyth Institute was founded in 1910 as a free dental clinic for the children of Boston through a generous gift from the Forsyth family. Between 1914 and the 1950s, more than 500,000 children received care at Forsyth. In the 1950s, realizing that dental diseases could not simply be "treated away," Forsyth's mission evolved to one of research into the causes and pathogenic mechanisms that underlie these conditions, and to the application of this knowledge to the development of better modes of disease prevention and treatment. Today Forsyth is recognized as a world leader in oral and craniofacial research, focusing on the disciplines of microbiology, immunology, bone and mineralized tissue biology, developmental biology, and clinical research.

Tohoku University, located in the city of Sendai, Miyagi Prefecture, is one of the foremost academic research institutions in Japan. Because the Tohoku Faculty of Dentistry, much like Forsyth, fosters an interdisciplinary approach to dental biomedicine, a sister relationship was established between the two institutions in 2005.

The Tohoku Faculty of Dentistry is to be commended for its long history of scientific scholarly activity and for contributing to our mutual efforts to understand basic biological processes, to elucidate host–pathogen interactions at the molecular level, and to develop cutting-edge technology and biomaterials for diagnostic and therapeutic applications. During the course of our two-day symposium, we heard many outstanding presentations of pioneering research activities by both the Tohoku and Forsyth scientists. It is in the spirit of our shared purpose that this book entitled Interface Oral Health Science 2009 has been prepared as a compendium that attests to the Forsyth–Tohoku collaboration. Its publication will be a milestone, leading us into the next century of dental and craniofacial sciences and service to the public.

Philip Stashenko
President and Chief Executive Officer
The Forsyth Institute
Boston, Massachusetts, USA

Acknowledgment

The Editors wish to acknowledge the following members of Tohoku University Graduate School of Dentistry and administrative members at The Forsyth Institute, who have contributed their expertise and time to the review of manuscripts submitted to *Interface Oral Health Science 2009*. These colleagues have provided the important assistance that made it possible for this monograph to publish critically reviewed papers in a timely manner.

Tohoku University Graduate School of Dentistry,

Teruko TAKANO-YAMAMOTO
Hidetoshi SHIMAUCHI
Minoru WAKAMORI
Satoshi FUKUMOTO
Shunji SUGAWARA
Takeyoshi KOSEKI
Kaoru IGARASHI
Masahiko KIKUCHI
Hiroyuki KUMAMOTO
Takahisa ANADA
Takashi TODA
Takuji YOSHIDA
Aya YAMADA
Ryo TOMIZUKA
Noriaki SHOJI
Shizuko SATO
Eiji NEMOTO
Kouki HATORI
Masahiro TSUCHIYA
Tadao KOBAYASHI
Mutsuo TAGUCHI
Tetsuo HAYASAKA
Akio MATSUMOTO
Hiromi YAMAZAKI
Megumi OTSUKI
Satoshi TAKAHASHI
Shinichi KIKUCHI
Takuichi SATO
Yoshitomo HONDA

The Forsyth Institute,

Catherine O'HARA
Ana RIVKIN
Kathleen M. MALONEY
Joe BUCHANAN

Contents

Plenary Lecture

Shear-stress-sensing and response mechanisms in vascular endothelial cells

Joji Ando and Kimiko Yamamoto

Abstract. Vascular endothelial cells (ECs) change their morphology, function, and gene expression in response to shear stress, a fluid mechanical force generated by flowing blood. This fact suggests that ECs recognize shear stress and transmit signals to the interior of the cell. Shear-stress-sensing and response mechanisms, however, have not been fully understood. We have demonstrated that ECs are capable of converting information regarding shear stress intensity into changes in intracellular Ca^{2+} concentration. The Ca^{2+} signaling is based on cell-surface ATP synthase-mediated ATP release and subsequent activation of an ATP-operated cation channel P2X4, which leads to a Ca^{2+} influx. Our studies using *P2X4*-deficient mice revealed that P2X4-mediated Ca^{2+} signaling of shear stress plays a crucial role in the homeostasis of the circulatory system, including the control of blood pressure, blood-flow-dependent vasodilation, and vascular remodeling, through endothelial nitric oxide production.

Key words. shear stress, endothelial cell, P2X purinoceptor, ATP, Ca^{2+}, ATP synthase

1 Introduction

Endothelial cells (ECs) lining blood vessels are constantly exposed to shear stress, a fluid mechanical force generated by flowing blood. A number of recent studies have revealed that ECs recognize changes in shear stress and transmit signals to the interior of the cell, which leads to cell responses that involve changes in cell

J. Ando (✉)
Laboratory of Biomedical Engineering, School of Medicine, Dokkyo Medical University, 880 Kita-kobayashi, Mibu, Tochigi 321-0293, Japan
e-mail: jo-ji@umin.ac.jp

K. Yamamoto
Laboratory of System Physiology, Department of Biomedical Engineering, Graduate School of Medicine, University of Tokyo, 7-3-1 Hongo, Bunkyo-ku, Tokyo 113-0033, Japan

T. Sasano et al. (eds.), *Interface Oral Health Science 2009*,
DOI 10.1007/978-4-431-99644-6_1, © Springer 2010

morphology and cell functions, including the production of a potent vasodilator, nitric oxide (NO), and an antithrombotic protein, thrombomodulin [1–3]. It has also become clear that shear stress regulates endothelial gene expression through transcription and/or mRNA stabilization [4–6]. Our DNA microarray analysis showed that approximately 3% of all genes examined in ECs showed some kind of response to shear stress, indicating that more than 600 genes are shear-stress-responsive [7]. These EC responses to shear stress are thought to play important roles in blood-flow-dependent phenomena, such as vascular tone control, angiogenesis, vascular remodeling, and atherogenesis. However, the precise mechanisms of the shear-stress-sensing are not yet completely understood. Here, we demonstrate the existence of ATP receptor-mediated Ca^{2+} signaling that occurs in ECs in response to shear stress and its physiological role in the vascular system.

2 Ca^{2+} Signaling of Shear Stress

Our previous studies demonstrated that Ca^{2+} signaling plays an important role in shear-stress-sensing and signal transduction [8, 9]. Human pulmonary artery ECs (HPAECs) that had been labeled with a fluorescent Ca^{2+} indicator, Indo-1, were exposed to controlled levels of shear stress in a parallel-plate-type flow chamber, and changes in the intracellular Ca^{2+} concentration were monitored. The intracellular Ca^{2+} concentration increased in a shear-stress-dependent manner (Fig. 1a). There is a good, almost linear, correlation between shear stress and Ca^{2+} concentration. This means that ECs can accurately convert information regarding shear stress intensity into changes in Ca^{2+} concentration. When extracellular Ca^{2+} was removed with EGTA, the shear-induced Ca^{2+} response completely disappeared, indicating that the response was due to an influx of extracellular Ca^{2+} across the cell membrane.

3 P2X4 Channels Mediate Ca^{2+} Influx in Response to Shear Stress

We found that P2X4, a subtype of ATP-operated cation channels known as P2X purinoceptors, plays a crucial role in the shear-stress-dependent Ca^{2+} influx [10, 11]. HPAECs were treated with antisense-oligonucleotides (AS-oligos) targeted to the P2X4 receptor or control scramble-oligos (S-oligos), and their Ca^{2+} responses were determined. A shear-stress-dependent Ca^{2+} response was seen in the cells treated with control S-oligos but not in those treated with AS-oligos (Fig. 1b). To further examine the role of P2X4 in flow-related Ca^{2+} signaling, we transfected P2X4 cDNA into human embryonic kidney (HEK) cells, which are basically insensitive to flow, and established cell lines that stably express P2X4 receptors. The

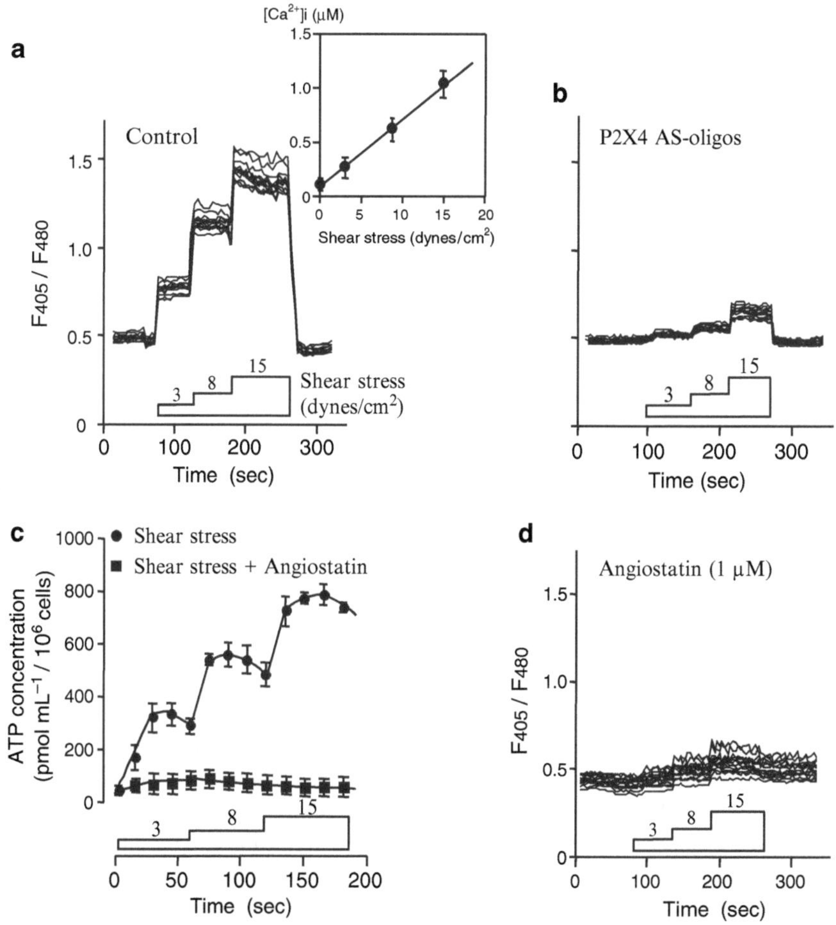

Fig. 1. Shear-stress-dependent Ca^{2+} influx via P2X4 channels. (**a**) Shear-stress-induced Ca^{2+} response. Intracellular Ca^{2+} concentrations ($[Ca^{2+}]i$) increased in a stepwise manner when cultured human pulmonary artery ECs (HPAECs) were exposed to stepwise increases in shear stress, and a linear relationship was found between the Ca^{2+} concentration and shear stress, indicating that ECs are capable of accurately converting information on shear stress into changes in Ca^{2+} concentration. The Ca^{2+} response was attributable to an influx of extracellular Ca^{2+} because it did not occur in the absence of extracellular Ca^{2+}. The ratio of the emitted light of the fluorescent Ca^{2+} indicator Indo-1/AM at 405 nm (F405) and 480 nm (F480) reflects $[Ca^{2+}]i$. (**b**) Involvement of P2X4 in the Ca^{2+}influx. Antisense-oligonucleotides (AS-oligos) targeted against P2X4 that knockout P2X4 expression in HPAECs markedly suppressed the shear-stress-dependent Ca^{2+} responses. (**c**) ATP release in response to shear stress. HPAECs released ATP in a shear-stress-dependent manner, and the ATP-releasing response was completely blocked by angiostatin, a membrane-impermeable ATP synthase inhibitor, suggesting the involvement of cell-surface ATP synthase in the shear-stress-induced ATP release. (**d**) Involvement of ATP release in shear-stress-dependent Ca^{2+} influx. A membrane-impermeable ATP synthase inhibitor, anigostatin, almost completely blocked the Ca^{2+} response to shear stress, suggesting that ECs have shear stress mechanotransduction mechanisms in which shear stress stimulates ECs to release ATP via cell-surface ATP synthase, which leads to P2X4 activation followed by a Ca^{2+} influx

control HEK cells showed no Ca^{2+} response when exposed to flow, whereas the HEK cells that stably expressed P2X4 exhibited a stepwise increase in Ca^{2+} concentrations in response to graded increments in shear stress. These findings suggest that P2X4 receptors have a 'shear-transducer' property through which shear stress signals are transmitted into the cell interior via the Ca^{2+} influx.

4 Shear-Stress-Induced ATP Release Via Cell-Surface ATP Synthase

Our recent study revealed that shear-stress-induced activation of P2X4 requires ATP, which is supplied in the form of endogenous ATP released by ECs [12]. We determined the amount of ATP released into the perfusate using a sensitive luciferase luminometric assay. HPAECs released ATP in response to shear stress, and the ATP release was dose-dependent (Fig1c). A membrane-impermeable ATP synthase inhibitor, angiostatin, and an antibody against ATP synthase markedly suppressed the shear-stress-dependent ATP release, which resulted in significant inhibition of the shear-stress-dependent Ca^{2+} response (Fig. 1d). This means that endogenously released ATP plays an important role in the shear-stress-induced activation of P2X4 receptors. These findings also indicated that ATP synthase is involved in the shear-stress-induced ATP release.

We found that HPAECs express ATP synthase on their cell surface [13]. The cell-surface ATP synthase is distributed in caveolae/lipid rafts and colocalized with caveolin-1, a marker protein of caveolae. Depletion of plasma membrane cholesterol with methyl-β cyclodextrin disrupted the lipid rafts and abolished the colocalization of ATP synthase with caveolin-1, which resulted in a marked reduction in shear-stress-induced ATP release. To further examine the role of caveolin-1 in the flow-induced ATP release, we used siRNA to specifically knock down expression of caveolin-1. Transfection of ECs with caveolin-1 siRNA resulted in a significant reduction in caveolin-1 protein expression and markedly inhibited the flow-induced ATP release. These results suggest that the localization and targeting of ATP synthase to caveolae/lipid rafts is critical for shear stress-induced ATP release. However, it remains unknown how shear stress activates cell-surface ATP synthase.

5 Vascular Physiological Roles of Ca^{2+} Signaling of Shear Stress

To gain insight into the roles of this shear-stress-sensing mechanism via P2X4 in vascular homeostasis, we generated a *P2X4*-knockout (KO) mouse [14]. The absence of P2X4 impaired the EC response to flow stimulation. When the pulmonary microvascular ECs of wild-type (WT) mice were exposed to flow, intracellular Ca^{2+} concentration increased stepwise in tandem with the increase in shear stress,

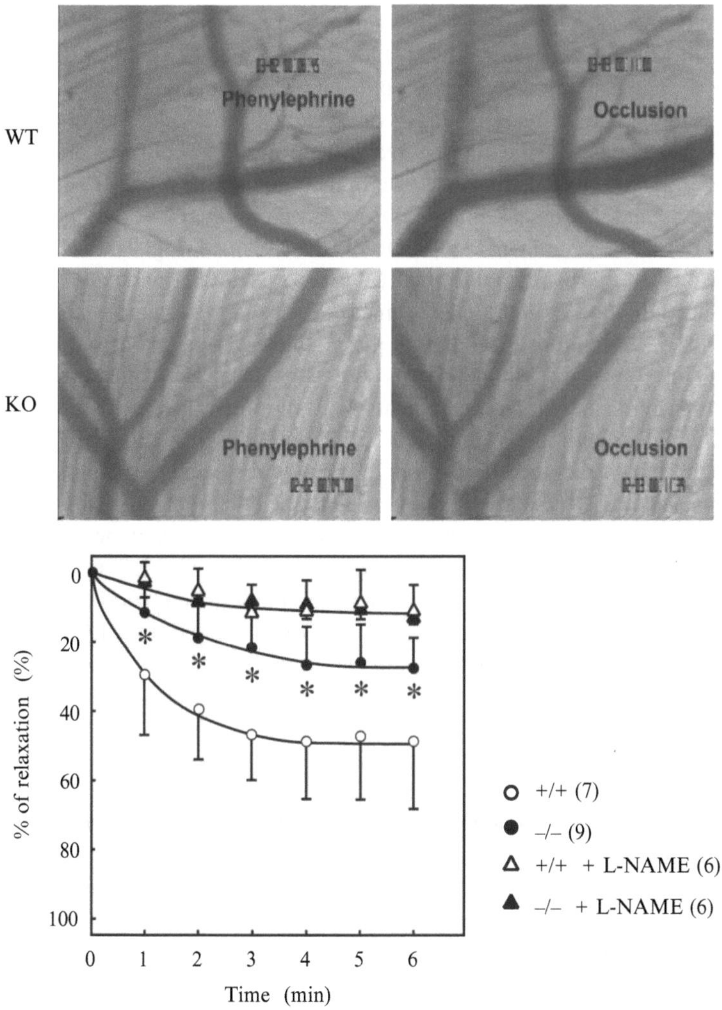

Fig. 2. Impaired flow-induced vasodilation in P2X4 knockout (KO) mice. *Top panel*, intravital microscopic images of vasodilator responses of cremaster muscle arterioles to an increase in blood flow. Arterioles were pre-constricted with phenylephrine. *Bottom panel*, results of a quantitative analysis of flow-induced vasodilation. The increase in blood flow caused marked vasodilation in the wild-type (WT) mice and much less prominent vasodilation in the KO mice. Blockade of NO synthesis with L-NAME markedly reduced the flow-induced vasodilation in both groups of mice, indicating that P2X4 plays an important role in the blood-flow-mediated vasodilation through endothelial NO production. Sample numbers are indicated in parentheses. * $p<0.01$ WT mice vs KO mice

whereas no flow-induced Ca^{2+} response occurred in the ECs of KO mice. Since increases in intracellular Ca^{2+} concentrations directly lead to the production of a potent vasodilator, NO, the ECs were examined for changes in NO production with a fluorescence indicator, diaminofluorescein (DAF-2). NO production by the ECs of

WT mice increased in response to flow, and the response was shear-stress-dependent. The ECs of KO mice, however, did not show flow-induced NO production, indicating that P2X4 channels are involved in endothelial NO production.

We examined the effects of *P2X4* deficiency on the endothelium-dependent vasodilator response in the murine cremaster muscle. Occlusion of one of the branches of an arteriole with a glass micropipette increases blood flow through the other branch, and the increase in blood flow caused marked vasodilation in the WT mice but much less prominent vasodilation in the KO mice (Fig. 2). Blockade of NO synthesis by N^G-nitro-L-arginine methyl ester (L-NAME) markedly reduced the flow-induced dilation in both types of mice. These findings suggest that the blood-flow-sensitive vasomotor mechanisms that regulate vascular tone are impaired in *P2X4* KO mice. A marked difference was observed in mean blood pressure determined by an intra-arterial catheter measurements, with significantly higher values recorded in the KO mice than in the WT mice (125 ± 8 mmHg vs 104 ± 10 mmHg).

Chronic changes in blood flow through large arteries induce structural remodeling of the vascular wall. Increases in blood flow cause enlargement of vessel diameter while decreases in blood flow have the opposite effect. To examine the role of the P2X4 receptor in flow-dependent vascular remodeling, the left external carotid artery of mice was ligated for 2 weeks. The ligation reduced blood flow in the left common carotid artery (LC), and we compared the diameter of the LC and the right common carotid artery (RC) histologically at the end of the 2-week period. Ligation of the left external carotid artery resulted in a significant reduction in lumen diameter in the LC in the WT mice but not in the KO mice. The absence of the blood-flow-induced change in diameter in the *P2X4* KO mice resembled the structural changes that occurred in eNOS-deficient mice [15], suggesting that P2X4 plays a critical role through endothelial NO production in controlling vascular structural adaptation to chronic changes in blood flow.

These results indicate that Ca^{2+} signaling of shear stress via endothelial P2X4 channels play an important role in the control of blood pressure, blood-flow-dependent vasodilation, and vascular remodeling, through endothelial NO production.

References

1. Korenaga R, Ando J, Tsuboi H et al (1994) Laminar flow stimulates ATP- and shear stress-dependent nitric oxide production in cultured bovine endothelial cells. Biochem Biophys Res Commun 198:213–219
2. Takada Y, Shinkai F, Kondo S et al (1994) Fluid shear stress increases the expression of thrombomodulin by cultured human endothelial cells. Biochem Biophys Res Commun 205:1345–1352
3. Davies PF (1995) Flow-mediated endothelial mechanotransduction. Physiol Rev 75:519–560
4. Korenaga R, Ando J, Kosaki K et al (1997) Negative transcriptional regulation of the VCAM-1 gene by fluid shear stress in murine endothelial cells. Am J Physiol 273:C1506–C1515
5. Kosaki K, Ando J, Korenaga R et al (1998) Fluid shear stress increases the production of granulocyte-macrophage colony-stimulating factor by endothelial cells via mRNA stabilization. Circ Res 82:794–802

6. Ando J, Korenaga R, Kamiya A (1999) Flow-induced endothelial gene regulation. In: Lelkes PI (ed) Mechanical Forces and the Endothelium. Harwood Academic Publishers, London, pp 111–126
7. Ohura N, Yamamoto K, Ichioka S et al (2003) Global analysis of shear stress-responsive genes in vascular endothelial cells. J Atheroscler Thromb 10:304–313
8. Ando J, Komatsuta T, Kamiya A (1988) Cytoplasmic calcium response to fluid shear stress in cultured vascular endothelial cells. In Vitro Cell Dev Biol 24:871–877
9. Ando J, Ohtsuka A, Korenaga R et al (1993) Wall shear stress rather than shear rate regulates cytoplasmic Ca^{2+} responses to flow in vascular endothelial cells. Biochem Biophys Res Commun 190:716–723
10. Yamamoto K, Korenaga R, Kamiya A et al (2000) P2X4 receptors mediate ATP-induced calcium influx in human vascular endothelial cells. Am J Physiol Heart Circ Physiol 279:H285–H292
11. Yamamoto K, Korenaga R, Kamiya A et al (2000) Fluid shear stress activates Ca^{2+} influx into human endothelial cells via P2X4 purinoceptors. Circ Res 87:385–391
12. Yamamoto K, Sokabe T, Ohura N et al (2003) Endogenously released ATP mediates shear stress-induced Ca^{2+} influx into pulmonary artery endothelial cells. Am J Physiol Heart Circ Physiol 285:H793–H803
13. Yamamoto K, Shimizu N, Obi S et al (2007) Involvement of cell surface ATP synthase in flow-induced ATP release by vascular endothelial cells. Am J Physiol Heart Circ Physiol 293:H1646–H1653
14. Yamamoto K, Sokabe T, Matsumoto T et al (2006) Impaired flow-dependent control of vascular tone and remodeling in P2X4-deficient mice. Nat Med 12:133–137
15. Rudic RD, Shesely EG, Maeda N et al (1998) Direct evidence for the importance of endothelium-derived nitric oxide in vascular remodeling. J Clin Invest 101:731–736

Symposium I
Novel Bioengineering

Cleft formation and branching morphogenesis of salivary gland: exploration of new functional genes

Takayoshi Sakai, Tomohiro Onodera, and Kenneth M. Yamada

Abstract. Epithelial branching morphogenesis is important to form many organs. Embryonic salivary glands provide an excellent model for clarifying the mechanisms of this phenomenon. As clefts form, epithelial cell–cell adhesions are converted to cell–matrix adhesions. Nevertheless, the mechanism of cleft formation is not well understood. Here, we describe a set of approaches being used to identify and characterize molecules necessary for branching morphogenesis. A combination of laser microdissection with T7-SAGE has been established as a gene discovery method for identifying candidate molecules that may be essential for early organ morphogenesis. Progress in understanding the mechanisms of salivary branching morphogenesis will provide novel approaches to future tissue engineering or regeneration of damaged salivary glands.

Key words. salivary gland, branching morphogenesis, cleft formation, molecular analysis, T7-SAGE

1 Introduction

Branching morphogenesis is an important developmental process that is required for the formation of a number of organs, including kidney, lung, pancreas, prostate, and salivary gland. Salivary epithelium undergoes repetitive cycles of branching

T. Sakai (✉)
Department of Oral-Facial Disorders, Division of Functional Oral Neuroscience, Osaka University Graduate School of Dentistry, 1-8 Yamadaoka, Suita, Osaka 565-0871, Japan
e-mail: sakai@dent.osaka-u.ac.jp

T. Onodera and K.M. Yamada
Laboratory of Cell and Developmental Biology, National Institute of Dental and Craniofacial Research, National Institutes of Health, Bethesda, MD 20892-4370, USA

T. Sasano et al. (eds.), *Interface Oral Health Science 2009*,
DOI 10.1007/978-4-431-99644-6_2,

to create an organ-specific pattern of clefts and buds. This process generates the large areas of epithelial surface needed to fulfill essential functions such as secretory activity and gas exchange, yet which can still be packed efficiently into a compact adult organ [1]. Many functional molecules that are present or change in quantity, as organs develop, have been identified recently using gene profiling techniques. However, little is known regarding how these molecules cooperate to regulate or to mediate the process [2].

Previously published studies have reported that the process of branching morphogenesis depends on epithelial–mesenchymal interactions, growth factors, Wnt signaling, and extracellular matrix (ECM) proteins [3]. It appears likely that different local patterns and degrees of outgrowth of buds and formation of clefts contribute to overall organ-specific branching patterns. This review will describe specific strategies to identify and characterize the patterns of gene expression in specific regions of tissues as well as methods to characterize the biological importance of specific molecules essential for branching morphogenesis. These strategies depend on the existence of functionally important differences in the levels and sites of expression of specific genes needed for morphogenesis.

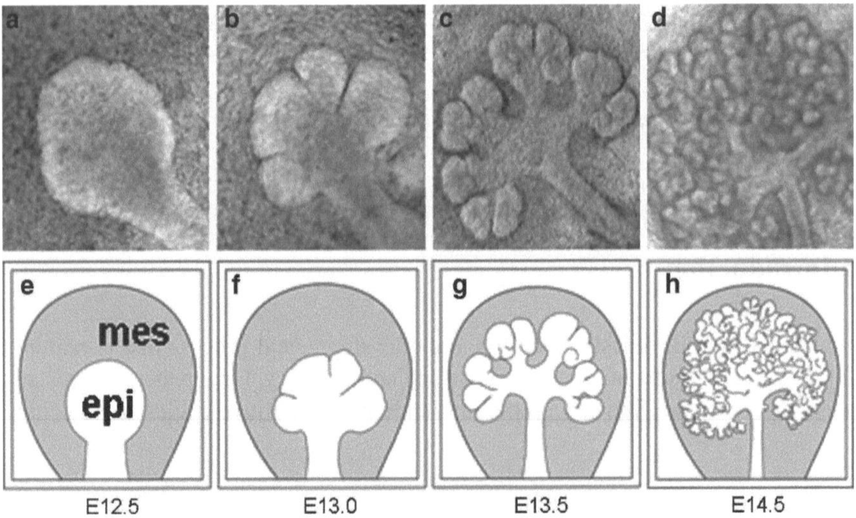

Fig. 1. Photographs and diagrams of epithelial branching morphogenesis of mouse embryonic salivary submandibular gland. Salivary glands from embryonic day 12.5 (E12.5) mice (**a**, **e**) generally exist as a single epithelial bud. The salivary epithelium is surrounded by mesenchyme, and a basement membrane is present around the epithelium, separating it from mesenchymal cells (**a**, **e**). As branching initiates, the gland has rounded buds separated by narrow, deep clefts at E13.0 (**b**, **f**). The epithelial bud progresses to a multilobed structure by E13.5 (**c**, **g**) and E14.5 (**d**, **h**). *epi* epithelium, *mes* mesenchyme

2 Branching Morphogenesis: Embryonic Submandibular Gland Development

The salivary gland starts extensive branching morphogenesis on embryonic day 12.5 (E12.5). Four pictures and diagrams depict a developing salivary gland in organ culture on a filter membrane. The branching process is initiated by the formation of shallow clefts in a single epithelial bud on E13.0 (Fig. 1b, f). These clefts deepen to subdivide the single bud into multiple smaller buds on E13.5 (Fig. 1c, g). Subsequent repetitive cycles allow the developing glands to branch into increasingly intricate three-dimensional patterns on E14.5 (Fig. 1d, h) and later [1].

3 A Strategy to Characterize Molecules Necessary for Branching Morphogenesis

Differential gene expression patterns within specific regions of a tissue can be identified using a global approach to quantify expression levels that begin with laser microdissection (LMD) [4] to isolate mRNA from potentially functionally distinct regions of tissue and is followed by amplification and serial analysis of gene expression (SAGE) [5]. The tissues for LMD consisted of E13 salivary glands cultured on membranes [6]. From salivary gland frozen tissue sections, regions corresponding to cleft and bud epithelia were laser-microdissected [7]. Profiles of gene expression in salivary epithelium from small amounts of cells were identified by a procedure termed T7-SAGE [8]. Here, we describe this current strategy and highlight a specific example of gene discovery in salivary branching morphogenesis using this combination of LMD and T7-SAGE.

4 Laser Microdissection

We were interested in identifying the genes that regulate cleft formation, which represents the initial step of branching morphogenesis. We wanted to identify genes differentially expressed in the specific cell populations immediately adjacent to the clefts versus the end buds of the salivary epithelium (Fig. 2). We used LMD of cryostat sections of multiple developing submandibular glands to isolate the pools of tissue that were used to prepare RNA from each region.

5 T7-SAGE

While there are many techniques available for global gene expression profiling, to profile gene expression in developing salivary glands, we developed a modification of the SAGE technique that we termed T7-SAGE [8]. SAGE is a gene profiling

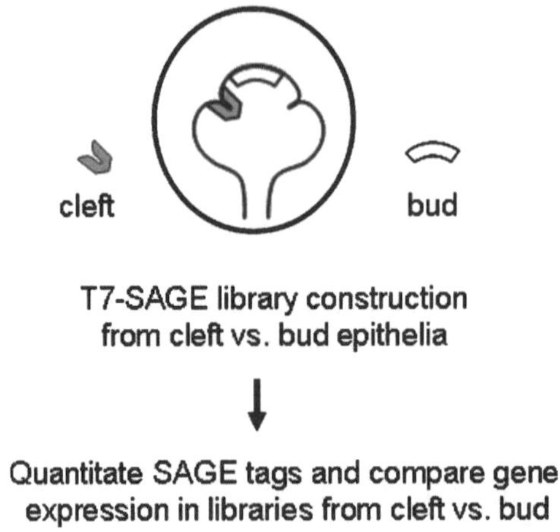

Fig. 2. Strategy for laser microdissection (LMD) and T7-SAGE. Salivary epithelial tissue isolated by LMD from cryostat sections is shown diagrammatically: cleft epithelium (*gray*) and bud epithelium (*white*). T7-amplified gene expression libraries were produced from pooled samples of tissues laser-microdissected from each region. The gene expression profiles from cleft and bud libraries were compared

technique that is unique in that it provides absolute transcript numbers in a digital format [5]. Even though microarrays have the advantages of being relatively easy to use and suitable for high-throughput applications, mRNA quantification is more accurate with SAGE than with microarrays [5]. A disadvantage of SAGE, however, is that it requires microgram quantities of starting poly(A)$^+$ mRNA, which prevents its use when mRNA is limited. Since we had very limited amounts of RNA from our samples, we created the T7 modification of SAGE by incorporating two cycles of high-fidelity T7-based RNA amplification as an initial amplification step [8]. We generated two SAGE libraries from each pool, sequenced greater than 20,000 SAGE ditags from each library, and compared the gene expression profiles derived from each library (Fig. 2).

6 The Essential Role of Fibronectin in Branching Morphogenesis

Unexpectedly, initial T7-SAGE results indicated that cleft epithelial cells expressed ECM protein fibronectin (FN) at levels much higher than bud epithelium. Certain other ECM proteins, such as collagen III, and basement membrane components, such as laminin and proteoglycans, are known to be required for salivary branching in general [9]. However, a regulatory role for FN had not previously been suggested in salivary morphogenesis. We were surprised to find that FN might be expressed specifically in cleft sites. To verify the T7-SAGE FN expression data, we confirmed the changes in gene expression using two methods. First, we performed quantitative reverse-transcription polymerase chain reaction (RT-PCR) using the original RNA preparations used to generate the SAGE libraries. RT-PCR confirmed that FN mRNA was in fact expressed 16-fold higher in cleft than end bud epithelial cells. To localize expression changes in the tissue itself, we performed in situ hybridization analysis on whole-mount salivary glands. In situ analysis confirmed not only that FN was expressed by the salivary gland epithelium but also that it was expressed at much higher levels in cleft compared to bud epithelium.

Immunostaining for FN revealed intense staining for fibrils of FN located in narrow clefts less than 3 μm wide in cultures of intact mesenchyme-containing salivary glands. This increased immunostaining complemented the RT-PCR and in situ data by showing expression of FN protein by the epithelial cells adjacent to clefts, whereas neighboring end bud epithelial cells were negative. We tested experimentally the role of the FN molecule by suppressing its function, examining whether inhibiting FN protein function blocked branching. We found that anti-FN function-blocking antibodies prevented salivary cleft formation and branching in a dose-dependent manner.

In addition, function-blocking antibodies against the β_1 and α_5 integrin subunits, which together comprise the classical FN receptor, inhibited salivary branching. Antibodies against the α_6 integrin laminin receptor subunit were also inhibitory, which had been previously shown [6]. Taken as a whole, our results indicated that both FN and FN receptor function are needed for cleft formation, the initial stage of salivary gland branching morphogenesis.

7 Other ECM Molecules

Searching for extracellular matrix proteins that were strongly differentially expressed in clefts versus end buds identified two more interesting differences. The laminin $\gamma 2$ chain is a subunit of the basement membrane protein laminin-5 and is expressed in epithelial cells of many embryonic tissues including skin, teeth, and collecting tubules of developing kidney [10]. Studies have also revealed that laminin $\gamma 2$ is expressed at the edges of malignant carcinomas, consistent with a possible role in inducing cell motility. Our initial T7-SAGE analyses revealed that laminin $\gamma 2$ is expressed highly in the epithelial end bud regions of E13 salivary glands; the relative number of SAGE tags (proportional to gene expression) in cleft sites was 4 compared to 22 in end buds. Our initial finding suggests that laminin-5 is involved selectively in some end bud function. Because it has been shown to regulate cell migration in vitro, laminin $\gamma 2$ might regulate the development of embryonic epithelium in various organs by stimulating cell motility; in the salivary gland, it might be involved in outward bud expansion.

Extracellular matrix proteins undergo degradation and turnover by matrix metalloproteinases (MMPs). Tissue inhibitors of metalloproteinase (TIMPs) are endogenous protein inhibitors of MMPs. Previous studies have shown that ECM remodeling results from a change in the balance between active MMPs and TIMPs [11], allowing TIMPs to regulate ECM degradation by MMPs. A recent study shows that inactivation of TIMP3 impairs lung branching morphogenesis [11]; in support of our study implicating FN expression in branching morphogenesis, the authors note that FN levels are abnormally low in their system lacking TIMP3, and branching morphogenesis is defective. Our initial T7-SAGE data revealed that TIMP3 is highly expressed in epithelial cleft regions compared to end buds of developing salivary gland (clefts: 27; end buds: 0). This increased expression of TIMP3 in the clefts of salivary epithelium undergoing branching morphogenesis could serve to protect locally expressed FN from degradation by MMPs. These findings support the concept that local regulation of expression and stability of specific ECM molecules play important roles in the mechanisms of branching morphogenesis of salivary and other tissues. In addition, the capacity to stimulate this complex process by the addition of a single purified protein, FN, suggests that tissue engineering with FN and other regulators may become practical.

Recently, we have been attempting to find novel regulators of branching morphogenesis. We started studies to identify and characterize new molecules that may be important for cleft formation. Further investigations using LMD and T7-SAGE will be useful for indentifying additional regulatory genes, such as growth factors, transcription factors, and other matrix proteins, to understand the complex process of branching morphogenesis in early development.

8 Conclusions

Following the procedures outlined above, we concluded that FN expression is required for cleft formation in glandular branching morphogenesis. These studies are significant from a technical standpoint in that we demonstrated that T7-SAGE gene expression analysis of laser-microdissected tissues provides new possibilities for the characterization of region-specific expression profiles in other complex, heterogeneous tissues. As outlined in this review, we believe that the combination of the amplification and profiling technique of T7-SAGE with inhibition and augmentation approaches similar to those that we have been using to address the mechanism of branching morphogenesis of the salivary gland will be broadly applicable to other important developmental and pathological processes in oral biology and other fields.

Acknowledgment This research was supported by the Intramural Research Program of the NIDCR, NIH, and Grant-in Aid for Scientific Research (B) in Japan Society for the Promotion of Science. We thank Yukinori Endo and Jill Harunaga for comments on this manuscript.

References

1. Davies JA (2002) Do different branching epithelia use a conserved developmental mechanism? Bioessays 24:937–948
2. Sakai T, Larsen M, Yamada KM (2005) Morphogenesis and branching of salivary glands: characterization of new matrix and signaling regulators. Oral Biosci Med 213:105–113

3. Hoffman MP, Kidder BL, Steinberg ZL et al (2002) Gene expression profiles of mouse submandibular gland development: FGFR1 regulates branching morphogenesis in vitro through BMP- and FGF-dependent mechanisms. Development 129:5767–5778
4. Kolble K (2000) The LEICA microdissection system: design and applications. J Mol Med 78:B24–B25
5. Velculescu VE, Zhang L, Vogelstein B et al (1995) Serial analysis of gene expression. Science 270:484–487
6. Larsen M, Hoffman MP, Sakai T et al (2003) Role of PI 3-kinase and PIP3 in submandibular gland branching morphogenesis. Dev Biol 255:178–191
7. Sakai T, Larsen M, Yamada KM (2003) Fibronectin requirement in branching morphogenesis. Nature 423:876–881
8. Sakai T, Larsen M, Yamada KM (2002) Current protocols in cell biology. Wiley, New York, pp 19.13.11–19.13.30
9. Nakanishi Y, Nogawa H, Hashimoto Y et al (1988) Accumulation of collagen III at the cleft points of developing mouse submandibular epithelium. Development 104:51–59
10. Lu W, Miyazaki K, Mizushima H et al (2001) Immunohistochemical distribution of laminin-5 gamma2 chain and its developmental change in human embryonic and foetal tissues. Histochem J 33:629–637
11. Gill SE, Pape MC, Khokha R et al (2003) A null mutation for tissue inhibitor of metalloproteinases-3 (Timp-3) impairs murine bronchiole branching morphogenesis. Dev Biol 261:313–323

Strategies underlying research in tooth regenerative therapy as a possible model for future organ replacement

Kazuhisa Nakao and Takashi Tsuji

Abstract. The ultimate goal of regenerative therapy is to develop fully functioning bioengineered organs that can replace lost or damaged organs after disease, injury, or aging. We have previously developed a three-dimensional culture system with the aim of reconstituting a bioengineered organ germ at an early developmental stage. The regeneration of a functional tooth unit is critical issue to achieving proper oral function, including mastication. Recently, we successfully demonstrated that our bioengineered tooth germ could develop a fully functioning tooth with sufficient hardness for masticatory potential, the ability to withstand mechanical stress in the maxillofacial region, and in which the innerved neural fibers had an adequate perceptive potential for noxious stimulations. Our results thus show that bioengineered tooth germ can develop a fully functioning regenerated tooth in vivo after engraftment and therefore that organ replacement regenerative therapy in this way is feasible.

Key words. regenerative therapy, tooth, organ germ method, bioengineered organ, transplantation

1 Introduction

Regenerative medicine is an anticipated clinical application in coming years [1–4], and stem cell transfer therapy (in which stem cells are removed and transferred to damaged organs and tissues of the same individual) has already begun to be developed [5]. The ultimate goal of regenerative medicine is to replace a defective organ with artificially regenerated organ that has full functionality [6–9]. In the dental field, therapies have

K. Nakao and T. Tsuji
Research Institute for Science and Technology, Tokyo University of Science, Noda, Chiba 278-8510, Japan

T. Tsuji (✉)
Graduate School of Industrial Science and Technology, Tokyo University of Science, Noda, Chiba 278-8510, Japan
and Organ Technologies Inc., Tokyo 101-0048, Japan
e-mail: t-tsuji@rs.noda.tus.ac.jp

T. Sasano et al. (eds.), *Interface Oral Health Science 2009*,
DOI 10.1007/978-4-431-99644-6_3, © Springer 2010

been established to replace teeth, which are ectodermal organs, with artificial implants that can provide some functions [10–12]. However, although these treatments can effectively substitute a lost tooth, transplantations of a natural tooth such as third molar are being attempted to provide more biological functionality [13]. For next-generation therapies, it is expected that tooth replacement will involve transplantation of a bioengineered tooth which has been constructed from stem cells [7,10,11].

2 The Strategies Underlying Current Research on Tooth Regenerative Therapy

In current research on whole-tooth regenerative therapies, one of the basic underlying strategies involves the transplantation of bioengineered tooth germ, which can then develop into a fully functional tooth [7,10,11]. Teeth arise from the tooth germ, which is induced by reciprocal epithelial-mesenchymal interactions in the developing embryo [14,15]. The epithelium and the mesenchyme differentiate into ameloblasts, which later become enamel and odontoblasts, respectively, which will form dentin. The mesenchyme also differentiates into dental pulp and into periodontal tissues, which will become cementum, alveolar bone, and periodontal ligament (PDL).

As described by many experts in the dental field, four major hurdles need to be overcome to enable the development of tooth regenerative therapy [7,10,11]. The first is the establishment of a more effective bioengineering method of producing three-dimensional organ germs from single cells. The second is the development of this bioengineered tooth germ in an adult oral environment. The third relates to how dental regenerative therapy may be optimized using the patient's own cells. Finally, the transplantation of a morphologically controlled regenerated tooth would be improved by more effective in vitro organ processing.

3 The Development of a Novel Bioengineered Organ Germ Method

Previously in our laboratory, we investigated the feasibility of developing a bioengineering cell processing method for three-dimensional organ germ using single cells. To precisely replicate the process of tooth organogenesis at early developmental stages, we employed a cell aggregation method using epithelial cells and mesenchymal cells isolated from cap stage tooth germ from the lower jaw of E14.5 mice. The epithelial and mesenchymal single cells were prepared using enzymatic treatments (Fig. 1a). Explants that reconstituted the cell compartmentalization between epithelial and mesenchymal cells at a low-cell density ($0.5–1 \times 10^8$ cells/ml), or that did not form cell compartmentalization at high-cell density (5×10^8 cells/ml), failed to generate a correct tooth structure. To reconstitute a bioengineered tooth germ with the correct cell compartmentalization between epithelial and mesenchymal-derived single cells, these cells were injected in turn at a high density (5×10^8 cells/ml) into

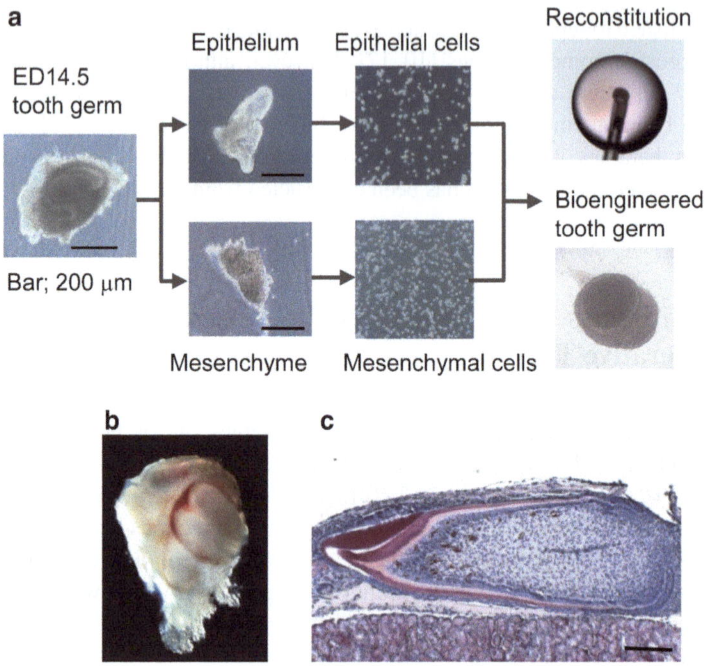

Fig. 1. Generation of a whole tooth using bioengineered tooth germ derived from dissociated single cells. (**a**) Schematic of the bioengineering technology used for the generation of a reconstituted tooth germ. (**b**) Representative phase contrast images showing a bioengineered tooth developed in a subrenal capsule environment for 14 days. (**c**) Histological analysis of the reorganized tooth germ under a subrenal capsule for 14 days. Scale bar, 250 μm

a collagen gel drop (Fig. 1a). Within 2 days of organ culture, a tooth germ was observed to form with the appropriate compartmentalization and cell–cell compaction. At 14 days after the transplantation into a subrenal capsule, this germ could successfully generate plural teeth in the alveolar bone of the mouse (Fig. 1b, c). These results emphasize that a high-cell density and a correct cell compartmentalization are essential for a bioengineered tooth to develop properly, and we termed our successful approach in this regard as a "bioengineered organ germ method" [16]. Our model improves the current understanding of the principles by which organ reconstitution can be achieved with tissues that have been bioengineered in vitro and increases the potential for bioengineered organ replacement in the future.

4 Eruption and Structure af a Bioengineered Tooth

We further investigated whether bioengineered molar tooth germ reconstituted using our novel method could erupt in an adult murine lost tooth transplantation model [17]. We extracted the first molar from the subject mice and allowed the cavity to

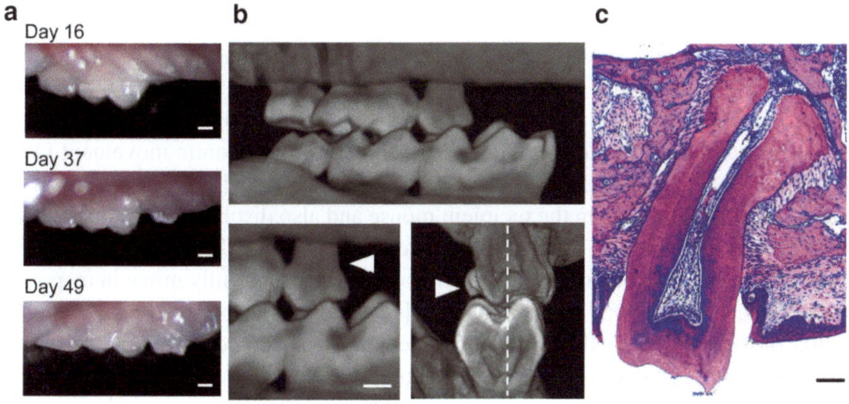

Fig. 2. Eruption and occlusion of a bioengineered tooth. (**a**) Oral photographs of the bioengineered tooth during eruption and occlusion processes. Scale bar, 200 μm. (**b**) MicroCT image of the occlusion of a bioengineered tooth (*arrowhead*). (**c**) Histological analysis of the bioengineered tooth in full occlusion. Scale bar, 100 μm

undergo repair by osteogenesis for 1 month. We then drilled a hole in the alveolar bone and transplanted our bioengineered molar tooth germ into this cavity in the correct orientation. At 16 days after this transplantation, eruption of the bioengineered tooth could not be observed (Fig. 2a). However, at 37 days post-transplantation, a cusp tip could be observed in the gingival area of the transplantation site, indicating an eruption of the bioengineered tooth. At 49 days post-transplantation and thereafter, this bioengineered tooth was observed to reach the occlusal plane and achieve opposing tooth occlusion (Fig. 2a, b). Following the achievement of occlusion, there was no excessive increase found in the tooth length at 120 days post-transplantation [17]. The bioengineered tooth also formed a correct structure comprising enamel, ameloblasts, dentin, odontoblasts, dental pulp, alveolar bone, and blood vessels (Fig. 2c). These results indicated that the tooth tissue structures of the bioengineered tooth were similar to those of a normal tooth [17].

5 Functional Bioengineered Tooth Replacement In an Adult Oral Environment

To develop a tooth regenerative method for possible future clinical applications, a bioengineered tooth must necessarily achieve full functionality, including sufficient masticatory performance [18], biomechanical cooperation with tissues in the oral and maxillofacial regions [19], and proper responsiveness via sensory receptors to noxious stimulations in the maxillofacial region [20]. Masticatory potential in particular is essential for achieving proper tooth function [18]. Significantly, the hardness of the mineralized tissues within our bioengineered tooth, as analyzed by the Knoop hardness test and including both enamel and dentin, was equivalent to that of a normal tooth. These results indicated an equivalent masticatory performance to a normal, mature tooth [17].

It has been established previously that alveolar bone remodeling is induced by the response of the PDL to mechanical stress such as orthodontic movement [19]. We found in this regard that our regenerated tooth achieved normal occlusion in harmony with other teeth in the recipient mouse and also displayed opposing cuspal contacts that maintained a proper occlusal vertical dimension between the opposing arches (Fig. 2b). The bioengineered tooth could also successfully move in response to mechanical stress as well as a normal tooth [17]. These findings indicate that the PDL of the bioengineered tooth successfully mediate bone remodeling in response to mechanical stress induced by the experimental orthodontic treatment.

The perception of noxious stimulations such as mechanical stress and pain is an important tooth function [20]. We found that the nerve fibers innervating both the pulp and the PDL of our bioengineered tooth had perceptive potential for nociceptive stimulation (Fig. 3) and could transduce these events to the central nervous system (the medullary dorsal horn) [17]. In these analyses, we observed that our bioengineered tooth germ developed into a fully functioning tooth with sufficient hardness for mastication and a functional responsiveness to mechanical stress in the maxil-

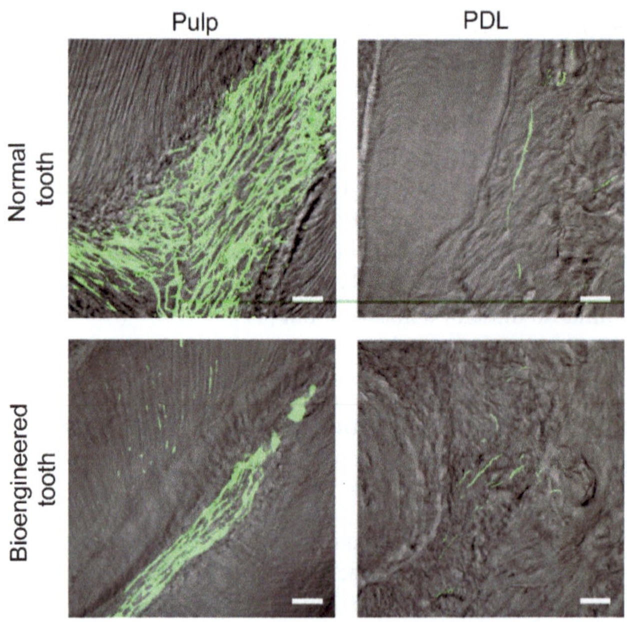

Fig. 3. Immunohistochemical analysis of nerve fibers entering the tissue of the bioengineered tooth. Nerve fibers in the pulp and periodontal ligament (PDL) in the normal and the bioengineered tooth were analyzed immunohistochemically using specific antibodies for neurofilament-H. Scale bar, 25 μm

lofacial region. We further confirmed that the neural fibers that has re-entered the pulp and PDL tissues of the bioengineered tooth had a proper level of perceptive potential in response to noxious stimulations such as orthodontic treatment and pulp stimulation. These findings further indicate that the bioengineered tooth generation techniques we developed can contribute to the rebuilding of a fully functional tooth.

6 Conclusions

We have provided the first description of a successful replacement of an entire and fully functioning organ in an adult body through the transplantation of bioengineered organ germ reconstituted by single cell manipulation in vitro. Our studies have therefore made a substantial contribution to the future development of bioengineering technology for organ replacement therapy. Further studies on the identification of available adult tissue stem cells for the reconstitution of a bioengineered tooth germ and the regulation of tooth shape will help to achieve the realization of tooth regenerative therapy in humans.

References

1. Brockes JP, Kumar A (2005) Appendage regeneration in adult vertebrates and implications for regenerative medicine. Science 310:1919–1923
2. Watt FM, Hogan BL (2000) Out of Eden: stem cells and their niches. Science 287:1427–1430
3. Langer RS, Vacanti JP (1999) Tissue engineering: the challenges ahead. Sci Am 280:86–89
4. Atala A (2005) Tissue engineering, stem cells and cloning: current concepts and changing trends. Expert Opin Biol Ther 5:879–892
5. Korbling M, Estrov Z (2003) Adult stem cells for tissue repair – a new therapeutic concept? N Engl J Med 349:570–582
6. Griffith LG, Naughton G (2002) Tissue engineering – current challenges and expanding opportunities. Science 295:1009–1014
7. Ikeda E, Tsuji T (2008) Growing bioengineered teeth from single cells: potential for dental regenerative medicine. Expert Opin Biol Ther 8:735–744
8. Purnell B (2008) New release: the complete guide to organ repair. Introduction. Science 322:1489
9. Lechler RI, Sykes M, Thomson AW et al (2005) Organ transplantation – how much of the promise has been realized? Nat Med 11:605–613
10. Sharpe PT, Young CS (2005) Test-tube teeth. Sci Am 293:34–41
11. Duailibi SE, Duailibi MT, Vacanti JP et al (2006) Prospects for tooth regeneration. Periodontol 2000 41:177–187
12. Voruganti K (2008) Clinical periodontology and implant dentistry, 5th edition. Br Dent J 205:216
13. Tsukiboshi M (1993) Autogenous tooth transplantation: a reevaluation. Int J Periodontics Restorative Dent 13(2):120–149
14. Pispa J, Thesleff I (2003) Mechanisms of ectodermal organogenesis. Dev Biol 262:195–205
15. Tucker A, Sharpe P (2004) The cutting-edge of mammalian development; how the embryo makes teeth. Nat Rev Genet 5:499–508

16. Nakao K, Morita R, Saji Y et al (2007) The development of a bioengineered organ germ method. Nat Methods 4:227–230
17. Ikeda E, Morita R, Nakao K et al (2009) Fully functional bioengineered tooth replacement as an organ replacement therapy. Proc Natl Acad Sci U S A 106:13475–13480
18. Manly RS, Braley LC (1950) Masticatory performance and efficiency. J Dent Res 29:448–462
19. Shimono M, Ishikawa T, Ishikawa H et al (2003) Regulatory mechanisms of periodontal regeneration. Microsc Res Tech 60:491–502
20. Byers MR, Narhi MV (1999) Dental injury models: experimental tools for understanding neuroinflammatory interactions and polymodal nociceptor functions. Crit Rev Oral Biol Med 10:4–39

Molecular basis for specification of the vertebrate head field

Akihito Yamamoto

Abstract. The network of regulatory factors involved in the development of the vertebrate head is an exquisitely tuned system of almost baroque complexity. During early embryogenesis, arrays of activating and inhibitory factors work to mold an initially unstructured clump of cells into an increasingly recognizable body with a distinct back, belly, head, and tail. The development of the head is enabled and guided by the activity of a region of cells called the organizer, which secretes inhibitors of multiple growth factor pathways that affect the axial orientation and germ layer formation of the embryo. In a sense, the organizer acts as a defense system against suppressors that frustrate the development of the nascent head. Here, I briefly discuss recent progress in understanding the molecular basis for specification of the head field.

Key words. vertebrate, head formation, Spemann–Mangold organizer

1 Introduction

During vertebrate development, a head is molded through a series of inductive signals that are followed by the coordinated movement and differentiation of multipotent progenitor cells. In the process, the dorsal gastrula or Spemann–Mangold organizer plays a prominent role in the early specification of the head. There is comprehensive published literature regarding the mechanism of establishment of the organizer [1], neural induction [2], and patterning [3]; and overviews of the entire process of gastrulation in a variety of organisms are available [4, 5]. In this review, I focus on the molecular basis of the organizer functions involved in specification of the head. The review is mostly based on studies of *Xenopus* embryos, since this organism has provided a number of common paradigms for vertebrate embryogenesis.

A. Yamamoto
Nagoya University School of Medicine, 65 Tsurumai-cho, Showa-ku,
Nagoya, Aichi 466-8550, Japan
e-mail: akihito@med.nagoya-u.ac.jp

T. Sasano et al. (eds.), *Interface Oral Health Science 2009*,
DOI 10.1007/978-4-431-99644-6_4, © Springer 2010

2 Axis Formation by the Dorsal Organizer

The work of Spemann and his colleagues, following the initial experiments carried out by Hilde Mangold [6], developed the organizer concept in which a small group of cells located in the dorsal mesoderm of the gastrula embryo produces marvelous factors that control the entire body patterning. In the initial experiment, they transplanted the organizer excised from early-pigmented newt dorsal gastrula mesoderm (Fig. 1a: open circle), which was just involuting though the dorsal blastopore lip, into the ventral side of a nonpigmented host of the same developmental stage; and the organizer generated a complete secondary axis (Fig. 1b). The grafted cells mainly differentiated into axial-mesoderm, i.e., prechordal plate and notochord. Most of the cells that formed the secondary axis were derived from nonpigmented ventral host cells that would normally participate in the formation of blood-lineage cells and not the axial structure. Thus, these organizer experiments prompted the realization that intercellular communication of early progenitor cells plays an important role in the proper formation of the body axis.

3 Region-Specific Induction by the Dorsal Organizer

Spemann also realized that induction by the organizer is region specific. Transplants of the organizer from the earliest cells to anterior mesendoderm (AME) of involutes induced a secondary head in the host, while transplantation of the organizer from late gastrulae induced an extra trunk (Fig. 1d, e). From these results Spemann predicted the existence of a head and trunk inducer, called a head and trunk organizer, respectively.

Extensive molecular cloning has now identified genes that are specifically expressed and function in the organizers. The majority of these genes encode the protein of signal inhibitors (Fig. 1c, f). The head organizer expresses the anti-bone morphogenetic proteins (BMPs) *chordin* [7], *noggin* [8], and *follistatin*; anti-Wnts *sfrps* (secreted frizzled related protein) [9] and *dickkopf1* (dkk1) [10]; anti-Nodal/Activin *lefty-1/antivin* [11]; multifunctional protein *cerberus* inhibiting Wnt, BMP, and Nodal/Activin [12, 13]; and *shisa* which suppresses both Wnt and fibroblast growth factor (FGF) [14]. In contrast, the trunk organizer expresses only three anti-BMPs.

4 Specification of the Head in Mice

In the early gastrula, mouse homologs of *cerberus*, *dkk-1*, and *lefty-1*, but not anti-BMPs, are expressed in the anterior visceral endoderm (AVE) which underlines presumptive head ectoderm (HE) (Fig. 1g) [15]. With the progress of gastrulation,

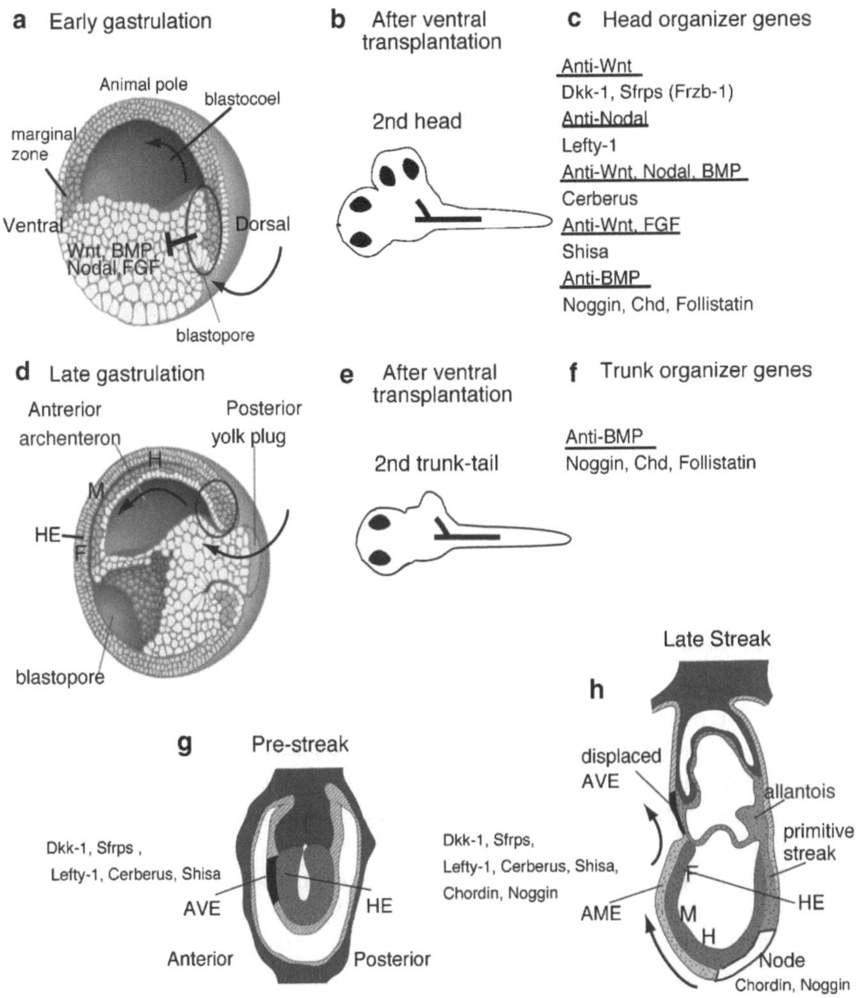

Fig. 1. Regional induction and early specification of head field in *Xenopus* and mouse. (**a**, **b**) In the early *Xenopus* gastrulae, head organizer (*open circle*) is located in the upper dorsal blastopore lip where cells start to involute and differentiate into the anterior mesendoderm (AME), and has the most potent head-inducing activity after transplantation into the ventral side. (**c**) Head organizer protects presumptive head ectoderm (HE) from multiple ventralizing and/or caudarizing factors, including Wnt, BMP, Nodal, and FGF. (**d**, **e**, **f**) In the late gastrulae, the last cells to involute, trunk organizer, induce a trunk-tail, and express anti-BMPs. These two types of organizer provides major driving forces to establish a regional identities of HE, fore-(F), mid-(M), hindbrain (H). (**g**) In the mouse pregastrulae, anterior visceral endoderm (AVE) underlines the HE. (**h**) During gastrulation, involuting cells emigrate from node and form AME that displaces AVE toward extraembryonic region. Genes are specifically expressed in the AVE, AME, or Node listed in (**g**, **h**). Modified with permissions from [4] for (**a**) and (**d**), and [5] for (**g**) and (**h**), Oxford University Press and Wiley-Blackwell, respectively

anteriorly migrating mesendoderm (AME) generated from the node (anterior end of the primitive streak) displaces the AVE and starts to express *cerberus*, *dkk-1*, and *left-1*, together with the BMP inhibitors *chordin* and *noggin* (Fig. 1h). Removal of AVE or AME from gastrula embryos suppresses head formation [16]. Thus, accumulated evidence shows that AVE and AME exhibit characteristics similar to the *Xenopus* head organizer and cooperatively specify the head field in mice. In contrast, the node expresses anti-BMPs alone and has trunk-inducing activity in transplantation analysis, suggesting that the node is a tissue equivalent to the *Xenopus* trunk organizer.

5 Role of Head Organizer Genes in Head Formation

BMPs ventralize embryos and inhibit axis as well as head formation [17]. Inhibition of BMPs in the ventral side results in generation of a secondary axis, which frequently lacks a head structure. Chordin, Noggin, and Follistatin prevent interaction of BMP ligands with their receptor by direct ligand binding. Simultaneous depletion of anti-BMPs together with antisense morpholino oligos leads to a catastrophic loss of neural tissue [18, 19]. In mice, compound mutants of both *chordin* and *noggin* show defects in formation of the head and anterior trunk but have relatively normal posterior trunk structures [20]. BMP inhibitors regulate dorsal–ventral axis formation and induction of neural tissues.

Wnt3a and Wnt8 are expressed in the posterior mesoderm, i.e., the lateral and ventral mesoderm of the marginal zone of *Xenopus* and mesoderm surrounding the streak in mice [21]. Overexpression of Wnts or components of the canonical β-Catenin/Wnt signal cascade suppresses head formation though caudalization of HE. Inhibition of the β-Catenin/Wnt signal enlarges head structure, but has no axis-inducing activity. Frzb-1 is a secreted form of the Wnt receptor Frizzled (Fz), which binds Wnt proteins in the extracellular space and prevents them from signaling [9]. Dickkopf-1 (Dkk-1) encodes a cysteine-rich secreted protein, which binds to Wnt coreceptor LRP5/6, and transmembrane protein Kremen. The trimolecular complex of LRP5/6, Dkk-1, and Kremen is endocytosed, resulting in the depletion of the coreceptor complex on the plasma membrane [22, 23]. Dkk-1 depletion in *Xenopus* and gene knockout in mice leads to a defect in the formation of the forebrain and prechordal plate [10]. Thus Wnt inhibitors protect the head field from caudalization and allow the establishment of a proper anterior–posterior body axis.

Nodal induces mesoderm and endoderm tissues and inhibits head formation. Increased amounts of Nodal induce dorsal tissue including the organizer, whereas low doses induce a ventral fate (blood) [24]. Low-dose inhibition of Nodal enlarges the head structure, but a high dose suppresses head formation by inhibiting the establishment of the organizer. Cerberus directly interacts with BMP, Wnt, and Nodal-ligand, with three different protein motifs in the extracellular space [13]. Antivin/Lefty-1 is a member of the TGF-β gene family which functions as competitive inhibitor of Nodal/Activin by interfering with its binding to type II receptors [25].

In mouse embryos, compound mutants of both *cerberus* and *lefty-1* induce an extra primitive streak [26]. Nodal inhibitors function to restrict mesoderm formation to generate proper body patterning.

Secretion of multisignal inhibitors is one of the major properties of the head organizer and is required to specify the head field. In *Xenopus*, inhibition of BMP signals alone induces a secondary trunk in the ventral side, while combinatorial inhibitions of BMP together with Wnt or Nodal induce a secondary head, mimicking head organizer activity [13, 27]. In mice, heterozygous mutants $dkk^{+/-}noggin^{+/-}$ fail to form a head [28]. Thus secreted inhibitors emanating from the head organizer function cooperatively and partially redundantly with each other to specify the head field during early embryogenesis.

6 Role of FGF Signal Inhibitor in Head Formation

FGFs were the molecules first proposed as a caudalizing factor in *Xenopus*. FGFs are expressed in the posterior mesoderm as Wnts in *Xenopus* and mouse embryos, and play an essential role in the induction of trunk-tail mesoderm. Activation of FGF-mitogen-activated protein kinase signaling abolishes head formation by inducing the hindbrain and spinal cord though the regulation of *Hox* gene expressions, while inhibition promotes an anterior neural fate [29, 30]. Shisa, a cell-autonomous inhibitor of both Wnt and FGF signals, is expressed in the HE as well as the head organizer [14]. The head organizer controls the expression of *shisa* in the HE, suggesting that Shisa functions as an ultimate lock system, preventing caudalization within the newly generated HE. Shisa is an endoplasmic reticulum (ER) protein that physically interacts with immature forms of Fz and FGF receptors within the ER, suppressing their protein maturation and transportation to the cell surface, and thereby preventing cells from receiving both Wnt and FGF signals. Knockdown of Shisa in *Xenopus* suppresses head formation by promoting transportation of Fz and FGF receptors to the cell surface. In mouse Shisa mutants, the process of head formation during gastrulation was normal, suggesting that redundant molecules or mechanisms might function together with Shisa to regulate the anterior–posterior patterning of HE in a cell-autonomous fashion.

References

1. De Robertis EM, Larrain J, Oelgeschlager M et al (2000) The establishment of Spemann's organizer and patterning of the vertebrate embryo. Nat Rev Genet 1:171–181
2. Stern CD (2006) Neural induction: 10 years on since the 'default model'. Curr Opin Cell Biol 18:692–697
3. Niehrs C (2004) Regionally specific induction by the Spemann-Mangold organizer. Nat Rev Genet 5:425–434
4. Slack JM (2005) Essential developmental biology, 2nd edn. Wiley-Blackwell, New Jersey

5. Wolpert L, Smith J, Jessell T et al (2006) Principles of development, 3rd edn. Oxford University Press, Northants
6. Spemann H, Mangold H (2001) Induction of embryonic primordia by implantation of organizers from a different species. Int J Dev Biol 45:13–38 (Reprinted from Archiv Mikroskopische Anatomie Entwicklungsmechanik 100:599–638, 1924)
7. Sasai Y, Lu B, Steinbeisser H et al (1994) Xenopus chordin: a novel dorsalizing factor activated by organizer-specific homeobox genes. Cell 79:779–790
8. Smith WC, Harland RM (1992) Expression cloning of noggin, a new dorsalizing factor localized to the Spemann organizer in Xenopus embryos. Cell 70:829–840
9. Leyns L, Bouwmeester T, Kim SH et al (1997) Frzb-1 is a secreted antagonist of Wnt signaling expressed in the Spemann organizer. Cell 88:747–756
10. Glinka A, Wu W, Delius H et al (1998) Dickkopf-1 is a member of a new family of secreted proteins and functions in head induction. Nature 391:357–362
11. Meno C, Saijoh Y, Fujii H et al (1996) Left-right asymmetric expression of the TGF beta-family member lefty in mouse embryos. Nature 381:151–155
12. Bouwmeester T, Kim S, Sasai Y et al (1996) Cerberus is a head-inducing secreted factor expressed in the anterior endoderm of Spemann's organizer. Nature 382:595–601
13. Piccolo S, Agius E, Leyns L et al (1999) The head inducer Cerberus is a multifunctional antagonist of Nodal, BMP and Wnt signals. Nature 397:707–710
14. Yamamoto A, Nagano T, Takehara S et al (2005) Shisa promotes head formation through the inhibition of receptor protein maturation for the caudalizing factors, Wnt and FGF. Cell 120:223–235
15. Robb L, Tam PP (2004) Gastrula organiser and embryonic patterning in the mouse. Semin Cell Dev Biol 15:543–554
16. Beddington RS, Robertson EJ (1999) Axis development and early asymmetry in mammals. Cell 96:195–209
17. De Robertis EM, Kuroda H (2004) Dorsal-ventral patterning and neural induction in Xenopus embryos. Annu Rev Cell Dev Biol 20:285–308
18. Kuroda H, Wessely O, De Robertis EM (2004) Neural induction in Xenopus: requirement for ectodermal and endomesodermal signals via Chordin, Noggin, beta-Catenin, and Cerberus. PLoS Biol 2:E92
19. Khokha MK, Yeh J, Grammer TC et al (2005) Depletion of three BMP antagonists from Spemann's organizer leads to a catastrophic loss of dorsal structures. Dev Cell 8:401–411
20. Bachiller D, Klingensmith J, Kemp C et al (2000) The organizer factors Chordin and Noggin are required for mouse forebrain development. Nature 403:658–661
21. Kiecker C, Niehrs C (2001) A morphogen gradient of Wnt/beta-catenin signalling regulates anteroposterior neural patterning in Xenopus. Development 128:4189–4201
22. Mao B, Wu W, Li Y et al (2001) LDL-receptor-related protein 6 is a receptor for Dickkopf proteins. Nature 411:321–325
23. Mao BY, Wu W, Davidson G et al (2002) Kremen proteins are Dickkopf receptors that regulate Wnt/beta-catenin signalling. Nature 417:664–667
24. Kimelman D (2006) Mesoderm induction: from caps to chips. Nat Rev Genet 7:360–372
25. Thisse C, Thisse B (1999) Antivin, a novel and divergent member of the TGFbeta superfamily, negatively regulates mesoderm induction. Development 126:229–240
26. Perea-Gomez A, Vella FD, Shawlot W et al (2002) Nodal antagonists in the anterior visceral endoderm prevent the formation of multiple primitive streaks. Dev Cell 3:745–756
27. Glinka A, Wu W, Onichtchouk D et al (1997) Head induction by simultaneous repression of Bmp and Wnt signalling in Xenopus. Nature 389:517–519
28. Barrantes ID, Davidson G, Grone HJ et al (2003) Dkk1 and noggin cooperate in mammalian head induction. Genes Dev 17:2239–2244
29. Cox WG, Hemmati-Brivanlou A (1995) Caudalization of neural fate by tissue recombination and bFGF. Development 121:4349–4358
30. Bottcher RT, Niehrs C (2005) Fibroblast growth factor signaling during early vertebrate development. Endocr Rev 26:63–77

Dental epithelium proliferation and differentiation regulated by ameloblastin

Satoshi Fukumoto, Aya Yamada, Tsutomu Iwamoto, and Takashi Nakamura

Abstract. The principal components of the enamel matrix that are synthesized by secretory ameloblasts can be classified into two major categories, amelogenin and nonamelogenis, which includes ameloblastin (AMBN) and enamelin. AMBN is an enamel matrix protein that regulates cell adhesion, proliferation, and differentiation of ameloblasts. In AMBN-deficient mice, ameloblasts are detached from the enamel matrix, continue to proliferate, and form a multiple cell layer; often, odontogenic tumors develop in the maxilla with age. AMBN had heparin-binding domains at the C-terminal half and that these domains were critical for AMBN binding to dental epithelial cells. Overexpression of full-length AMBN protein inhibited proliferation of human ameloblastoma cells, but overexpression of heparin-binding-domains-deficient AMBN protein had no inhibitory effect. AMBN promotes cell binding through the heparin-binding sites and plays an important role in preventing odontogenic tumor development by suppressing cell proliferation and maintaining differentiation phenotype.

Key words. ameloblasts, enamel matrix, ameloblastin, heparin-binding domain, odontogenic tumor

1 Ameloblast Differentiation

The extracellular matrix (ECM) plays a critical role in tissue development and homeostasis by mediating cell growth, migration, differentiation, apoptosis, and gene expression [1, 2]. Tooth development is regulated by sequential and

S. Fukumoto (✉), A. Yamada, T. Iwamoto, and T. Nakamura
Division of Pediatric Dentistry, Department of Oral Health and Development Sciences, Tohoku University Graduate School of Dentistry, 4-1 Seiryo-machi, Aoba-ku, Sendai 980-8575, Japan
e-mail: fukumoto@mail.tains.tohoku.ac.jp

T. Sasano et al. (eds.), *Interface Oral Health Science 2009*,
DOI 10.1007/978-4-431-99644-6_5, © Springer 2010

reciprocal interaction between the neural crest-derived mesenchyme and oral ectoderm [3]. Although the precise molecular mechanisms that mediate the interactions between epithelia and mesenchyme remain unknown, the basement membrane between epithelial and mesenchymal cells has been reported to play a critical role in the induction and maintenance of cellular differentiation phenotypes. The basement membrane is a specialized ECM that mainly consists of various forms of laminin, type IV collagen, perlecan, and nidogen, and provides a scaffolding for epithelium and signaling for cellular differentiation in a variety of tissues through interactions between specific cellular receptors.

Oral ectoderm differentiates to four different cell types, outer enamel epithelium, stellate reticulum, stratum intermedium, and inner enamel epithelium. Inner enamel epithelium polarizes and differentiates to enamel-matrices-secreted ameloblasts. Inner enamel epithelium was supported by tooth-specific basement membrane. Laminin alpha5 subunit-containing laminin-10/11 is the major laminin in the tooth germ basement membrane. We have examined the role of laminin alpha5 (Lama5) in tooth development using laminin alpha5-null mouse primary dental epithelium and tooth germ organ cultures [4]. Lama5-null mice develop a small tooth germ with defective cusp formation and have reduced proliferation of dental epithelium. Also, cell polarity and formation of the monolayer of the inner dental epithelium are disturbed. In normal mice, integrin alpha6beta4, a receptor for laminin alpha5, is strongly localized at the basal layer of the epithelium, whereas in mutant mice, integrin alpha6beta4 is expressed around the cell surface. In primary dental epithelium culture, laminin-10/11 promotes cell growth, spreading, and filopodia-like microspike formation. This promotion is inhibited by anti-integrin alpha6 and beta4 antibodies and by phosphatidylinositol 3-kinase inhibitors and dominant negative Rho-GTPase family proteins Cdc42 and Rac. In organ culture, anti-integrin alpha6 antibody and wortmannin reduce tooth germ size and shape. Our studies demonstrate that laminin alpha5 is required for the proliferation and polarity of basal epithelial cells and suggest that the interaction between laminin-10/11-integrin alpha6beta4 and the phosphatidylinositol 3-kinase-Cdc42/Rac pathways plays an important role in determining the size and shape of tooth germ.

Laminin 5 regulates anchorage and motility of epithelial cells through integrins alpha6beta4 and alpha3beta1, respectively. Targeted disruption of the Lama3 gene, which encodes the alpha3 subunit of laminin 5 and other isoforms, showed developmental functions that are regulated by adhesion to the basement membrane [5]. In homozygous null animal, profound epithelial abnormalities were detected that resulted in neonatal lethality, consistent with removal of all alpha3-laminin isoforms from epithelial basement membranes. This mouse showed abnormalities in ameloblast differentiation in developing mutant incisors indicating that events downstream of adhesion are affected in mutant animals. These results indicate that laminin 5 has an important role in regulating ameloblast differentiation, enamel matrices gene expression, and survival of dental epithelium.

2 Amelogenesis Imperfecta Resulting from Mutation in Enamel Matrix

Amelogenesis imperfecta is a clinically and genetically diverse group of conditions caused by mutations in genes critical for normal enamel formation. Mutations in the amelogenin (AMEL), enamelin (ENAM) genes are associated with specific amelogenesis imperfecta types having X-linked, autosomal dominant, and autosomal recessive modes of inheritance [6]. The generation of animal models provides an important resource to study normal and abnormal enamel development. Generation of mice lacking expression of genes associated with enamel formation provides a potentially useful tool for understanding biomineralization of enamel and the pathogenesis of the different amelogenesis imperfecta types [7]. AMEL is the most abundant ECM protein in developing enamel. AMELs are encoded by two single copy genes on chromosome Xp22.3–p22.1 and on chromosome Yp11. Mutations in the X chromosome AMEL gene (AMELX) cause a variety of changes in the AMEL protein and are associated with amelogenesis imperfecta phenotypes ranging from hypoplastic to hypomineralized enamel [8]. ENAM is a relatively low-abundance matrix protein in developing enamel and is encoded by the ENAM gene which is located within a cluster of genes critical to biomineralization on chromosome 4q21 [9]. Mutations in ENAM cause amelogenesis imperfecta types characterized by localized pitted enamel or generalized thin enamel [10]. Ameloblastin (AMBN) gene is located on chromosome 4q21. AMBN overexpression in transgenic mice results in abnormal enamel crystallite formation and enamel rod morphology. Further, AMBN-null mice showed severe enamel hypoplasia [11]. These findings suggest that enamel crystallite formation and rod morphology are influence by the temporal and spatial expression of AMBN and imply that the AMBN gene locus may be involved in the etiology of a number of undiagnosed hereditary amelogenesis imperfecta.

3 Expression of Enamel Matrix During Tooth Development

The major secretory proteins synthesized by ameloblasts can be divided into two categories, AMEL (90%) and non-AMEL (10%). AMBN (5%) and ENAM (1%) are well-known members of the non-AMEL proteins. AMBN was identified by three groups and reported to code for an anionic protein that is rich in proline, glycine, and leucine. Immunodetection analysis shows that AMEL is expressed in the ECM prior to AMBN [12]. At the presecretory stage, when the basement membrane is still present between ameloblasts and odontoblasts, only AMEL is detectable in secretory granules. Typically, the dispersed presence of AMEL can be observed in the ECM prior to the disappearance of the basement membrane. AMBN is not detectable extracellularly at this early stage but later appears as patches of electron-dense matrix in the dentin mantle. Based on the relative amounts of enamel proteins' mRNA, identified by real-PCR, the ratio of each gene product was calculated after normalization to the amount of the GAPDH signal [13]. The ratio of AMEL mRNA for secretory ameloblasts, maturation ameloblasts, young odontoblast, and mature odontoblast samples was 4096:2:64:1. The ratios for ENAM and AMBN mRNA in these same tissues were similar to those of AMEL, although their values in the secretory amelobast layer were lower. These results indicate that enamel matrix proteins were highly expressed in secretory ameloblast, and little in odontoblasts.

4 Structure of Ameloblastin and Regulatory Mechanism of Proliferation by Ameloblastin

AMBN has a VTKG motif, which is a potential thrombospondin-like cell adhesion domain, also known as a heparin-binding domain [11]. In addition, AMBN has positively charged lysine (K), arginine (R), and histidine (H) amino acids-rich sequences in the middle- and C-terminal regions, and a KRH-rich motif has been proposed as a heparin-binding domain. We found that heparin and heparan sulfate inhibited dental epithelial cell adhesion to full-length AMBN but not laminin 10/11 [14]. These findings suggest that the heparin-binding domains are involved in dental epithelium cell adhesion to AMBN. Many ECM proteins bind to cells through integrins or calcium-dependent cell adhesion molecules, and this binding is inhibited in the presence of EDTA. EDTA inhibited cell adhesion to laminin 10/11, which has integrin-binding regions. However, the inhibitory effect of EDTA on dental epithelial cell adhesion to AMBN was less effective than that of laminin10/11. These results suggest that integrin-independent cell adhesion is important for cell binding of AMBN.

We previously found using AMBN-null mice and cell culture that AMBN is an adhesion molecule for ameloblasts and required for maintaining a single ameloblast cell layer attached to the enamel matrix and the differentiation state of ameloblasts [11]. We found that recombinant AMBN serves as a cell adhesion molecule. This binding activity of AMBN is specific to dental epithelium and not to other cell types. Although the primary culture of dental epithelium contains mixed cell types, we are able to distinguish preameloblasts and ameloblasts by immunostaining marker proteins and correlating their binding activity to AMBN. Using these assays, we demonstrate that ameloblasts preferentially bind to AMBN, but not to fibronectin or laminin 10/11. In contrast, preameloblast attachment shows the opposite substrate specificity, i.e., preferential adhesion to fibronectin laminin 10/11. We also demonstrate that AMBN inhibits proliferation of AMBN-null ameloblasts in culture. This inhibitory activity is not observed in nonameloblast cell types suggesting that cell type specific interactions between ameloblast and AMBN are involved. Thus, the cell-binding function of AMBN may explain the abnormal phenotypes of ameloblasts observed in the mutant mice. We identify the heparin-binding domains of AMBN and demonstrate that these domains play a critical role of AMBN binding to dental epithelial cells.

AMBN inhibits proliferation of human ameloblastoma cells. This inhibition is accompanied with the induction of p21 and p27 and ENAM and the reduction of Msx2. Those results implicate that AMBN regulates cell proliferation and differentiation through cellular signaling induced by the AMBN interaction with cells. Our finding that AMBN cell binding is mediated through the heparin-binding domains suggests that AMBN interacts with heparan sulfate cell surface receptors.

Epithelial odontogenic tumors are histologically related to the remnants of odontogenic epithelium, which includes the dental lamina, enamel organ, and Hertwig's epithelial root sheath. Actively growing dental lamina is present within the jaws for a considerable time after birth and because of the widespread presence of odontogenic epithelium, some tumors may arise from residues of those cells. AMBN is expressed by differentiated ameloblasts and also in forming Hertwig's epithelial root sheath cells, and can be used as a marker of their migration. Previous immunohistochemical studies have attempted to investigate the differentiation of neoplastic cells in odontogenic tumors; however, it was reported that AMBN, AMEL, and ENAM were not expressed in ameloblastoma cells. AMBN-null mice develop odontogenic tumors of dental epithelium origin in addition to severe enamel hypoplasia. Further, because the ameloblasts disappear after eruption, tooth enamel is never replaced or repaired and odontogenic epithelium almost completely disappears when tooth formation is completed in those mice. However, it is known that discrete clusters of odontogenic epithelial cells remain in the periodontal ligament as the epithelial cell rests of Malassez (ERM). Although the function of ERM cells is still unclear, it is considered that a number of odontogenic tumors arise from them. Recently, it was reported that ERM cells express AMBN, but not AMEL or ENAM. It was also reported that AMBN gene mutations are associated with odontogenic tumors, including ameloblastomas [15, 16]. We showed that human ameloblastoma cell line, AM-1, did not express AMBN, but overexpression of AMBN suppresses proliferation of AM-1 cells. Taken together, we speculate that AMBN functions as an odontogenic tumor suppressor. Msx2, a homeobox-containing transcription factor, was previously shown to be expressed in undifferentiated ameloblasts, while it is downregulated in differentiated ameloblasts [11]. In AMBN-null ameloblasts, an abnormal upregulation of Msx2 was observed, suggesting that AMBN inhibits the expression of Msx2 in normal tooth development. Our finding that AMBN transfection dramatically reduced Msx2 expression supports the notion that AMBN negatively regulates Msx2 expression. It has been suggested that Msx homeobox genes inhibit differentiation through upregulation of cyclin D1. In AMBN-transfected AM-1 cells, the cyclin-dependent kinase inhibitors p21 and p27 were upregulated. Thus, downregulation of Msx2 and upregulation of p21 and p27 by AMBN expression likely cause reduced proliferation of AM-1 cells. Further, the overexpression of AMBN lacking three heparin-binding domains did not inhibit proliferation of AM-1 cells, suggesting the crucial role of the heparin-binding domains of AMBN for the inhibition of AM-1 proliferation. It is conceivable that AMBN induces cellular signaling for these cellular changes by its interaction with AM-1 cells.

References

1. Damasky CH, Werb Z (1992) Signal transduction by integrin receptors for extracellular matrix: cooperative processing of extracellular information. Curr Opin Cell Biol 4:772–781
2. Lin CQ, Bissell MJ (1993) Multi-faceted regulation of cell differentiation by extracellular matrix. FASEB J 7:737–743
3. Thesleff I, Hurmerinta K (1981) Tissues interactions in tooth development. Differentiation 18:75–88
4. Fukumoto S, Miner JH, Ida H et al (2005) Laminin alpha5 is required for dental epithelium growth and polarity and the development of tooth bud and shape. J Biol Chem 281:5008–5016
5. Ryan MC, Lee K, Miyashita Y et al (1999) Targeted disruption of the LAMA3 gene in mice reveals abnormalities in survival and late stage differentiation of epithelial cells. J Cell Biol 145:1309–1323
6. Wright JT (2006) The molecular etiologies and associated phenotypes of amelogenesis imperfecta. Am J Med Gent A 140:2547–2555
7. Gibson CW, Yuan ZA, Hall B et al (2001) Amelogenin-deficient mice display an amelogenesis imperfecta phenotype. J Biol Chem 276:31871–31875
8. Wright JT, Hart PS, Aldred MJ et al (2003) Relationship of phenotype and genotype in X-linked amelogenesis imperfecta. Connect Tissue Res 44(Suppl 1):72–78
9. Hu JC, Yamakoshi Y (2003) Enamelin and autosomal-dominant amelogenesis imperfecta. Crit Rev Oral Biol Med 14:387–398
10. Rajpar MH, Harley K, Laing C et al (2001) Mutation of the gene encoding the enamel-specific protein, enamelin, causes autosomal-dominant amelogenesis imperfecta. Hum Mol Genet 10:1673–1677
11. Fukumoto S, Kiba T, Hall B et al (2004) Ameloblastin is a cell adhesion molecule required for maintaining the differentiation state of ameloblasts. J Cell Biol 167:973–983
12. Nanci A, Lavoie ZP, Kunikata M et al (1998) Comparative immunochemical analyses of the developmental expression and distribution of ameloblastin and amelogenin in rat incisors. J Histochem Cytochem 46:911–934
13. Nagano T, Oida S, Ando H et al (2003) Relative levels of mRNA encoding enamel proteins in enamel organ epithelia and odontoblasts. J Dent Res 82:982–986
14. Sonoda S, Iwamoto T, Nakamura T et al (2009) The critical role of heparin binding domains of ameloblastin for dental epithelium cell adhesion and ameloblastoma proliferation. J Biol Chem 284:27176–27184
15. Perdigao PF, Gomez RS, Pimenta F (2004) Ameloblastin gene (AMBN) mutations associated with epithelial odontogenic tumors. Oral Oncol 40:841–846
16. Toyosawa S, Fujiwara T, Ooshima T et al (2000) Cloning and characterization of the human ameloblastin gene. Gene 256:1–11

Symposium II
Mechanobiology

Stress fiber and the mechanical states in a living endothelial cell

Masaaki Sato

Abstract. We have assumed that stress fibers play a critical role in transmitting intracellular forces to mechanosensing sites. Magnitude of preexisting tension in a single stress fiber of endothelial cells was estimated on the basis of measurements of its preexisting stretching strain and tensile properties. Cultured endothelial cells expressing fluorescently labeled actin were treated with detergents to extract stress fibers. One end of a stress fiber was then dislodged from the substrate by using a microneedle, resulting in a shortening of the stress fiber due to a release of preexisting tension. Assuming that the shortened stress fibers reached its nonstress state, preexisting stretching strain was determined to be 0.24 on average. A tensile test of the isolated single stress fiber was conducted with a pair of cantilevers. The magnitude of the preexisting tension was estimated as around10 nN on average in living endothelial cells.

Key words. actin filament, endothelial cell, mechanical property, mechanotransduction, stress fiber

1 Introduction

Mechanical forces transmitted via cytoskeleton in adherent vascular cells are important for activation of proteins localized at focal adhesions, plasma membrane, intercellular junctions, and so on, which induce the downstream signaling related to gene expression and protein synthesis [1–3]. Hayakawa et al. [4] directly showed that mechanical forces were transmitted from cell surface to focal adhesions through stress fibers.

M. Sato
Graduate School of Biomedical Engineering, Tohoku University,
6-6-01 Aramaki-aoba, Aoba, Sendai 980-8579, Japan
e-mail: sato-m@bme.tohoku.ac.jp

T. Sasano et al. (eds.), *Interface Oral Health Science 2009*,
DOI 10.1007/978-4-431-99644-6_6, © Springer 2010

To elucidate the mechanical pathway, it is of importance to understand the cell mechanical structure, in other words, how the structure of the adherent cell is constructed and from which subcellular structural components. Here, we focus on stress fibers as structural components in endothelial cells. Stress fibers are composed of actin filaments grouped together with myosin, vinculin, and other actin-binding proteins to form a thick fiber with a diameter of several hundred nanometers [5–7]. Stress fibers often develop in mechanical stress-imposed condition and run transversely across the cytoplasm.

In the present paper, we review the mechanobiology of stress fibers of endothelial cells. Preexisting strain of stress fibers was first investigated after being chemically and mechanically isolated from the cells and the substrate [8]. Tensile tests of the isolated stress fibers were then conducted to measure the force required for keeping the preexisting strain [8, 9]. From both these data, the mechanical states of stress fibers in a living cell were estimated [8].

2 Materials and Methods

2.1 Cell Culture

Freshly excised bovine thoracic aortas were obtained from a local slaughterhouse. Endothelial cells were isolated according to the reported technique [10]. Cells were cultured in Dulbecco's modified Eagle medium (DMEM, Invitrogen) supplemented with 10% fetal bovine serum (FBS, JRH Biosciences) and 1% each of penicillin and streptomycin. Cells were used at passages 5–12. Cells were cotransfected with pEGFP-actin vector (GFP-actin, Clontech) and pdsFP593-focal adhesion targeting vector (RFP-FAT; FAT, an amid acid sequence at C-terminus of focal adhesion kinase) using a liposomal method. At 24 h after passage, cells were incubated with a mixture of the DNA plasmids, Lipofectamin (Invitrogen), and Plus Reagent (Invitrogen) in a serum-free medium (Opti-MEM, Invitrogen) for 5 h. After the mixture was replaced with FBS-containing DMEM, cells were incubated in a humidified 5% CO_2 atmosphere at 37°C overnight. Before experiments, cells expressing GFP-actin and RFP-FAT were incubated with a buffer A (10 mM imidazole (Wako), 100 mM KCl, and 2 mM EGTA (Wako), pH 7.2) containing 25 μg/ml saponin (Wako) for 8 min at 37°C to remove intracellular ATP and cations, which induce actomyosin-based contraction [7].

2.2 Isolation of Stress Fiber

Culture medium was washed with a PBS containing 1 μg/ml each of leupeptin and pepstatin and kept at 4°C. To extract stress fibers according to the reported technique [7], cells were treated with a low-ionic-strength extraction solution

(2.5 mM triethanolamine and 1 μg/ml each of leupeptin and pepstatin in distilled water) for 20–40 min, the PBS with 0.05% NP-40 (pH 7.2) for 5 min, and the PBS with 0.05% Triton X-100 (pH 7.2) for 5 min. Extracted stress fibers were then washed gently with the PBS to remove the detergents.

2.3 Evaluation of Preexisting Strain of Stress Fiber

The dish with the cells was placed on a stage of an inverted microscope (IX-70, Olympus). The stress fiber extraction treatments were performed while the dish was being fixed on the stage. After the extraction, the PBS were replaced with the buffer A. Oxygen-removal reagents (2.3 mg/ml glucose, 0.018 mg/ml catalase, and 0.1 mg/ml glucose oxidase) were added to reduce photobleach. By manipulating a fine glass needle made of a capillary tube, one of the focal adhesions visualized by RFP-FAT was carefully dislodged from the substrate surface. Fluorescence images were acquired by a CCD camera to examine changes in the length of stress fiber. The lengths were determined by manually tracing stress fiber images from one end to the other with NIH image software.

2.4 Tensile Test of Stress Fiber

A carbon fiber with approximately 7 μm in diameter was attached to the tip of a rigid glass rod with epoxy resin. The carbon fiber–glass rod, referred to hereafter as cantilever, was used for tensile tests of the isolated stress fibers [8, 9]. Tensile tests were carried out at room temperature (20°C) on an inverted microscope (IX-71, Olympus, Japan) fixed on a vibration-free table. Immediately prior to tests, the tips of both cantilevers were thinly coated with epoxy resin (Araldite, Vantico, Japan). Under illumination from a halogen and a mercury light, the cantilevers were positioned on a targeted single stress fiber and moved toward the end of the stress fiber by using two hydraulic micromanipulators equipped on both sides of the microscope. Here, one cantilever was placed vertically to the specimen and the other in parallel. Great care was taken not to apply any tension to the specimen before tests. Tensile tests were initiated while controlling the position of the parallel cantilever with a piezoelectric actuator connected to the base glass rod to stretch the specimen at 0.02/s strain rate with a program written in LabVIEW programming language (National Instruments, USA). Leverage was applied at the base part of the parallel cantilever to increase its maximum extension. The deflection of cantilever and the displacement of specimen were calculated by using NIH image software from the images taken at 5 s intervals through a digital CCD camera, yielding force–displacement relation.

3 Results and Discussion

The cells were almost confluent at experiments, and many stress fibers were localized around the cell periphery. The cell membrane and cytoplasmic constituents, including the nucleus, were removed after the chemical treatments, and stress fibers were

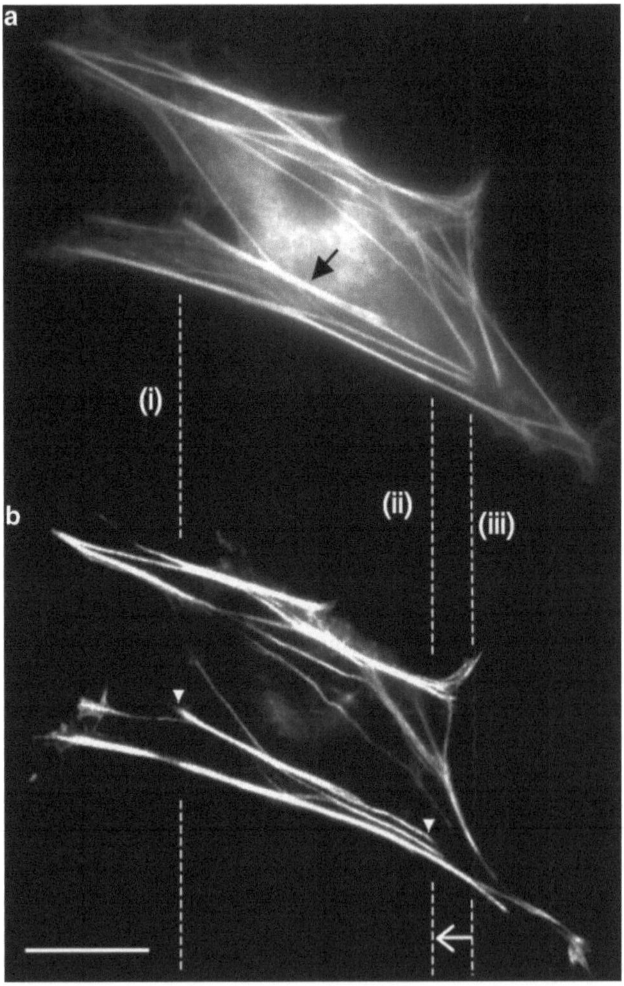

Fig. 1. Preexisting strain of stress fibers [8]. GFP-actin images of a cultured endothelial cell before (**a**) and after (**b**) the treatments of the chemical extraction and the glass needle manipulation. After the right end of a stress fiber (*arrow* in (**a**); each end, *arrowheads* in (**b**)) had been dislodged from the substrate, it was displaced from a place on the *line (iii)* to another on the *line (ii)*. Scale bar = 10 μm

extracted. To confirm whether stress fibers carry preexisting tension, we observed shape changes after the extracted stress fiber had been detached at one end from the substrate. After being detached owing to the glass needle manipulation, the stress fibers shrank somewhat like a recoil of an elastic material as shown in Fig. 1. The shrink is attributable to a sudden release of preexisting tension in the stress fiber. Assuming that the detached stress fiber finally reached its nonstress state, we examined the magnitude of the shortening. The ratio of initial length (before detachment) to nonstress length (after detachment) was 0.82 ± 0.11 (mean ± SD, $n = 11$). If we define preexisting strain as ((initial length) − (nonstress length))/(nonstress length), stress fibers had a preexisting stretching strain of 0.24 ± 0.18 before extraction.

The chemically extracted stress fibers were scraped off from the substrate before the tensile test. The isolated single stress fibers were stretched in the tensile tests from the nonstress state up to a strain of >1.0 across the preexisting strain level (i.e., 0.24 on average). The flexible cantilever was gradually bent since tensile load was given via the stress fiber. Force–strain relations were then obtained as shown in Fig. 2. Initial length of the specimen was $10.3 \pm 2.8\,\mu m$. In a higher stretching strain range of >0.1, tensed stress fibers detached at one end from either of the cantilevers.

The principal purpose of the present study is to evaluate the magnitude of the tension in single stress fibers for better understanding of the cell structure. The strategy was to first identify preexisting stretching strain of a single stress fiber by making it free from surrounding mechanical constraints (i.e., the cell membrane,

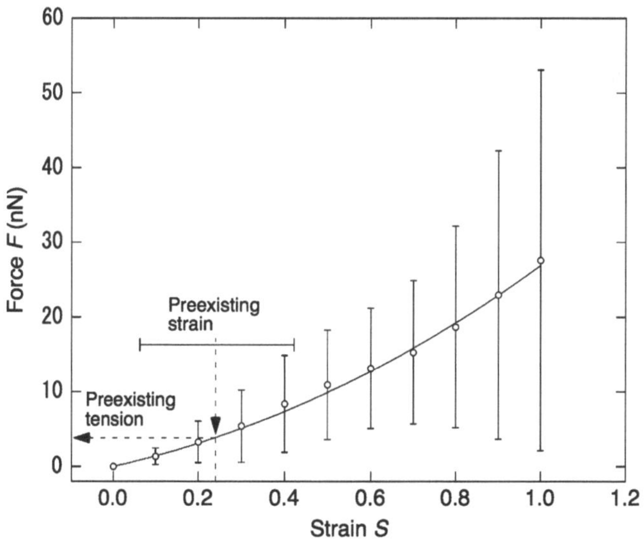

Fig. 2. Relationship between tensile force and stretching strain [8]. Vertical *solid bars* indicate standard deviation ($n=6$). A curve was obtained by the least-squares regression for the mean plots

the other cytoplasmic constituents, and the substrate) to observe resultant shortening, then to measure its tensile force–strain relation, and lastly to examine the average tensile force required for keeping preexisting strain to evaluate preexisting tension. The result showed that preexisting tension was estimated around 4 nN on average as shown in Fig. 2. Tan et al. [11] measured traction force of adherent ECs applied to the substrate at single focal adhesion sites to obtain around 10-nN traction force, the order of which is comparable to that of the estimated preexisting tension in single stress fibers. In contrast, actin microfilament, which is a major component of stress fibers, can bear a tensile force of at most 600 pN [12] that would be insufficient for bearing the traction force. Hence, the quantitative comparison suggests that the principal component responsible for the traction force or the mechanical integrity at the cell bottom is most likely to be "bundled" actin filaments.

Since the diameter of the stress fiber is of submicron order of the magnitude, it was difficult to directly measure the diameter of individual stress fibers from the phase-contrast or fluorescence microscopy during the tensile tests. The diameter was therefore evaluated from a separate experiment with electron microscopy to investigate the order of its average value although diameters of individual specimens cannot be specified [8]. The Young's modulus of the stress fibers was determined to be 287 kPa assuming a uniform circle cross-section with the average diameter (0.25 μm), and was also evaluated at the preexisting strain level (i.e., $S = 0.24$) to be 408 kPa. The Young's modulus of the stress fiber was almost three orders of magnitude smaller than that of its principal component, actin filament, which has around 1 GPa Young's modulus according to a previous report [13]; however, the mechanism of the difference remains unclear.

Acknowledgments This work was supported financially in part by the Grant-in-Aid for Scientific Research (Scientific Research A #17200030 and Specially Promoted Research #20001007) by the MEXT, Japan and the Mitsubishi Foundation.

References

1. Davies PF (1995) Flow-mediated endothelial mechanotransduction. Physiol Rev 75:519–560
2. Ingber DE (1997) Tensegrity: the architectural basis of cellular mechanotransduction. Annu Rev Physiol 59:575–599
3. Wang N, Suo Z (2005) Long-distance propagation of forces in a cell. Biochem Biophys Res Commun 328:1133–1138
4. Hayakawa K, Tatsumi H, Sokabe M (2008) Actin stress fibers transmit and focus force to activate mechanosensitive channels. J Cell Sci 121:496–503
5. Satcher RL Jr, Dewey CF Jr (1996) Theoretical estimates of mechanical properties of the endothelial cell cytoskeleton. Biophys J 71:109–118
6. Furukawa R, Fechheimer M (1997) The structure, function, and assembly of actin filament bundles. Int Rev Cytol 175:29–90
7. Katoh K, Kano Y, Fujiwara K (2000) Isolation and in vitro contraction of stress fibers. Methods Enzymol 325:369–380

8. Deguchi S, Ohashi T, Sato M (2005) Evaluation of tension in actin bundle of endothelial cells based on preexisting strain and tensile properties measurements. Mol Cell Biomech 2:125–134
9. Deguchi S, Ohashi T, Sato M (2006) Tensile properties of single stress fibers isolated from cultured vascular smooth muscle cells. J Biomech 39:2603–2610
10. Shasby DM, Shasby MW (1987) Effect of albumin concentration on endothelial albumin transportation in vitro. Am J Physiol 253:H654–H661
11. Tan JL, Tien J, Pirone DM et al (2003) Cells lying on a bed of microneedles: an approach to isolate mechanical force. Proc Natl Acad Sci USA 100:1484–1489
12. Tsuda Y, Yasutake H, Ishijima A et al (1996) Torsional rigidity of single actin filaments and actin-actin bond breaking force under torsion measured directly by in vitro micromanipulation. Proc Natl Acad Sci USA 93:12937–12942
13. Kojima H, Ishijima A, Yanagida T (1994) Direct measurement of stiffness of single actin filaments with and without tropomyosin by *in vitro* nanomanipulation. Proc Natl Acad Sci USA 91:12962–12966

Transient receptor potential channels and mechanobiology

Minoru Wakamori

Abstract. Mechanotransduction is a fundamental process converting mechanical force into electrical and chemical responses, in which mechanosensitive channels are thought to play crucial roles. They are expressed in a variety of cells, including hair cells, baroreceptors, muscle spindle, and bone cells. Recent molecular biological analyses have revealed that several members of transient receptor potential (TRP) channels may play important roles in detection of mechanical stimuli. Twenty-seven *trp*-related genes in human have been identified to date. TRP proteins can be classified into six subfamilies: TRPC (Canonical), TRPV (Vanilloid), TRPM (Melastatin), TRPP (Polycystin), TRPML (Mucolipin), and TRPA (Ankyrin). Among the TRP superfamily TRPC1, TRPC6, TRPV1, TRPV2, TRPP1/TRPP2, TRPM4, TRPM7, and TRPA1 channels mediate mechanosensation. TRPC1, TRPC5, TRPC6, TRPV2, TRPV4, TRPM3, and TRPM7 channels are involved in osmosensation. I will review mechanosensitive channels and discuss their activation mechanism.

Key words. mechanotransduction, channel, mechanosensor, transient receptor potential, TRP

Mechanotransduction is a fundamental process converting mechanical force into electrical and chemical responses, in which mechanosensitive channels are thought to play crucial roles. Many intracellular proteins including actin fiber, membrane-spanning integrin, and extracellular matrix play important roles in mechanotransduction. Ca^{2+}-permeable stretch-activated channel participates in mechanotransduction as well. In this review, I'd like to focus on ionic channels.

M. Wakamori
Laboratory of Molecular Pharmacology and Cell Biophysics, Department of Oral Biology, Tohoku University Graduate School of Dentistry, Sendai 980-8575, Japan
e-mail: wakamori@m.tains.tohoku.ac.jp

T. Sasano et al. (eds.), *Interface Oral Health Science 2009*,
DOI 10.1007/978-4-431-99644-6_7, © Springer 2010

1 Classification of Ionic Channels

Ionic channels are elementary excitable elements in the cell membranes of nerve, muscle, and other tissues. Recently, with infusion of new techniques of biochemistry, anatomy, pharmacology, and physiology, we can now recognize increasingly wide roles for them in nonexcitable cells, including sperm, white blood cells, and endocrine glands. Two apparatuses that characterize the channels are permeation and gating. On the basis of their gating properties, the channels are classified into four groups. The most well-known channel group is voltage-gated ion channels. The voltage-dependent Na^+ and K^+ currents were recorded in the squid giant axon by Prof. Hodgkin and Prof. Huxley [1]. They introduced the physicochemical analysis into biology and proposed the Hodgkin–Huxley model to explain the action potential. At that time, the putative channels were given the same names as the permeability components. However, only about 25 years ago the late Prof. Numa in Kyoto University provided much molecular evidence that voltage-gated Na^+ channel had the pore and the voltage sensor [2–4]. The second group is ligand-gated ion channels which are activated by binding with agonists such as glutamate, acetylcholine (ACh), serotonin (5-HT), ATP, γ-aminobutyric acid (GABA), and glycine. The second group is divided into two subgroups by the permeation. Glutamate, ACh, 5-HT, and ATP are excitatory neurotransmitters whose receptors permeate cations, Na^+, K^+, and Ca^{2+}. However, GABA and glycine are inhibitory neurotransmitters and induce anion (Cl^-) currents. The third group is mechanosensitive or stretch-activated ion channels. The last group is receptor-operated channels. Some studies are just beginning to elucidate functions of the mechanosensitive channels and the receptor-operated channels.

2 Mechanosensitive Channels

Mechanosensitive channels are expressed in a wide range of cells, including hair cells in the ear, baroreceptor in carotid body, chondrocytes and osteocytes in bone, endothelial cells in blood vessel, gastrointestinal tract, skeletal muscles, and periodontal membranes in tooth sockets. Originally mechanosensitive ion channels were supposed to have mechanical sensor in the channel. But now, other activation mechanisms are also proposed. Four models of channel gating by mechanical stimuli are proposed. In the first model, force is delivered to the channel by surface tension or bending the lipid bilayer, causing hydrophobic mismatch which favors opening. Channel opening decreases the energy stored in the membrane. The second is the tether model. The tether binds to channel protein directly and specific accessory proteins such as intracellular cytoskeletal elements and/or extracellular matrix molecules. The tethers directly convey the stimulus force to the channel protein in order to induce its conformational change. In contrast, in the third model, tethers are indirect. Tethers convey mechanical force to an accessory protein and in consequence induce its conformational change. This signal is communicated to pore-forming subunits. Finally in the fourth model, mechanosensing protein is more distant and communicates with the channel by generating secondary signals such as diffusible second-messenger molecules or activation of kinases. In this model, the channel is considered mechanically sensitive, but not mechanically gated, because the gate is regulated by diffusible second messenger(s).

3 Mechanosensitive Channels

As I mentioned, many kinds of activation mechanisms are proposed, but molecular identity of the mechanically gated and mechanically sensitive ion channels has been elusive. Possible candidates are transient receptor potential (TRP) channels. The first identified TRP channel is *Dolosophila* TRP channel expressed in retina. Its photoreceptors fail to generate the Ca^{2+}-dependent sustained phase of receptor potential and to induce subsequent Ca^{2+}-dependent adaptation to light [5, 6]. Therefore, the receptor potential is transient, which is the origin of the name of TRP channel. Up to now, there are at least 33 TRP channel genes in mammals. They are subdivided into six subfamilies, TRPM, TRPC, TRPV, TRPP, TRPML, and TRPA, on the bases of sequence similarity. The most familiar TRP channel is TRPV1, which is called as capsaicin receptor. TRPV1 channel is activated in a polymodal manner by capsaicin, which is an irritant of hot pepper, as well as acid and noxious heat. Therefore, TRPV1 monitors extracellular environmental condition of the cells. Other TRP channels are also involved in the transduction of a wide variety of other sensations, with roles in vision, olfaction, taste, chemosensation, and thermosensation. For example, TRPC5 channel monitors extracellular Ca^{2+} concentration [7] and is activated by nitric oxide through cysteine S-nitrosylation [8]. TRPM2 monitors redox state of the cell [9]. TRPV1, TRPV2, TRPV3, TRPV4, TRPM8, and TRPA1 channels are sensitive to temperature, although their response range is different [10]. These TRP channels are named as thermo-TRP. Further, TRPC1, TRPC5, TRPC6, TRPV2, TRPV4, TRPM3, and TRPM7 channels are sensitive to osmostimuli and called as osmo-TRP [10]. In addition, TRPC1, TRPC6, TRPV1, TRPV2, TRPM4, TRPM7, TRPP2, and TRPA1 channels are sensitive to mechanostimuli, and called as mechano-TRP [10]. TRPC2, TRPC6, TRPV2, and TRPM7 channels are sensitive to both osmostress and mechanostress [10]. The borderline between osmosensitivity and mechanosensitivity is poorly defined. Some of the osmo- and mechano-TRP channels have specific domain to form a spring [11]. PKD1 and TRPC1 channels have a coiled-coil domain in C-termini. TRPA1, TRPV4, and TRPC1 channels have ankyrin repeat domains in N-termini. Because long ankyrin repeats are curved like a spring, it has been suggested that these ankyrin repeats might form the biophysically defined gating spring. But the gating mechanism is not estab-

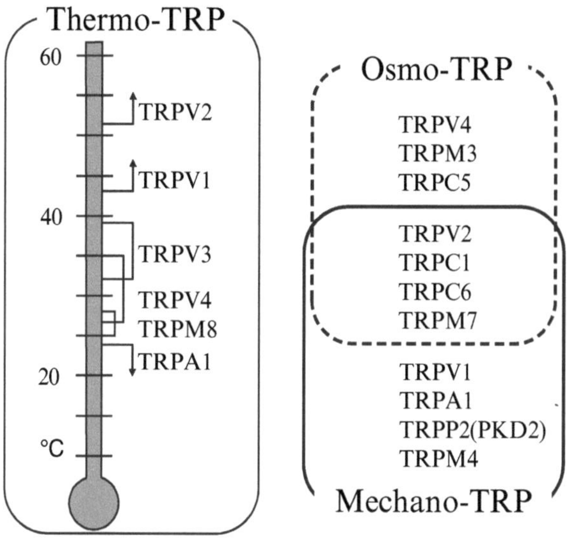

Fig. 1. Thermo-, osmo-, and mechano-TRP channels

lished yet. It will be interesting to dissect the molecular chain which conveys force to the transmembrane domains of a TRP channel and to understand how that force causes a conformational change to open the pore. However, we still have a long way to go.

References

1. Hodgkin AL, Huxley AF (1952) The components of membrane conductance in the giant axon of Loligo. J Physiol 116:473–496
2. Noda M, Ikeda T, Suzuki H et al (1986) Expression of functional sodium channels from cloned cDNA. Nature 322:826–828
3. Numa S, Noda M (1986) Molecular structure of sodium channels. Ann N Y Acad Sci 479:338–355
4. Stumer W, Conti F, Suzuki H et al (1989) Structural parts involved in activation and inactivation of the sodium channel. Nature 339:597–603
5. Fein A, Payne R, Corson DW et al (1984) Photoreceptor excitation and adaptation by inositol 1, 4, 5-trisphosphate. Nature 311:157–160
6. Ranganathan R, Malicki DM, Zuker CS (1995) Signal transduction in *Drosophila* photoreceptors. Annu Rev Neurosci 18:283–317
7. Okada T, Shimizu S, Wakamori M et al (1998) Molecular cloning and functional characterization of a novel receptor-activated TRP Ca^{2+} channel from mouse brain. J Biol Chem 273:10279–10287
8. Yoshida T, Inoue R, Morii T et al (2006) Nitric oxide activates TRP channels by cysteine S-nitrosylation. Nat Chem Biol 2:596–607

9. Hara Y, Wakamori M, Ishii M et al (2002) LTRPC2 Ca^{2+}-permeable channel activated by changes in redox status confers susceptibility to cell death. Mol Cell 9:163–173
10. Venkatachalam K, Montell C (2007) TRP channels. Annu Rev Biochem 76:387–417
11. Christensen AP, Corey DP (2007) TRP channels in mechanosensation: direct or indirect activation? Nat Rev Neurosci 8:510–521

Molecular mechanisms of the response to mechanical stimulation during chondrocyte differentiation

Ichiro Takahashi, Taisuke Masuda, Kumiko Kohsaka, Fumie Terao, Takahisa Anada, Yasuyuki Sasano, Teruko Takano-Yamamoto, and Osamu Suzuki

Abstract. The differentiation of mesenchymal cells and the metabolism of skeletal tissues are regulated by multiple factors, such as growth factors, cytokines, and the interaction between the extracellular matrices (ECMs) and mechanical stress. Bone and cartilage are tissues that support body movement, which consist of tissue-specific ECM and specifically differentiated cells, such as osteocytes and chondrocytes, in each tissue. Cartilage contributes to bone growth under mechanical compressive loading and buffers mechanical stress during joint action. Thus, mechanical stress could be an important regulatory factor in the differentiation of chondrocytes from mesenchymal stem cells and/or their metabolism. In this chapter, we review the intracellular signal transduction that occurs through the mitogen-activated protein kinase pathway in response to mechanical stimulation during the differentiation of chondrocytes from mesenchymal stem cells.

I. Takahashi: Presently belonging to Section of Orthodontics, Kyushu University Faculty of Dental Science having moved from Division of Orthodontics and Dentofacial Orthopedics, Tohoku University Graduate School of Dentistry after presentation to IOHS.

I. Takahashi, K. Kohsaka, F. Terao, and T. T.-Yamamoto
Division of Orthodontics and Dentofacial Orthopedics, Tohoku University Graduate School of Dentistry, 4-1 Seiryo-machi, Aoba-ku, Sendai 980-8575, Japan

I. Takahashi (✉)
Section of Orthodontics, Kyushu University Faculty of Dental Science, 3-1-1 Maidashi, Higashi-ku, Fukuoka 812-8582, Japan
e-mail: takahashi@dent.kyushu-u.ac.jp

T. Masuda, T. Anada, and O. Suzuki
Division of Craniofacial Function Engineering, Tohoku University Graduate School of Dentistry, 4-1 Seiryo-machi, Aoba-ku, Sendai 980-8575, Japan

Y. Sasano
Division of Craniofacial Development and Regeneration, Tohoku University Graduate School of Dentistry, 4-1 Seiryo-machi, Aoba-ku, Sendai 980-8575, Japan

T. Sasano et al. (eds.), *Interface Oral Health Science 2009*,
DOI 10.1007/978-4-431-99644-6_8, © Springer 2010

Key words. chondrocytes, mesenchymal stem cell, differentiation of mesenchymal cells, MAPK, signal transduction

1 Introduction

Bone, cartilage, and tendon are tissues that support voluntary movement during muscular function in animals. Bony pillars in the cancellous bone in the epiphysis of long bones show a honeycombed architecture that sustains the mechanical stress generated by voluntary muscular actions and body weight bearing. These structures are formed by complex interactions between the biomechanical environment and the cells composing the supporting tissues. Chondrocytes generate cartilaginous tissues by producing specific extracellular matrix (ECM) molecules, which are replaced by calcified cancellous bone produced by osteoblasts and osteocytes after mineralization of the cartilaginous ECM. Consequently, calcified cancellous bone is remodeled under the control of mechanical stress exerted on the bone-cartilage complex, which consists of cartilage and the joints of long bones. This multistep sequential process of bone formation is defined as endochondral bone formation. During this process, various types of epigenetic and environmental factors such as growth factors, cytokines, and mechanical factors influence cell growth, proliferation, metabolism, and cytodifferentiation in different phenotypes and lineages of cells.

In the present review, we attempt to introduce the roles of mechanical stress in the cytodifferentiation and regulation of metabolism in various types of cells and discuss the intracellular signaling pathways involved in mechanical stress loading on chondrocytes differentiating from mesenchymal stem cells.

2 General Mechanoreaction of Cells

Recently, the cellular responses to a variety of mechanical stresses have been investigated at the molecular level and millisecond order by focusing on channel activities, cytoskeletal architecture, membrane deformation, and cell-ECM adhesion [1–3]. Of the various types of cells, fibroblasts, vascular endothelial cells, and myofibroblasts are the most common cell types being investigated with regard to their mechanical stress response [4–7]. These cells are exposed to cyclical tensile and/or shear stress during muscular function and/or fluid flow such as that in the blood vessels. Osteoblasts and chondrocytes are of interest to researchers [8] because they exist in a unique three-dimensional context, i.e., embedded in or attached to the quite specific and complex ECM of bone and/or cartilage. Therefore, in previous studies, fibroblasts and vascular endothelial cells were commonly used as model cells for the analysis of mechanical stress responses [6, 8, 9].

The molecular mechanism of mechanical stress response involves mechanosensing, by which the cells recognize mechanical stress or strain; mechanotransduction, in which cells translate mechanical stimulation into a chemical reaction and transfer this signal through the cytosol to the nucleus; and mechanoresponse, in which the cells express genes and generate final products in order to respond to the mechanical stimulation. The mechanosensing step is usually considered as the step when cells sense the deformation of the cell membrane or cytoskeleton, or recognize a distraction, deformation, and/or detachment of the cell-ECM adhesion and/or activate calcium ion channels through a putative mechanoreceptor. Cell-ECM adhesion through integrins and stretch-activated calcium ion channels is considered as a mechanosensing component in combination with the actin cytoskeleton. Recently, mechanical deformation of the lipid bilayered cell membrane and energy transfer to channel proteins has also been focused on as part of the molecular mechanics of mechanosensing machinery [1]. Thus, calcium ion influx and cell-ECM adhesion-mediated signal transduction are considered as signal transduction pathways downstream of mechanical stimulation.

Integrin-mediated cell-ECM adhesion complex is a putative mechanosensor, and the pathways activated downstream of focal adhesion complex are strong candidates for mechanotransduction pathways. Indeed, a recent study revealed conformational changes in integrin-related molecules after mechanical stimulation [2], which led to the activation of downstream signals mediated by the mitogen-activated protein kinase (MAPK) pathway and/or small GTPases [8, 9]. Previous studies indicated that the extracellular signal-regulated kinase (ERK) or p38-MAPK pathway is activated under fluid shear stress in endothelial cells [10, 11]. In addition, the Rho and Rock pathways are also activated in fibroblasts under shear stress loading [8]. Consequently, cell-type specific mechanoreactions result from differences in the signaling molecules expressed in specific types of cell, and their downstream gene regulation is dependent upon the phenotype of the cell. In many cases, the metabolism and/or proliferative activities of cells are regulated to maintain or remodel the structure of the tissues and/or organs under mechanical stimulation without changing cell phenotypes. In some cases, mechanotransduction actually controls the phenotypes of cells by means of differentiation or dedifferentiation to another phenotype.

3 Differentiation of Chondrocytes

Chondrocytes generate quite specific ECM macromolecules such as type II collagen and aggrecan after they have differentiated from mesenchymal cells as described earlier [11–13]. During embryonic development, mesenchymal stem cells congregate along with fibronectin and tenascin in the location where future long bone will be formed in the immature tissue-specific ECM. Once cellular condensation begins using these ECM molecules as a scaffold, the cells in the aggregation begin to express cell–cell adhesion molecules such as N-cadherin and N-CAM [14, 15].

After cellular condensation, mesenchymal cells start to differentiate into chondrocytes by firstly expressing the transcription factors *Sox-5*, *-6*, and *-9*, which regulate the gene expression of the phenotypic genes, *Col2a1* and *aggrecan* [13]. By producing these cartilage-specific ECM macromolecules, cell shape changes from ovoid to round, and polarity is obtained by arranging the positions of the nucleus and a lipid storage area in the cytosol, and the intercellular space is enlarged through the progression of chondrocyte differentiation. Further hypertrophic differentiation continues as endochondral bone formation progresses during growth. Chondrocytes terminally differentiate into hypertrophic chondrocytes expressing the *Col10a1* gene and induce calcification of the cartilaginous matrix, which is later replaced by bone.

These stepwise differentiation processes of chondrocytes progress under the regulation of growth factors and hormones. Fibroblast growth factors (FGFs) 8 and 10 enhance the proliferation of mesenchymal stem cells during the initiation of limb bud formation prior to chondrocyte differentiation, and FGF1 and 2 enhance chondrogenesis after the initial differentiation of chondrocytes [16]. Bone morphogenetic proteins (BMP) 2, 4, and 7 induce and promote the chondrogenesis of mesenchymal stem cells, even before cell condensation. Downstream of BMP and FGF are Smad and ERK, which send signals to the nucleus. BMP activate their own serine-threonine kinase receptors, and FGF promote the activities of the ERK signaling pathway. In the case of bone- or cartilage-derived mesenchymal cells, ERK phosphorylates the linker region of Smad so as to inhibit BMP signaling, while ERK synergistically upregulate the activity of BMP in tooth-forming cells. On the other hand, Indian hedgehog (Ihh) and Parathyroid-hormone-related protein (PTHrP) regulate the hypertrophy of chondrocytes. While PTHrP enhances the differentiation of mature chondrocytes to hypertrophic chondrocytes, Ihh inhibits the progression to hypertrophy [17]. Once the expression profile of these growth factors is imbalanced, the rate and the amount of chondrogenesis and endochondral bone formation are affected, resulting in alterations in the size and shape of bones. Thus, the environmental factors surrounding cartilage differentiation have considerable impact on the resulting body shape and size.

4 Mechanobiology of Chondrocytes

4.1 Mechanoresponse of Chondrocytes

The chondrocytes in articular cartilage are fully differentiated and function as the metabolic machinery of cartilage by maintaining the integrity of the ECM so as to support joint function. Therefore, the mechanoresponse of chondrocytes in healthy articular cartilage is limited to the range of the physiological fluctuation in the expression of phenotypic genes and/or to the range of tissue remodeling necessary to maintain joint function including the regulation of cell proliferative activity. In general, the mechanoresponse of chondrocytes under physiological mechanical loading is considered to affect the expression level of type II collagen and/or aggrecan and their metabolic enzymes, such as matrix metalloproteinases (MMP) and aggrecanases, and inhibitors, such as tissue inhibitor of metalloproteinases (TIMP) [18]. These genes and proteins are physiologically expressed in chondrocytes, and their expression levels may be affected by alterations in the mechanical load placed on cartilage as a way of regulating and maintaining tissue integrity. Indeed, cyclic compressive stress exerted on cartilage was found to upregulate the expression of the aggrecan gene without changing tissue phenotype [18–21].

On the other hand, nonphysiological levels of mechanical loading drastically change the phenotypes of chondrocytes. Pathological levels of shear stress destroy the articular cartilage in knee joints, and chondrocytes are not capable of maintaining their viability in these circumstances [22]. In the initial stage of osteoarthritis, aggrecanases are expressed and secreted into the synovial fluid, which is followed by the secretion and activation of MMP and chondrocyte cell death. Tensile stress inhibits the differentiation of chondrocytes. In particular, mechanical stretching caused the replacement of the midpalatal suture cartilage with bone in rodent animal model experiments. In our previous studies, expansive stress induced the expression and accumulation of focal adhesion complex-related proteins, such as paxillin and vinculin in undifferentiated mesenchymal cells in the midpalatal suture cartilage [23]. At the same time, the cytoskeletal configuration was altered to produce stress fibers in these cells, which were directed to become chondrocytes. Thus, differentiating chondrocytes can be dedifferentiated or have their maturation inhibited.

4.2 Mechanotransduction in Chondrocytes

Based on the results of the previous studies described earlier, integrin-mediated cell-ECM adhesion complex is a putative mechanosensor, and its downstream signaling pathways are strong candidates for mechanotransduction pathways that act during chondrogenesis. Indeed, a recent study revealed the structural changes that occur in integrin-related molecules after mechanical stimulation lead to the activation of downstream signaling mediated by the MAPK pathway and/or small GTPases such as Cdc42, Ras, and Rho [8].

In our studies, the following results were obtained by using micromass culture in combination with a stepwise mechanical stretch culture system. Chondrogenic differentiation was inhibited at day 4 after stretch stimulation had started in an in vitro stretched micromass culture system. The number of cartilaginous nodules and their area were reduced in the stretched culture compared with a nonstretched culture, and increases in *Col2a1* gene expression were also inhibited in the nonstretched culture. Thus, the chondrogenic differentiation of rat limb bud cells was inhibited by stepwise stretch stimulation.

The phosphorylation of ERK was transiently and directly upregulated and peaked at 1.0 h after stretch stimulation, while MAPK, p38-MAPK, and JNK were not activated. In addition, the gene expression level of ERK-1/2 did not change throughout the experimental period according to semiquantitative RT-PCR analysis. Taken together, mechanical stretching directly activates ERK-1/2, but not JNK or p38 MAPK. The phosphorylation of ERK-1/2 even increased under the inhibition of protein production, which can be interpreted as showing that the signal was transferred to the nucleus after the cell had sensed it. MEK1/2 and MEK-1 inhibitors rescued chondrogenic nodule formation from mechanical-stretch-induced inhibition. Consequently, MEK inhibitors prevented *Col2a1* gene expression from being inhibited by stretch stimulation. Thus, it is considered that the ERK pathway is directly involved in the mechanoresponse of chondrocytic differentiation in limb bud cells.

5 Summary

In the present review, we focused on the mechanotransduction and mechanoresponse of chondrocytes during chondrogenic differentiation of mesenchymal stem cells. We have described in previous studies that when cell-ECM adhesion through the RGD peptide on integrin molecules was blocked, the inhibition of chondrogenesis from mesenchymal cells was abrogated [24]. Therefore, mechanosensing machinery must make up part of the molecular structure of the focal adhesion complex. In combination with membrane channel activities related to membrane deformation, structural changes in integrin molecules could be involved as described earlier. In addition, different modes and magnitudes of mechanical stress lead to different patterns of mechanoresponse in certain cell phenotypes and induce not only metabolic activities, but also phenotypic changes in cells. Since mechanotransduction and the FGF signaling pathway share the MAPK pathway, which has crosstalk with the BMP signaling pathway [25], mechanical stress may regulate the organogenesis of skeletal organs.

Molecular mechanobiology is a rapidly expanding field of life science that explores the molecular mechanisms of the mechanosensing, mechanotransduction, and mechanoresponse of cells. Further progress in this field will contribute to clarifying the molecular mechanisms of the initiation and progression of joint, vascular, musculoskeletal, and cardiac diseases.

References

1. Phillips R, Ursell T, Wiggins P et al (2009) Emerging roles for lipids in shaping membrane-protein function. Nature 459:379–385
2. Friedland JC, Lee MH, Boettiger D (2009) Mechanically activated integrin switch controls $\alpha_5\beta_1$ function. Science 323:642–644
3. Kaazempur Mofrad MR, Abdul-Rahim NA, Karcher H et al (2005) Exploring the molecular basis for mechanosensation, signal transduction, and cytoskeletal remodeling. Acta Biomaterialia 1:281–293

4. Yamaki K, Harada I, Goto M et al (2009) Regulation of cellular morphology using temperature-responsive hydrogel for integrin-mediated mechanical force stimulation. Biomaterials 30:1421–1427
5. Tzima E (2006) Role of small GTPases in endothelial cytoskeletal dynamics and the shear stress response. Circ Res 98:176–185
6. Li C-H, Xu Q-G (2007) Mechanical stress-initiated signal transduction in vascular smooth muscle cells in vitro and in vivo. Cell Signal 19:881–891
7. Wei W-C, Lin H-H, Shen M-R et al (2008) Mechanosensing machinery for cells under low substratum rigidity. Am J Physiol Cell Physiol 295:C1579–C1589
8. Weyts FA, Li YS, van Leeuwen J et al (2002) ERK activation and alpha v beta 3 integrin signaling through Shc recruitment in response to mechanical stimulation in human osteoblasts. J Cell Biochem 87:85–92
9. Ali MH, Mungai PT, Schumacker PT (2006) Stretch-induced phosphorylation of focal adhesion kinase in endothelial cells: role of mitochondrial oxidants. Am J Physiol Lung Cell Mol Physiol 291:L38–L45
10. Sun HW, Li CJ, Chen HQ (2007) Involvement of integrins, MAPK, and NF-kappaB in regulation of the shear stress-induced MMP-9 expression in endothelial cells. Biochem Biophys Res Commun 353:152–158
11. von der Mark K, von der Mark H (1977) The role of three genetically distinct collagen types in endochondral ossification and calcification of cartilage. J Bone Joint Surg 59:458–464
12. Silbermann M, Reddi AH, Hand AR et al (1987) Further characterisation of the extracellular matrix in the mandibular condyle in neonatal mice. J Anat 151:169–188
13. Barna M, Niswander L (2007) Visualization of cartilage formation: insight into cellular properties of skeletal progenitors and chondrodysplasia syndromes. Dev Cell 12:931–941
14. Oberlender SA, Tuan RS (1994) Spatiotemporal profile of N-cadherin expression in the developing limb mesenchyme. Cell Adhes Commun 2:521–537
15. Tavella S, Raffo P, Tacchetti C et al (1994) N-CAM and N-cadherin expression during in vitro chondrogenesis. Exp Cell Res 215:354–362
16. Bobick BE, Thornhill TM, Kulyk WM (2007) Fibroblast growth factors 2, 4, and 8 exert both negative and positive effects on limb, frontonasal, and mandibular chondrogenesis via MEK-ERK activation. J Cell Physiol 211:233–243
17. Chung UI, Schipani E, McMahon AP et al (2001) Indian hedgehog couples chondrogenesis to osteogenesis in endochondral bone development. J Clin Invest 107:295–304
18. Mitani H, Takahashi I, Onodera K et al (2006) Comparison of age-dependent expression of aggrecan and ADAMTSs in mandibular condylar cartilage, tibial growth plate, and articular cartilage in rats. Histochem Cell Biol 126:371–380
19. Ikenoue T, Trindade MC, Lee MS et al (2003) Mechanoregulation of human articular chondrocyte aggrecan and type II collagen expression by intermittent hydrostatic pressure in vitro. J Orthopedic Res 21:110–116
20. Wang X, Mao JJ (2002) Accelerated chondrogenesis of the rabbit cranial base growth plate by oscillatory mechanical stimuli. J Bone Miner Rese 17:1843–1850
21. Kim YJ, Grodzinsky AJ, Plaas AH (1996) Compression of cartilage results in differential effects on biosynthetic pathways for aggrecan, link protein, and hyaluronan. Arch Biochem Biophys 328:331–340
22. Yamada S, Saeki S, Takahashi I et al (2002) Diurnal variation in the response of mandible to orthopedic force. J Dent Res 81:711–715
23. Takahashi I, Onodera K, Sasano Y et al (2003) Effect of stretching on gene expression of β1 integrin and Focal adhesion kinase and chondrogenesis through cell–extracellular matrix interactions. Eur J Cell Biol 82:182–192
24. Onodera K, Takahashi I, Sasano Y et al (2005) Stepwise mechanical stretching inhibits chondrogenesis through cell-matrix adhesion mediated by integrins in embryonic rat limb bud mesenchymal cells. Eur J Cell Biol 84:45–58
25. Kretzschmar M, Doody J, Massagué J (1997) Opposing BMP and EGF signalling pathways converge on the TGF-beta family mediator Smad1. Nature 389:618–622

Recruitment of masseter motoneurons by spindle Ia inputs and its modulation by leak K^+ channels

Youngnam Kang, Hiroki Toyoda, Mitsuru Saito, and Hajime Sato

Abstract. The slow-closing phase of the mastication cycle plays a major role in the mastication of foods. However, the neuronal mechanism underlying the slow-closing phase remains unknown. During the slow-closing phase, isometric contraction of jaw-closing muscles is developed through the recruitment of jaw-closing motoneurons (MNs). It is well established that motor units are recruited depending on the order of sizes or input resistances (IRs) of MNs, which is known as the size principle. TASK1/3 channels are recently found to be the molecular correlates of the IR, and also found to be expressed in the masseter MNs. The orderly recruitment of masseter MNs may be modified by the activity of TASK1/3 channels. In this chapter, we discuss the synaptic mechanisms underlying the orderly recruitment of masseter MNs that occurs during the slow-closing phase, together with the mechanism for the modulation of the orderly recruitment of motor units.

Key words. isometric contraction, orderly recruitment, muscle spindle, masseter motoneuron, TASK channel

1 Recruitment of Masseter Motoneurons Caused by Ia Inputs During the Slow-Closing Phase

During the slow-closing phase of the mastication cycle, the length of the jaw-closing muscles remains almost constant. Therefore, the slow-closing phase can be regarded as the isometric contraction. It is well established that during isometric contraction, motor units are recruited depending on the order of sizes or input resistances (IRs) of motoneurons (MNs), which is known as the size principle [1]. It has also been

Y. Kang (✉), H. Toyoda, M. Saito, and H. Sato
Department of Neuroscience and Oral Physiology, Osaka University Graduate School of Dentistry, 1-8 Yamadaoka, Suita, Osaka 565-0871, Japan
e-mail: kang@dent.osaka-u.ac.jp

T. Sasano et al. (eds.), *Interface Oral Health Science 2009*,
DOI 10.1007/978-4-431-99644-6_9, © Springer 2010

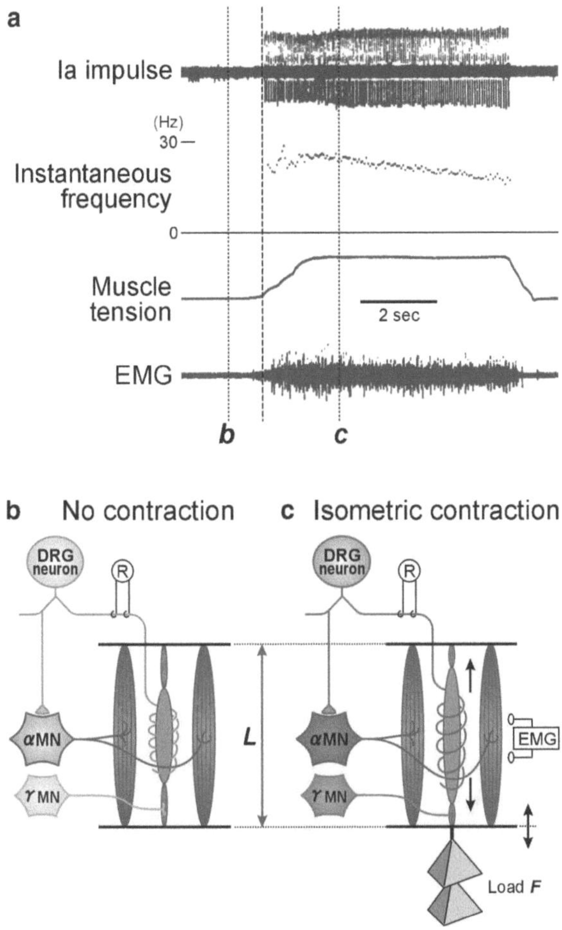

Fig. 1. α–γ linkage. (**a**) During voluntary isometric contraction of human lumbrical muscles, a constant discharge activity of spindle Ia afferent fibers was produced as soon as EMG activity and muscle tension were increased. (**b**) Resting state. (**c**) During isometric contraction, muscle length (*L*) is kept constant, indicating that γ-motoneuron (*γ-MN*) activates Ia sensory endings through the contraction of intrafusal fibers. *DRG* dorsal root ganglion (Adapted from [2])

reported that when the isometric tension of human lumbrical muscle is increased, Ia discharge is evoked by the activity of γ-MNs and is maintained constant throughout the contraction (Fig. 1, [2]). Since an activation of stretch-reflex pathway can cause the rank-ordered recruitment of motor units [3,4], it is possible that the rank-ordered recruitment of motor units during isometric contraction is at least partly caused by spindle discharges that are produced by the activity of γ-MNs (Fig. 2). Then, the activity of γ-MNs is crucial for the generation of the isometric contraction during the slow-closing phase [2]. In contrast to the role in limb movement, the role of γ-MNs is considered to be very special in jaw-closing movement because of the difference

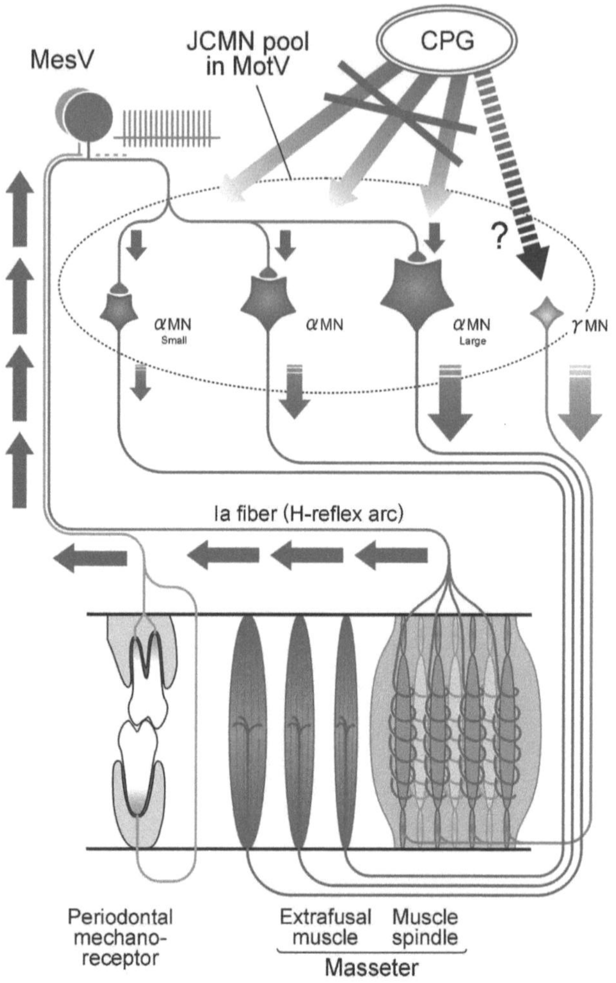

Fig. 2. A hypothesized model showing that the isometric contraction during the slow-closing phase is developed by the recruitment of masseter MNs through the activity of Ia inputs. *JCMN* jaw-closing motoneuron, *MotV* trigeminal motor nucleus, *CPG* central pattern generator

in the stretch-reflex circuit between large limb muscle and jaw-closing masseter muscle (Fig. 3). In masseter muscle, the number of intrafusal fibers included in single muscle spindle was found to be extremely large up to 36 (Fig. 3a) [5], while the number of synapses between Ia afferents and α-MNs is much smaller (Fig. 3a) [6,7], in comparison with the limb muscle (Fig. 3b) [8,9]. Thus, it is likely that in limb muscle, the spatial summation of Ia-EPSPs would easily activate α-MNs, while in masseter muscle, the temporal summation of Ia-EPSPs would be required to activate α-MNs. Therefore, it can be hypothesized that the quasi-isometric contraction during the slow-closing phase is developed by the recruitment of masseter α-MNs that can be caused by the activity of γ-MN through Ia inputs.

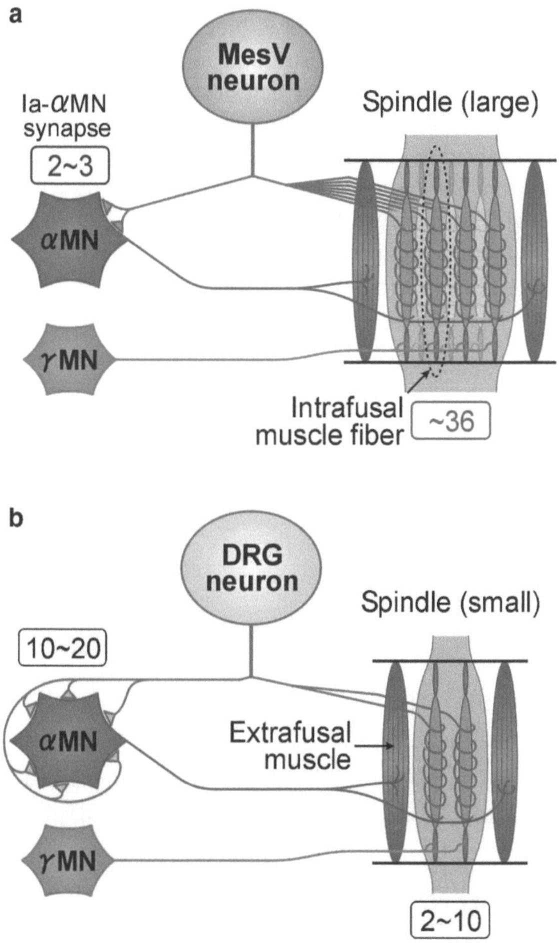

Fig. 3. (**a**, **b**) Differences in the stretch reflex circuit between masseter (**a**) and limb (**b**) muscles. A muscle spindle in masseter muscles contains many intrafusal fibers (up to 36, in human), while a spindle in limb muscles contains a few fibers. In contrast, the number of synaptic connections between a single Ia fiber and an α-motoneuron (*α-MN*) innervating masseter muscles is much smaller than that innervating limb muscles. *MesV* mesencephalic trigeminal nucleus

2 Possible involvements of leak K^+ channels, TASK1/3, in IR-ordered recruitment

What is the molecular mechanism critical for rank-order recruitment? It is known that leak K^+ currents play an essential role in determining the IR. Among several leak K^+ channels, TWIK-related acid-sensitive K^+ (TASK) channels are known to be responsible for neuronal leak K^+ currents (see reviews [10,11]). Since it is already known that TASK1 and TASK3 channels are strongly expressed in trigeminal MNs [12], it is likely that TASK channels play an important role in rank-order recruitment during the slow-closing phase. Nevertheless, it is not clear yet how differentially TASK1/3 channels are distributed in MNs depending on their sizes.

TASK channels are inhibited by many neuromodulators that activate G_q-coupled receptors and by local anesthetics and proton [10,11]. By contrast, endogenous neuromodulators activating TASK channels in neurons remained unknown, although TASK channels are activated by general anesthetics such as halothane and sevoflurane [10,11]. However, we have recently found that TASK1-like leak K^+ currents in basal forebrain cholinergic neurons were activated by nitric oxide (NO)

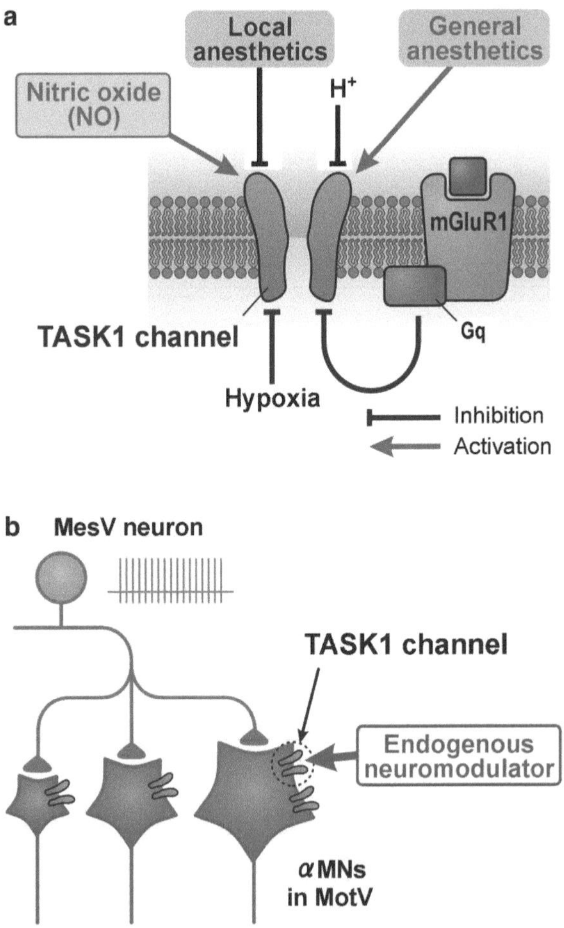

Fig. 4. (**a**, **b**) Modulation of leak K^+ (TASK) channels. (**a**) TASK1 channels are known to be inhibited by many neuromodulators in addition to local anesthetics and H^+, and activated by volatile anesthetics. We recently found that nitric oxide, one of endogenous neuromodulators, activates TASK1 channels in the basal forebrain cholinergic neurons. (**b**) TASK1 channels could be modulated by endogenous neuromodulators, affecting the order and extent of recruitment

signaling through cGMP/cGMP-dependent protein kinase (PKG) transduction pathway [13,14]. Cholinergic neurons located in the pedunculopontine and laterodorsal tegmentum nuclei and the ventromedial medullary reticular formation are known to be a source of nitrergic input to the trigeminal motor nucleus (MotV) [15,16]. Therefore, it is possible that NO can modulate TASK channels expressed in the MotV, thereby affecting the order and extent of the recruitment of masseter MNs (Fig. 4).

References

1. Henneman E (1991) The size principle and its relation to transmission failure in Ia projections to spinal motoneurons. Ann N Y Acad Sci 627:165–168
2. Vallbo AB (1970) Discharge patterns in human muscle spindle afferents during isometric voluntary contractions. Acta Physiol Scand 80:552–566
3. Bawa P, Binder MD, Ruenzel P et al (1984) Recruitment order of motoneurons in stretch reflexes is highly correlated with their axonal conduction velocity. J Neurophysiol 52:410–420
4. Calancie B, Bawa P (1985) Firing patterns of human flexor carpi radialis motor units during the stretch reflex. J Neurophysiol 53:1179–1193
5. Eriksson PO, Butler-Browne GS, Thornell LE (1994) Immunohistochemical characterization of human masseter muscle spindles. Muscle Nerve 17:31–41
6. Dessem D, Donga R, Luo P (1997) Primary- and secondary-like jaw-muscle spindle afferents have characteristic topographic distributions. J Neurophysiol 77:2925–2944
7. Yabuta NH, Yasuda K, Nagase Y et al (1996) Light microscopic observations of the contacts made between two spindle afferent types and alpha-motoneurons in the cat trigeminal motor nucleus. J Comp Neurol 374:436–450
8. Redman S, Walmsley B (1983) The time course of synaptic potentials evoked in cat spinal motoneurones at identified group Ia synapses. J Physiol 343:117–133
9. Redman S, Walmsley B (1983) Amplitude fluctuations in synaptic potentials evoked in cat spinal motoneurones at identified group Ia synapses. J Physiol 343:135–145
10. Bayliss DA, Sirois JE, Talley EM (2003) The TASK family: two-pore domain background K^+ channels. Mol Interv 3:205–219
11. Lesage F (2003) Pharmacology of neuronal background potassium channels. Neuropharmacology 44:1–7
12. Talley EM, Lei Q, Sirois JE et al (2000) TASK-1, a two-pore domain K^+ channel, is modulated by multiple neurotransmitters in motoneurons. Neuron 25:399–410
13. Kang Y, Dempo Y, Ohashi A et al (2007) Nitric oxide activates leak K^+ currents in the presumed cholinergic neuron of basal forebrain. J Neurophysiol 98:3397–3410
14. Toyoda H, Saito M, Sato H et al (2008) cGMP activates a pH-sensitive leak K^+ current in the presumed cholinergic neuron of basal forebrain. J Neurophysiol 99:2126–2133
15. Pose I, Fung S, Sampogna S et al (2005) Nitrergic innervation of trigeminal and hypoglossal motoneurons in the cat. Brain Res 1041:29–37
16. Travers JB, Yoo JE, Chandran R et al (2005) Neurotransmitter phenotypes of intermediate zone reticular formation projections to the motor trigeminal and hypoglossal nuclei in the rat. J Comp Neurol 488:28–47

Symposium III
Biomaterial Interface

Implant interface to bone tissue: biomimetic surface functionalization through nanotechnology

Ichiro Nishimura

Abstract. The living cell can interact with inorganic materials. This process is believed to regulate the behaviors of cells, as well as to generate unique hybrid structures of biological molecules and minerals such as bone. Nanotechnology has emerged with a novel promise of incorporating biological systems by developing materials and processing features in the scope of cells and biomolecules. The mechanism of biomineralization has been replicated in nanotechnology, which resulted in a new array of materials. Recently, three-dimensional bio-structures in micrometer and nanometer scales have been shown to exhibit unexpectedly robust biological responses. Although the mechanistic role of micro- and nanoscale structures in biology is not yet fully elucidated, nanotechnology-based surface functionalization has already been engineered as commercially viable products. This review will highlight the recent development of biomimetic nanotechnology in endosseous implant.

Key words. nanotechnology, surface topography, implant, bone, biomimetic technology

1 Introduction

In 1959, Richard Feynman presented his vision of a new technology at the annual meeting of the American Physical Society: "Many of the cells are very tiny, but they are very active; they manufacture various substances; they walk around; they wiggle; and they do all kinds of marvelous things – all on a very small scale" [1]. The efficient biological system has been a source of inspiration for many engineers.

I. Nishimura (✉)
The Weintraub Center for Reconstructive Biotechnology, Division of Advanced Prosthodontics, Biomaterials and Hospital Dentistry, UCLA School of Dentistry, Box 951668, CHS B3-087, Los Angeles, CA 90095-1668, USA
e-mail: inishimura@dentistry.ucla.edu

T. Sasano et al. (eds.), *Interface Oral Health Science 2009*,
DOI 10.1007/978-4-431-99644-6_10,

From the inception, rapidly emerging nanotechnology has always captured a novel promise of mimicking a wide range of biological systems.

To date, the process of biomineralization has been frequently mimicked to create nano-scale materials. The biomineralization process generally involves the synthesis of protein-based organic matrix, which facilitates the scaffold for inorganic growth and nucleation [2]. Although silver ions have been known for antibacterial activity, some microorganisms exhibit resistance against the noble metals. Silver-resistant *Pseudomonas stutzeri* has been shown to incorporate silver ions and organize nanoparticles within the cell wall [3]. This process was modulated by putative peptides, with a preferential enrichment of proline and hydroxyl-containing amino acid residues, suggesting that the organic matrix may guide the crystalline patterns of inorganic nanoparticles.

2 Bone: A Hybrid Structure of Organic Matrix and Crystalline Hydroxyapatite

The major component of bone tissue is collagenous and noncollagenous extracellular matrix (ECM) and inorganic crystalline hydroxyapatite (HA). Bone biomineralization utilizes the peptide–organic ion interaction. The nucleation, growth, and development of mineral crystals occur in the ECM (Fig. 1a) through the inorganic ion interaction primarily within the collagen fibers [4]. Other ECM molecules also participate in the biomineralization process for regulating the size, shape, and orientation of the crystalline HA [5]. The mineralized bone surface provides the critical homing site for osteoblasts and osteoclasts.

When the substrate materials are soaked in acellular aqueous solutions with different ion concentrations and pH, crystalline HA mimicking bone biomineralization can be precipitated [6]. This method omits the prerequisite ECM molecules; however, crystalline HA grown on synthetic materials has been shown to stimulate osteoblast viability and function [7–9]. Recently, a new material composed of reconstituted type I collagen and biomimetically precipitated nanocrystalline HA has been developed. When human bone marrow-derived stromal cells were cultured on this collagen-HA membrane, osteoblastic differentiation was promoted [10].

We have investigated the surface of calivarial bone harvested from type IX collagen null mutant mice, postulating that collagen IX may have a role in bone matrix organization. Numerous well-like structures were found on the surface of normal mouse bone (Fig. 1b), while the bone surface of collagen IX null mutant mouse showed relatively smooth topography (Fig. 1c). Further investigations revealed that bone resorbing osteoclasts widely spread over the mutant bone surface lacking the micro-topography, resulting in increased bone resorption and the progressive development of a severe form of age-dependent osteoporosis [11]. It is highly conceivable that this hybrid structure of bone organic matrix and crystalline HA may have a significant biological effect in bone generation and regeneration.

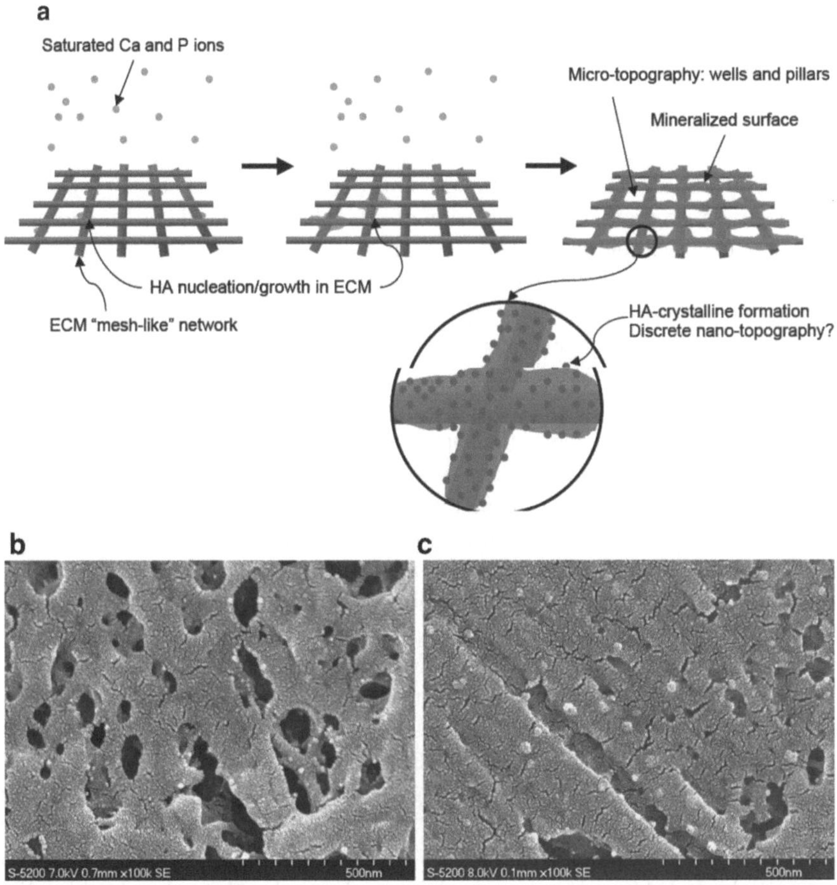

Fig. 1. (**a**) Schematic diagram of bone ECM biomineralization. The scaffold structure determined by ECM network plays a critical role in determining the micro-topography of bone. Furthermore, HA-crystalline formation gives rise to the discrete nano-topography. Both micro-topography and discrete nano-topography may contribute to the regulatory mechanism of bone remodeling. (**b**) Normal (wild-type) mouse bone surface exhibited the characteristic wells and pillars. (**c**) The reduced micro-topography in mutant mouse bone exhibited much smoother surface, which was found to be more susceptible to osteoclast bone resorption. (Reproduced from [11] with permission of the American Society for Bone and Mineral Research)

3 Biomimetic Surface Functionalization

The roughened implant surface with isotorphic micro-topography has been shown to improve implant and bone integration (osseointegration) [12–14], percutaneous implant integration [15], and breast implant tissue integration [16]. These improvements are due to increased adhesion of connective tissue cells onto roughened surfaces. For example, an acid-treated titanium implant exhibits roughened surface topography, which contributes to the molecular and cellular reaction by the wound healing tissue, leading to advantageous bone-implant integration [17].

It was noted that the surface of mineralized bone showed discrete nanoscale topography of 10–20 nm possibly composed of crystalline HA [11, 18]. Carbon nanofibers whose aspect ratio and physical dimensions are similar to that of crystalline HA selectively promote cellular adhesion of osteoblasts in vitro [19]. Furthermore, nanophase ceramics and titania with 100 nm grain sizes have also been associated with increased in vitro adhesion of osteoblasts [20] and chondrocytes [21] as well as increased function of osteoclasts [22]. Although future studies are necessary, the discrete nano-topography generated by HA-crystalline formation on bone surface may play a significant biological role.

To further improve the degree of osseointegration, we tested our biomimetic surface functionalization concept: a combination of micro-topography and discrete nano-topography with HA-crystalline formation. The use of biomimetic processing did not seem practical because of the lack of prior art on the protein-based biomineralization for the titanium substrate. Instead, we designed the sol–gel-based nanoparticle application. Titanium substrates were treated with 3-amino-propyltriethoxysilane, to which HA nanoparticles (20 nm) were deposited and chemically bonded to TiO_2. The HA deposition rate was linearly related to the treatment time, and HA nanoparticles were deposited up to 50% of the substrate surface (Fig. 2a). As the result, the discrete deposition of HA nanoparticles generated novel 20–40 nm nano-topography to Ti substrate with smooth (turned) or roughened by double acid etching (DAE) micro-topography (Fig. 2b, c) [23]. This simple and versatile nanotechnology-based surface modification exhibited a unique discrete nanoscale topography resembling the mineralized bone surface (Fig. 2c) and showed unexpectedly robust biological effect. The deposition of HA nanoparticles to DAE surface increased the mechanical withstanding load for 129 and 782% as compared to the control DAE and turned implants, respectively. This technology has been commercialized rather quickly and started serving the dental community and patients. Furthermore, the new implant can provide a novel clue to understand the biological mechanism of bone remodeling and metabolic bone diseases.

4 Conclusions

Biomimetic nanotechnology has been developed in two general areas: [1] material processing mimics biological mechanisms and [2] the end product mimics biological tissues. Both approaches appear to give rise to novel materials that have distinct benefits for biomedical applications. Elucidation of biological mechanisms is far from complete. The evaluation of nanotechnology-based biomaterials may also provide unique opportunities for understanding the cellular and molecular mechanisms.

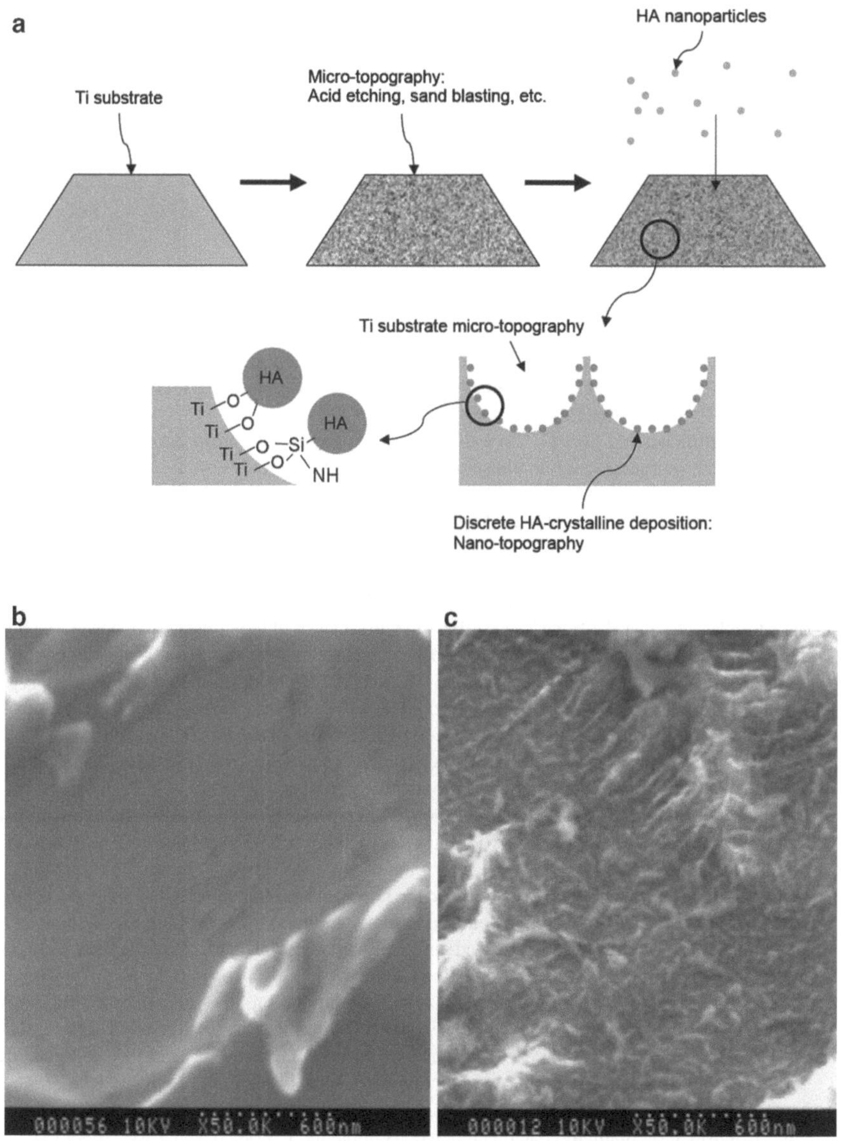

Fig. 2. (**a**) An application of nanotechnology for HA nanoparticle deposition on the surface of titanium substrate with predisposing micro-topography. The resultant surface exhibited biomimetic topography: a combined micro-topography and discrete nano-topography. This new titanium implant showed unexpectedly robust biological response that osseointegration was significantly enhanced. (Reproduced from [24] with permission of Quintessence Publishing Co Inc.) (**b**) Field emission SEM images of the titanium implant surface treated with double acid etching. (**c**) Double acid etching treated titanium implant was further modified with HA nanoparticles, which generated a novel discrete nano-topography resembling bone surface. (Reproduced from [23] with permission of IOP Publishing Ltd.)

Acknowledgments The studies presented were supported in part by NIH/NIDCR R01 DE10870, NIH/NIA UCLA Claude Pepper Center, the NIH/NIDCR SBIR program (R43 DE14927), Biomet 3i and Sumitomo Chemical Corp. This investigation was conducted in part in a facility constructed with the support from Research Facilities Improvement Program NIH/NCRR C06 RR014529.

References

1. Drexler KE (1992) Nanosystems: molecular machinery, manufacturing, and computation. Wiley, Hoboken, NJ
2. Lowenstam HA (1981) Minerals formed by organisms. Science 211:1126–1131
3. Klaus T, Joerger R, Olsson E et al (1999) Silver-based crystalline nanoparticles, microbially fabricated. Proc Natl Acad Sci USA 96:13611–13614
4. Lee DD, Glimcher MJ (1991) Three-dimensional spatial relationship between the collagen fibrils and the inorganic calcium phosphate crystals of pickerel (*Americanus americanus*) and herring (*Clupea harengus*) bone. J Mol Biol 217:487–501
5. Wiesmann HP, Meyer U, Plate U et al (2005) Aspects of collagen mineralization in hard tissue formation. Int Rev Cytol 242:121–156
6. Kokubo T, Ito S, Huang ZT et al (1990) Ca, P-rich layer formed on high-strength bioactive glass-ceramic A-W. J Biomed Mater Res 24:331–343
7. Matsuoka H, Akiyama H, Okada Y et al (1999) In vitro analysis of the stimulation of bone formation by highly bioactive apatite- and wollastonite-containing glass-ceramic: released calcium ions promote osteogenic differentiation in osteoblastic ROS17/2.8 cells. J Biomed Mater Res 47:176–188
8. Chou YF, Huang W, Dunn JC et al (2005) The effect of biomimetic apatite structure on osteoblast viability, proliferation, and gene expression. Biomaterials 26:285–295
9. Oh SH, Finones RR, Daraio C et al (2005) Growth of nano-scale hydroxyapatite using chemically treated titanium oxide nanotubes. Biomaterials 26:4938–4943
10. Bernhardt A, Lode A, Boxberger S et al (2008) Mineralised collagen-an artificial, extracellular bone matrix-improves osteogenic differentiation of bone marrow stromal cells. J Mater Sci Mater Med 19:269–275
11. Wang CJ, Iida K, Egusa H et al (2008) Trabecular bone deterioration in col9a1+/- mice associated with enlarged osteoclasts adhered to collagen IX-deficient bone. J Bone Miner Res 23:837–849
12. Thomas KA, Cook SD (1985) An evaluation of variables influencing implant fixation by direct bone apposition. J Biomed Mater Res 19:875–901
13. Bowers KT, Keller JC, Randolph BA et al (1992) Optimization of surface micromorphology for enhanced osteoblast responses in vitro. Int J Oral Maxillofac Implants 7:302–310
14. Qu J, Chehroudi B, Brunette DM (1996) The use of micromachined surfaces to investigate the cell behavioural factors essential to osseointegration. Oral Dis 2:102–115
15. Chehroudi B, Gould TR, Brunette DM (1992) The role of connective tissue in inhibiting epithelial downgrowth on titanium-coated percutaneous implants. J Biomed Mater Res 26:493–515
16. Barone FE, Perry L, Keller T et al (1992) The biomechanical and histopathologic effects of surface texturing with silicone and polyurethane in tissue implantation and expansion. Plast Reconstr Surg 90:77–86
17. Ogawa T, Nishimura I (2006) Genes differentially expressed in titanium implant healing. J Dent Res 85:566–570
18. Palin E, Liu H, Webster TJ (2005) Mimicking the nanofeatures of bone increases bone-forming cell adhesion and proliferation. Nanotechnology 16:1828–1835
19. Price RL, Ellison K, Haberstroh KM et al (2004) Nanometer surface roughness increases select osteoblast adhesion on carbon nanofiber compacts. J Biomed Mater Res A 70:129–138
20. Webster TJ, Ergun C, Doremus RH et al (2000) Specific proteins mediate enhanced osteoblast adhesion on nanophase ceramics. J Biomed Mater Res 51:475–483

21. Kay S, Thapa A, Haberstroh KM et al (2002) Nanostructured polymer/nanophase ceramic composites enhance osteoblast and chondrocyte adhesion. Tissue Eng 8:753–761
22. Webster TJ, Ergun C, Doremus RH et al (2001) Enhanced osteoclast-like cell functions on nanophase ceramics. Biomaterials 22:1327–1333
23. Nishimura I, Huang Y, Butz F et al (2007) Discrete deposition of hydroxyapatite nanoparticles on titanium implant with predisposing substrate microtopography accelerated osseointegration. Nanotechnology 18:245101 (9 pp)
24. Lin A, Wang CJ, Kelly J et al (2009) The role of titanium implant surface modification with hydroxyapatite nanoparticles in progressive early bone-implant fixation in vivo. Int J Oral Maxillofac Implants 24:808-816

Interface affinity between apatites and biological tissues

Masayuki Okazaki

Abstract. To develop a new biodegradable scaffold biomaterial, synthesized CO_3Ap was mixed with neutralized collagen gel and lyophilized into sponges. X-ray diffraction and FT-IR analyses, together with chemical analysis, indicated that synthesized CO_3Ap had crystallinity and a chemical composition similar to bone. SEM observation showed that the CO_3Ap-collagen sponge had a suitable pore size for cell invasion. When these sponge-frame complexes with rh-BMP2 were implanted beneath the periosteum cranii of rats, sufficient new bone was created at the surface of the periosteum cranii after 4 weeks' implantation. Furthermore, when a CO_3Ap-collagen sponge containing the SVVYGLR peptide was implanted into a tissue defect created in a rat tibia, the migration of numerous vascular endothelial cells, as well as prominent angiogenesis inside the graft, could be detected after 1 week. These CO_3Ap-collagen sponges with highly functional modifications are expected to be used as hard-tissue scaffold biomaterials for the therapeutic purpose of rapid healing.

Key words. interface affinity, apatites, biological tissues, BMP, angiogenesis

1 Introduction

Biological apatites contain several %(w/w) CO_3^{2-} ions. For many years, it was speculated that bone mineral is composed of calcium phosphate and calcium carbonate $CaCO_3$. However, LeGeros clarified that the inorganic composition of hard tissues, such as bone and teeth, is based on carbonate apatites [1]. In general, CO_3^{2-} ions can substitute into both the PO_4^{3-} position and OH^- position. It is said that apatite synthesized in an aqueous system contains CO_3^{2-} ions in the PO_4^{3-} position. When ions substitute into these positions, the *a*-axis dimension decreases. The crystallinity of CO_3apatites decreases with increases in CO_3^{2-} content, while the solubility increases [2].

M. Okazaki
Department of Biomaterials Science, Graduate School of Biomedical Sciences, Hiroshima University, Hiroshima 734-8553, Japan
e-mail: okazakix@hiroshima-u.ac.jp

T. Sasano et al. (eds.), *Interface Oral Health Science 2009*,
DOI 10.1007/978-4-431-99644-6_11, © Springer 2010

Hydroxyapatite and CO_3apatites with different carbonate contents were synthesized [3], mixed with atelocollagen, and made into sponge scaffolds [4]. The scaffolds were implanted into the femur bone sockets of male New Zealand white rabbits. Histological observation suggested that a CO_3Ap-collagen scaffold with carbonate content similar to that of human bone had optimal bone formation ability.

2 New Concept for Biological Adhesion

Classical adhesion theory, the so-called "wettability," has helped to explain adhesion of general materials using the concepts of hydrophilic and hydrophobic properties. However, biological adhesion cannot be explained only by classical adhesion theory. Cell adhesion starts from adhesion of a protein, followed by protein exchange, and finally, cell adhesion receptors target molecules or ligands of materials. Here, we would like to introduce "interface affinity" instead of "wettability" as a new concept for biological adhesion.

The interface affinity (wettability) is affected by morphology, surface mobility, surface composition, electrical charge, etc. [5] (Fig. 1). Since hydroxyapatite is an ionic crystal, the interface affinity is strongly related to the surface composition and electrical charge. The carbonate content may affect osteoblast behavior.

3 Effect of Mg^{2+} Ions on Bone Formation

Recently, adhesion molecules such as those of the integrin family were examined in terms of cell structure and function. Divalent ions affect cell adhesion in relation to the integrin molecule as an adhesion molecule at the cell surface. It has been

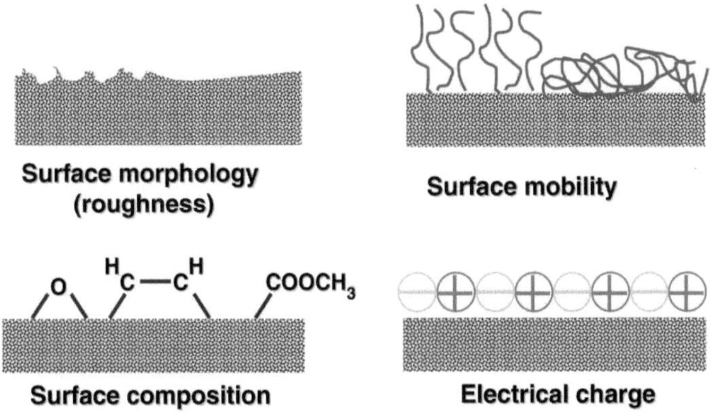

Fig. 1. Interface affinity affected by various surface properties of biomaterials

reported, especially, that Zn^{2+} and Mg^{2+} ions promote cell adhesion [6]. Integrins are crucially important receptor proteins because they are the main way through which the cells both bind and respond to the extracellular matrix. They are composed of two noncovalently associated transmembrane glycoprotein subunits called α and β, both of which contribute to the binding of the matrix protein (Fig. 2). The binding of integrins to their ligands depends on extracellular divalent cation, reflecting the presence of three or four divalent-cation-binding domains in the large extracellular part of the α chain [7].

Mg^{2+} ions also play some roles in cell adhesion. Thus, magnesium seems to be an important factor even in controlling in vivo bone metabolism since it plays a part in both bone formation and resorption [8]. Mg^{2+} ions may contribute to the bone metabolism of osteoclasts and osteoblasts' action with integrins at their cell surfaces.

Recently, scaffold biomaterials have become the focus in tissue engineering field. In a continuation of those studies, functionally graded CO_3apatite containing Mg, $FGMgCO_3Ap$, producing a negative gradient of magnesium concentration from the surface toward the core, was synthesized [9].

Four weeks after implantation into rabbit femurs, both the $FGMgCO_3Ap$-collagen composite (Fig. 3) [10] and the CO_3Ap-collagen composite showed clear bone formation, although the control hole with no implantation appeared to also have been covered with a thinner layer of new bone. The bone density of the $FGMgCO_3Ap$-collagen composite was higher than that of the CO_3Ap-collagen composite.

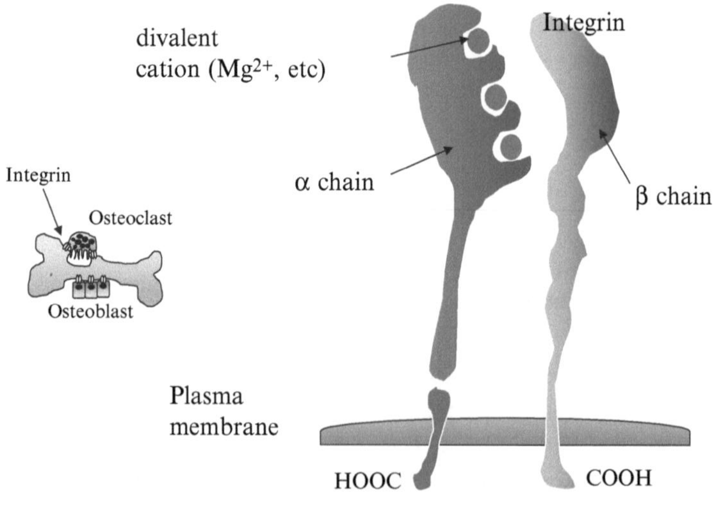

Fig. 2. Integrin with two noncovalently associated transmembrane glycoprotein subunits

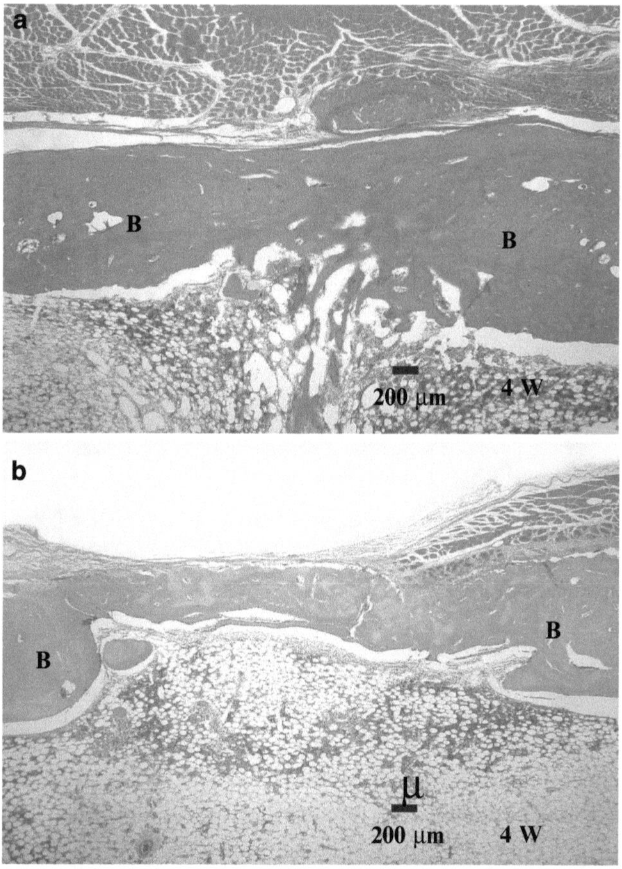

Fig. 3. Bone formation of FGMgCO_3Ap-collagen composite (**a**) and control (**b**) after 4 weeks of implantation into rabbit femurs

4 BMP Modification

Cytokines, such as BMP2 [11] and BMP7, have been successfully used to create and promote new bone growth using fixing biomaterials [12]. We investigated the acceleration of bone formation with rh-BMP2 using frame-reinforced CO_3Ap-collagen sponge scaffolds. To develop a new biodegradable scaffold biomaterial reinforced with a frame, synthesized CO_3Ap was mixed with neutralized collagen gel, and the CO_3Ap-collagen mixtures were lyophilized into sponges in a porous HAp-frame ring. X-ray diffraction and FT-IR analyses, together with chemical analysis, indicated that synthesized CO_3Ap had crystallinity and a chemical composition similar to bone. SEM observation showed that the CO_3Ap-collagen sponge had a suitable pore size for cell invasion. In proliferation and differentiation experiments with osteoblasts, ALP and OPN activity was clearly detected.

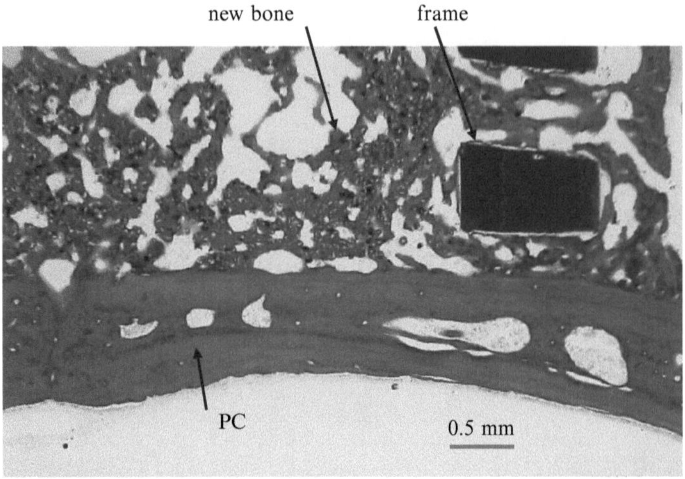

Fig. 4. Bone formation with a reinforced CO_3Ap-collagen sponge with rh-BMP after 4 weeks of implantation beneath the periosteum cranii of rats

When these sponge-frame complexes with rh-BMP2 were implanted beneath the periosteum cranii of rats, adequate new bone was created at the surface of the periosteum cranii after 4 weeks' implantation (Fig. 4) [13]. These reinforced CO_3Ap-collagen sponges with rh-BMP2 are expected to be used as hard-tissue scaffold biomaterials for the therapeutic purpose of rapid healing.

5 SVVYGLR Modification

Regeneration of the vessels that supply oxygen and nutrients to cells is essential to allow defective tissues to regenerate and biomaterials to engraft and express sufficient function. Blood vessels form a crucial lifeline for the maintenance and growth of bone in addition to providing hybrid functions to hard-tissue scaffold materials.

Recently, the novel binding sequence Ser-Val-Val-Tyr-Gly-Leu-Arg (SVVYGLR) has been identified as an amino acid sequence in osteopontin (OPN) that is involved in angiogenesis [14, 15]. This motif might be important in pathological conditions, as SVVYGLR is adjacent to the RGD sequence in OPN and is exposed by thrombin cleavage.

To modify angiogenesis properties of CO_3Ap-collagen sponges, OPN-derived peptide SVVYGLR was synthesized with high purity [16]. When the CO_3Ap-collagen sponges with the synthetic motif SVVYGLR peptide were implanted beneath the back skin of the rat, new blood vessels were dramatically induced in

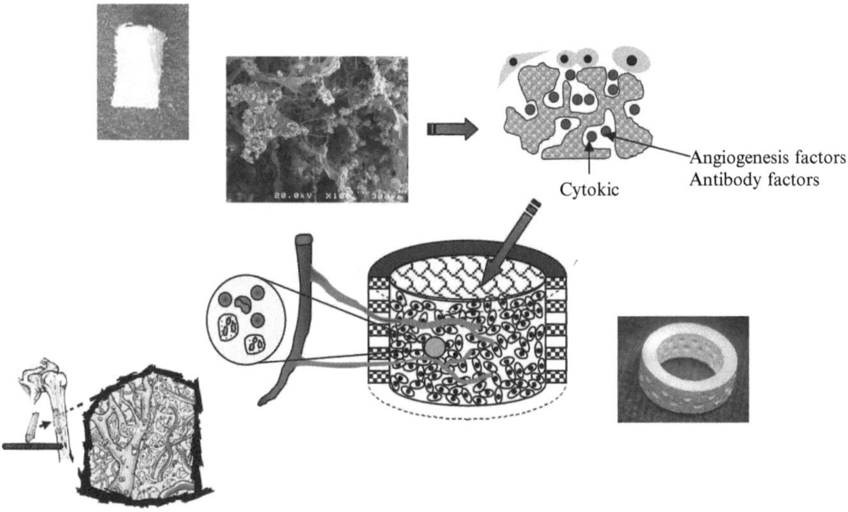

Fig. 5. Schema of highly functional CO_3Ap-collagen scaffold biomaterial

1 week [17], while no blood vessels were observed in the control sponge without SVVYGLR. SMA staining also indicated that smooth muscular actin of blood vessel was stained red for the sponge with SVVYGLR. These results suggest that the CO_3Ap-collagen sponge, combined with SVVYGLR, is a useful high-quality scaffold biomaterial that contributes to angiogenesis and bone formation.

6 Summary

CO_3Ap-collagen scaffold biomaterials have potentially useful therapeutic applications. They can be utilized as a rapid healing biomaterial by emphasizing the interface affinity with the adhesion motif and cytokines such as growth factor BMP or angiogenesis factor SVVYGLR (Fig. 5).

References

1. LeGeros RZ (1967) Apatite crystallites: effects of carbonate on morphology. Science 155:1409–1411
2. Okazaki M, Moriwaki Y, Aoba T, Doi Y et al (1981) Solubility behavior of CO_3apatites in relation to crystallinity. Caries Res 15:477–483
3. Yokota R, Hayashi H, Hirata I et al (2006) Detailed consideration of physicochemical properties of CO_3apatites as biomaterials in relation to carbonate content using ICP, X-ray diffraction, FT-IR, SEM and HR-TEM. Dent Mater J 25:597–603

4. Matsuura A, Kubo T, Doi K et al (2009) Bone formation ability of carbonate apatite-collagen scaffolds with different carbonate contents. Dent Mater J 28:234–242
5. Fuse Y, Hirata I, Kurihara H et al (2007) Cell adhesion and proliferation patterns on mixed self-assembled monolayers carrying various ratios of hydroxyl and methyl groups. Dent Mater J 26:814–819
6. Lange TS, Bielinsky AK, Kirchberg K et al (1994) Mg^{2+} and Ca^{2+} differentially regulate β_1 integrin-mediated adhesion of dermal fibroblasts and keratinocytes to various extracellular matrix proteins. Exp Cell Res 214:381–388
7. Albert B, Bray D, Lewis J et al (1994) Molecular biology of the cell, 3rd edn. Garland Publishing, New York, pp 949–1010
8. Serre CM, Papillard M, Chavassieux P et al (1998) Influence of magnesium substitution on a collagen-apatite biomaterial on the production of a calcifying matrix by human osteoblasts. J Biomed Mater Res 42:626–633
9. Yamasaki Y, Yoshida Y, Okazaki M et al (2002) Synthesis of functionally graded $MgCO_3$apatite accelerating osteoblast adhesion. J Biomed Mater Res 62:99–105
10. Yamasaki Y, Yoshida Y, Okazaki M et al (2003) Action of $FGMgCO_3Ap$-collagen composite in promoting bone formation. Biomaterials 24:4913–4920
11. Urist MR (1965) Bone: formation by autoinduction. Science 150:893–899
12. Service RF (2000) Bone remodeling and repair (News) – tissue engineerings build new bone. Science 289:1498–1500
13. Hirata I, Nomura Y, Ito M et al (2007) Acceleration of bone formation with BMP2 in frame-reinforced carbonate apatite-collagen sponge scaffolds. J Artif Organs 10:212–217
14. Yokosaki Y, Matsuura N, Sasaki T et al (1999) The integrin a9b1 bind to a novel recognition sequence (SVVYGLR) in the thrombin-cleaved amino-terminal fragment of osteopontin. J Biol Chem 274:36328–36334
15. Hamada Y, Nokihara K, Okazaki M et al (2003) Angiogenic activity of osteopontin- derived peptide SVVYGLR. Biochem Biophys Res Commun 310:153–157
16. Hamada Y, Yuki K, Okazaki M et al (2004) Osetpontin-derived peptide SVVYGLR induces angiogenesis in vivo. Dent Mater J 23:650–655
17. Hamada Y, Egusa H, Kaneda Y et al (2007) Synthetic osteopontin-derived peptide SVVYGLR can induce neovascularization in artificial bone marrow scaffold biomaterials. Dent Mater J 26:487–492

Biological reactions on titanium surface electrodeposited biofunctional molecules

Takao Hanawa

Abstract. Surface modification is an important and predominant technique for obtaining biofunction and biocompatibility in metals for biomedical use including dentistry. One approach is the immobilization of biofunctional molecules on the metal surface to control the adsorption of proteins and adhesion of cells, platelets, and bacteria. In particular, the immobilization of poly(ethylene glycol) (PEG) to a titanium surface with electrodeposition are effective to inhibit the adsorption of proteins, adhesion of platelet, and formation of biofilm. This technique is applied to conventional metals and biomolecules that have electric charges. On the other hand, when the peptides, which accelerate cell adhesion, are immobilized to titanium through PEG electrodeposited, bone formation and soft tissue adhesion may be improved.

Key words. titanium, PEG, peptide, electrodeposition, biofunction

1 Introduction

Abrupt technological evolution on ceramics and polymers make it possible to apply these materials to medical devices the last three decades. In particular, excellent biocompatibility and biofunction of ceramics and polymers are expected to show excellent properties as biomaterials; in fact, many devices consisting of metals have been substituted by those consisting of ceramics and polymers. In spite of this event, over 70% of implant devices in medical field, including dentistry, still consist of metals, and this share is currently maintained because of their high strength, toughness, and durability. Metallic biomaterials cannot be replaced with ceramics or polymers at present.

A disadvantage of using metals as biomaterials is that they are typically artificial materials and have no biofunction. To add biofunction to metals, surface modification is necessary because biofunction cannot be added during manufacturing processes

---

T. Hanawa
Institute of Biomaterials and Bioengineering, Tokyo Medical and Dental University, 2-3-10 Kanda-surugadai, Chiyoda-ku, Tokyo 101-0062, Japan
e-mail: hanawa.met@tmd.ac.jp

T. Sasano et al. (eds.), *Interface Oral Health Science 2009*,
DOI 10.1007/978-4-431-99644-6_12, © Springer 2010

such as melting, casting, forging, and heat treatment. Surface modification is a process that changes a material's surface composition, structure, and morphology, leaving the bulk mechanical properties intact. In addition, metals with biofunctions have been required in the recent past. For example, stents are placed at stenotic blood vessels for dilatation, and blood compatibility or prevention of adhesion of platelets is necessary. In guide wires and guiding catheters, lubrication in the blood vessels is important for proper sliding and insertion. If metals are used as sensing devices, the control of cell adhesion is necessary. Infection due to biofilm formation on implant devices must be inhibited. For these purposes, the fundamental property is to control the adsorption of proteins, and adhesion of cells, platelets, and bacteria.

When a metallic material is implanted into a human body, immediate reaction occurs between its surface and the living tissues. In other words, immediate reaction at this initial stage straightaway determines and defines a metallic material's tissue compatibility. With surface modification, tissue compatibility of surface layer could be improved. For these purposes, many techniques for surface modification of metals are attempted on a research stage and some of them are commercialized.

In this chapter, biofunctionalization of metals using biofunctional molecules are reviewed. In particular, advantages of electrodepostion of functional molecules to titanium surface are demonstrated.

2 Immobilization of PEG to Metals with Electrodeposition

The immobilization of biofunctional polymers on noble metals such as gold is usually conducted by using the bonding –SH or –SS– group; however, this technique can only be used for noble metals. The adhesion of platelets and adsorption of proteins, peptides, antibodies, and DNA is controlled by modifications of the above technique. On the other hand, poly(ethylene glycol) (PEG) is a biofuctional molecule on which adsorption of proteins is inhibited. Therefore, immobilization of PEG to metal surface is an important event to biofunctionalize the metal surface. No successful one-stage technique for the immobilization of PEG to base metals has ever been developed. In this section, immobilization of PEG, which modified both terminals or one terminal with amine bases onto titanium surface using electrodeposition will be explained.

Both terminals of PEG were terminated with $–NH_2$ (B-PEG; PEG1000 Diamine, NOF Corporation, Japan), and only one terminal was terminated with $–NH_2$ (O-PEG; SUNBRIGHT MEPA-10H, NOF Corporation, Japan). The chemical structures of the PEGs are shown in Fig. 1. The molecular weights of all PEGs were about 1,000. These terminated PEGs were dissolved in a 0.3-mol L^{-1} NaCl solution with a concentration of 2 mass%. In the solution, the $–NH_2$ terminal was dissociated and charged as $–NH_3^+$. The pH of the solution with B-PEG was 11.2 and that of the solution with O-PEG was 11.0. The resultant solution was used as an electrolyte for electrodeposition at 310 K. A commercially pure titanium disk with grade 2 was metallographically polished and ultrasonically rinsed in acetone and deionized

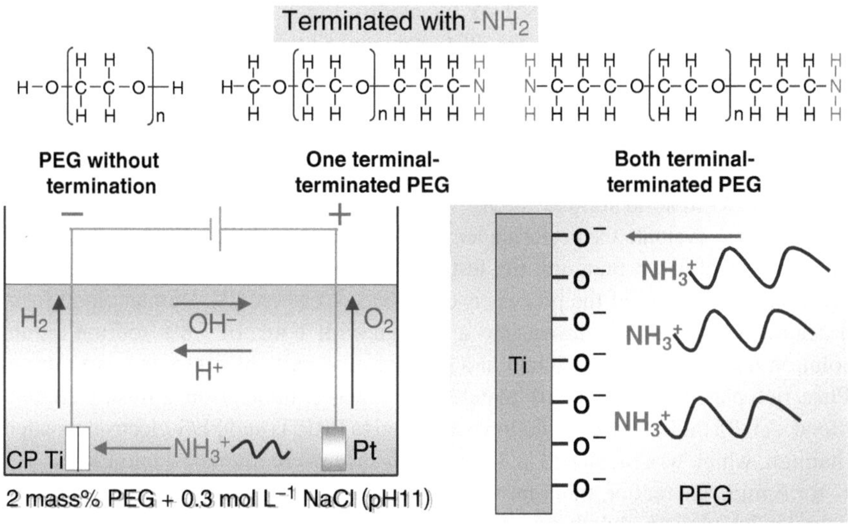

Fig. 1. PEG molecules were terminated with amine bases at one terminal or both terminals. Amine bases dissociate and are positively charged in aqueous solution and electrically attracted to titanium surface with cathodic charge, and eventually PEG molecules are immobilized

water. The titanium disk was fixed in a polytetrafluoroethylene holder that was insulated from the electrolyte except for an opening made for electrodeposition. The cathodic potential was charged from open circuit potential to −0.5 V vs. SCE with a sweep rate of 0.1 $V s^{-1}$ and maintained at this potential for 300 s. During charging, the terminated PEGs were electrically migrated to the titanium cathode and deposited on it as shown in Fig. 1. For comparison, titanium was immersed in the electrolyte containing B-PEG for 2 and 24 h without any electric charge at 310 K. After electrodeposition, specimens were rinsed in deionized water and dried with a stream of nitrogen gas (99.9%).

The thicknesses of the PEG deposition layers, in other words, the amount of deposited PEG, is the largest in this order: 24 h-immersion B-PEG, electrodeposition of B-PEG for 300 s, electrodeposition of O-PEG for 300 s, and 2 h-immersion B-PEG. This indicated that electrodeposition was more effective than immersion for the deposition of PEG on the titanium surface. However, the PEG layer increased after a 24-h immersion, indicating that the charged terminals of PEG attracted electrostatically titanium surface that is covered by titanium oxide with a large number of hydroxyl groups.

The bonding manner of PEG to titanium surface is significant to design PEG-immobilized materials, while characterization techniques for the determination of immobilization manner of PEG are little. Immobilization manner of PEG was characterized using X-ray photoelectron spectroscopy (XPS) with angle-resolved technique and glow discharge optical emission spectroscopy (GD-OES). As a result, not only electrodeposition, but also immersion led to the immobilization of PEG onto titanium surface. However, more terminated amines combined with titanium oxide as an ionic NH–O by electrodeposition, while more amines randomly existed as NH_3^+ in the PEG molecule by immersion. Moreover, the difference of amine termination led to different bonding manner, U-shape in PEG-terminated both terminals, and brush in PEG-terminated one terminal. Schematic illustration of immobilization manners of PEG molecules are shown in Fig. 2. Characterization with XPS and GD-OES is useful to determine immobilization mode of PEG to solid surface [1, 2].

In order to evaluate the performance for the inhibition of the adhesion of platelets in PEG-immobilized titanium, the test was conducted according to the following procedure; the detail of the process is described elsewhere [3]. Human blood from a healthy volunteer was drawn into a syringe with 1 mL of 3.8% sodium citrate solution used as an anticoagulant at a ratio of nine parts blood to one part citrate. Plate-rich plasma (PRP), 1×10^5 platelets μL^{-1}, was obtained from a freshly citrated blood. A 0.25 mol L^{-1} $CaCl_2$ solution was added to PRP. Ti and PEG-electrodeposited titanium, which was incubated at 310 K in advance, were immersed into PRP at 310 K for 5 min. Thereafter, titanium was rinsed with PBS(−), fixed with 2% glutaraldehyde, dehydrated, and observed through a scanning electron microscope. Platelet adhesion is inhibited on PEG-electrodeposited titanium surface (Fig. 3a), while platelets adhered on untreated Ti surface and fibrin network is formed on it (Fig. 3b). Bacteria (*Streptococcus mutans* MT8148) adhered to an untreated titanium surface (Fig. 4a), while bacterial adhesion was inhibited on a PEG-electrodeposited titanium surface (Fig. 4b).

This electrodeposition technique is applied to other conventional metals and biomolecules, which have electric charges. In addition, this technique is useful for complex surface morphology.

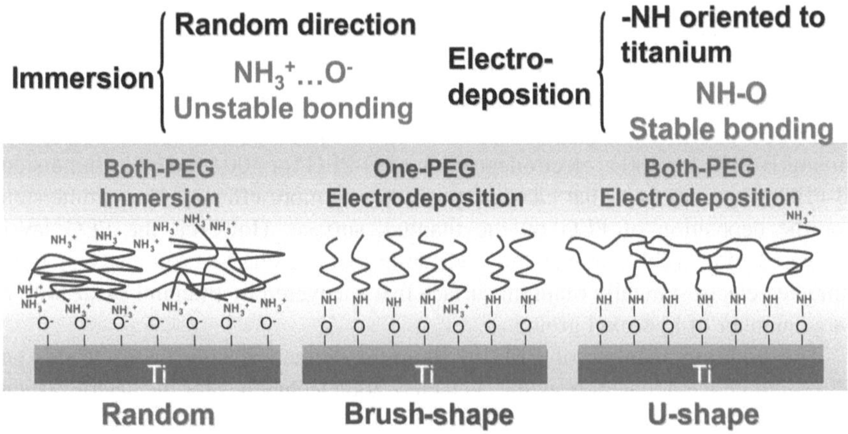

Fig. 2. Schematic model of immobilized manners of PEG to titanium surface with immersion and electrodeposition

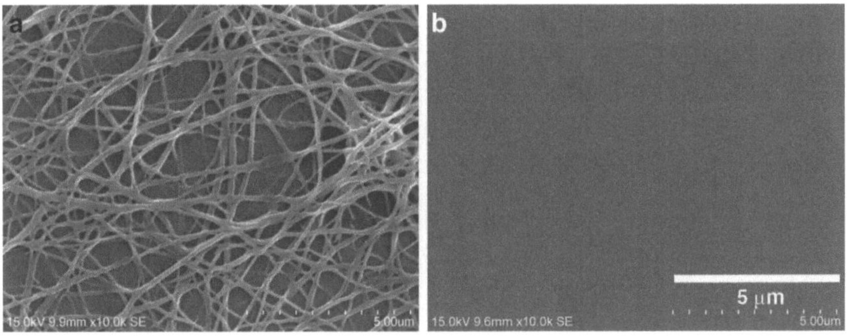

Fig. 3. Platelets adhered on untreated Ti surface and fibrin network is formed on it (**a**), while platelet adhesion is inhibited on PEG-electrodeposited Ti surface (**b**)

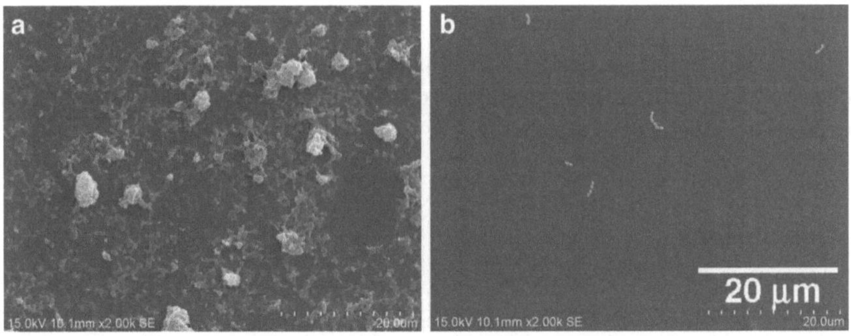

Fig. 4. Bacteria (*S. mutans* MT8148) adhered to an untreated Ti surface (**a**), while bacterial adhesion was inhibited on a PEG-electrodeposited Ti surface (**b**)

3 Immobilization of Biomolecules

Peptides containing Arg-Gly-Asp (RGD) accelerate cell attachment and extension of bone cells on Ti [4]. RGD is a peptide known to involve cell adhesion, which is involved in many extracellular matrix proteins [5]. Bone formation is accelerated by immobilizing RGD on a Ti surface [6]. Peptides with terminal cysteine residues were immobilized on maleimide-activated oxides [7–9].

To immobilize RGD to the electrodeposited PEG on Ti, PEG with an $-NH_2$ group and a –COOH group (NH_2–PEG–COOH) must be employed. One terminal group, $-NH_2$, is required to bind stably with a surface oxide on a metal. On the other hand, the other terminal group, –COOH, is useful to bond biofunctional molecules, such as RGD, [10] as shown in Fig. 5. This RGD/PEG/Ti surface accelerated calcification by MC3T3-E1 cell [11].

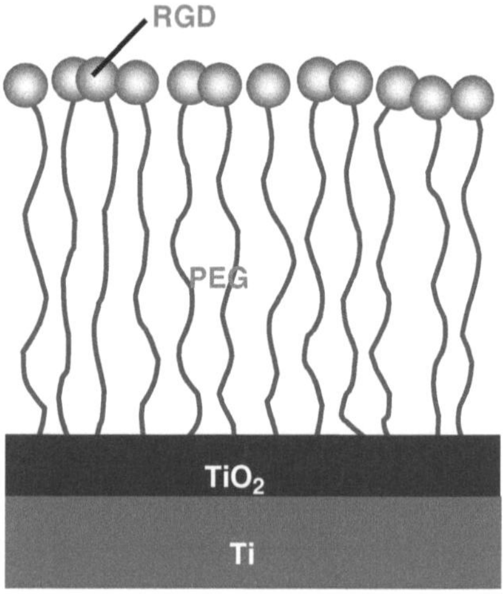

Fig. 5. Schematic structure of RGD immobilization on titanium through PEG electrodeposited

GRGDSP peptide is coated with chloride activation technique to enhance adhesion and migration of osteoblastic cells [12]. The expression levels of many genes in MC3T3-E1 cells are altered.

4 Conclusions

Metallic materials are widely used in medicine not only for dental and orthopedic implants but also as cardiovascular devices and for other purposes. Biomaterials are always used in close contact with living tissues. Therefore, interactions between material surfaces and living tissues must be well-controlled. Metal surface may be biofunctionalized by various techniques, such as immobilization of biofunctional molecules. These techniques make it possible to apply metals to a scaffold in tissue engineering.

References

1. Tanaka Y, Doi H, Iwasaki Y et al (2007) Electrodeposition of amine-terminated poly(ethylene glycol) to titanium surface. Mater Sci Eng C27:206–212
2. Tanaka Y, Doi H, Kobayashi E et al (2007) Determination of immobilization manner of amine-terminated poly(ethylene glycol) electrodeposited to titanium surface with XPS and GD-OES. Mater Trans 48:287–292
3. Tanaka Y, Kurashima K, Saito H, et al (2009) In vitro short term platelet adhesion on various metals. J Artf Org 12:182–186
4. Rezania A, Thomas CH, Branger AB et al (1997) The detachment strength and morphology of bone cells contacting materials modified with a peptide sequence found within bone sialoprotein. J Biomed Mater Res 37:9–19
5. Pierschbacher MD, Ruoslahti E (1984) Cell attachment activity of fibronectin can be duplicated by small synthetic fragments of the molecule. Nature 309:30–33

6. Schliephake H, Scharnweber D, Dard M et al (2002) Effect of RGD peptide coating of titanium implants on periimplant bone formation in the alveolar crest. An experimental pilot study in dogs. Clin Oral Implant Res 13:312–319
7. Xiao SJ, Textor M, Spencer ND et al (1998) Covalent attachment of cell-adhesive, (Arg-Gly-Asp)-containing peptides to titanium surfaces. Langmuir 114:5507–5516
8. Xiao SJ, Textor M, Spencer ND et al (1997) Immobilization of the cell-adhesive peptide Arg-Gly-Asp-Cys (RGDC) on titanium surfaces by covalent chemical attachment. J Mater Sci Mater Med 8:867–872
9. Rezania A, Johnson R, Lefkow AR et al (1999) Bioactivation of metal oxide surfaces. 1. Surface characterization and cell response. Langmuir 15:6931–6939
10. Tanaka Y, Saito H, Tsutsumi Y et al (2009) Effect of pH on the interaction between zwitterion and titanium oxide. J Colloid Interface Sci 330:138–143
11. Oya K, Tanaka Y, Saito H et al (2009) Calcification by MC3T3–E1 cells on RGD peptide immobilized on titanium through electrodeposited PEG. Biomaterials 30:1281–1286
12. Yamanouchi N, Pugdee K, Chang WJ et al (2008) Gene expression monitoring in osteoblasts on titanium coated with fibronectin-derived peptide. Dent Mater J 27:744–750

Effect of Young's modulus in metallic implants on atrophy and bone remodeling

Mitsuo Niinomi and Tomokazu Hattori

Abstract. The Young's modulus equal to that of the cortical bone can be achieved at the <100> direction of a single crystal of Ti–29Nb–13Ta–4.6Zr (TNTZ). The fatigue life of TNTZ can be sufficiently improved by keeping its Young's modulus low enough by short-time aging after severe cold rolling. Silane coupling treatment highly improves the strength of porous titanium and poly(methyl methacrylate) composite by keeping its Young's modulus just equal to that of the cortical bone. TNTZ with low Young's modulus inhibits the bone atrophy and enhances bone remodeling.

Key words. Ti–26Nb–13Ta–4.6Zr, low Young's modulus, porous titanium, porous titanium and PMMA composite, fatigue life, mechanical biocompatibility

1 Introduction

The main metallic biomaterials are stainless steels, cobalt (Co) alloys, and titanium (Ti) and its alloys. Among these biomaterials, the biocompatibility of Ti and its alloys is the highest. Because Ti alloys exhibit excellent biocompatibility and have high corrosion resistance and specific strength, which is the ratio of density to strength (density/strength), the demand for Ti alloys as biomaterials has increased, and extensive research and development on the use of Ti alloys for biomedical applications is being carried out. Among the current practical Ti alloys available for

M. Niinomi (✉)
Department of Biomaterials Science, Institute for Materials Research, Tohoku University, Sendai 980-8574, Japan
e-mail: niinomi@imr.tohoku.ac.jp

T. Hattori
Department of Materials Science and Engineering, Faculty of Science and Technology, Meijo University, Nagoya 468-8502, Japan

T. Sasano et al. (eds.), *Interface Oral Health Science 2009*,
DOI 10.1007/978-4-431-99644-6_13, © Springer 2010

biomedical applications, Ti–6Al–4V ELI is the most widely used. Ti–6Al–4V ELI was initially used for aerospace applications and then for surgical applications. It was found that vanadium (V) present in Ti–6Al–4V ELI was toxic for surgical applications; however, no problems have been encountered. Therefore, Ti–6Al–7Nb and Ti–5Al–2.5Fe, where V in Ti–6Al–4V ELI is replaced with Nb or Fe, which are nontoxic elements that act as β-stabilizing elements, similar to V, have been developed [1].

Recently, researchers are involved in metallic biometals, focusing on the development of low-modulus β-type Ti alloys composed of nontoxic and allergy-free elements such as Nb, Ta, and Zr [2]. Composing of nontoxic and allergy-free elements is considered to enhance biological biocompatibility. Low modulus is considered to enhance mechanical biocompatibility because it is considered to inhibit stress shielding between bone and implant. Finally, many low-modulus β-type Ti alloys, whose Young's moduli are relatively close to that of the cortical bone (10–30 GPa), have been developed. Further, researchers have attempted to develop Ti alloys that have functionality as well as biological and mechanical biocompatibility for biomedical applications. For example, the authors have developed Ti–29Nb–13Ta–4.6Zr (TNTZ) [3], which satisfies both biological and mechanical biocompatibilities [4]. In the broad sense, fatigue strength, fretting fatigue strength, wear properties, strength, ductility, and functionalities such as superelasticity and shape memory effect may be collectively referred to as mechanical biocompatibilities. These properties of TNTZ can be controlled by microstructural control through thermomechanical treatment, surface treatment, etc. On the other hand, mechanical biocompatibility such as Young's modulus of TNTZ with living tissue should be evaluated in vivo.

Mechanical biocompatibilities of TNTZ will be mainly described in this chapter.

2 Lowering Young's Modulus of Titanium Alloy Similar to That of Cortical Bone

It is desirable for Young's moduli of biomaterials to be equal to that of the cortical bone (10–30 GPa) because if the former is considerably greater than the latter, bone resorption and lack of bone remodeling occur. The smallest Young's modulus reported in bulk Ti alloys to date is around 40 GPa as shown in Fig. 1 [5]. It seems to be difficult to lower Young's modulus of bulk Ti alloys below 40 GPa. However, it is expected to achieve smaller Young's modulus than 40 GPa by controlling crystal orientation because the anisotropy of mechanical properties of Ti–Nb–Ta–Zr system alloy is significantly large [6]. Therefore, the single crystal of biomedical β-type Ti alloy, TNTZ, showing Young's modulus around 60 GPa was made, and then its orientation dependence of Young's modulus was evaluated. The result has been reported as shown in Fig. 2 [7]. The smallest Young's modulus is obtained to be 35 GPa at a direction of <100>. This Young's modulus is almost equal to the greatest level of Young's modulus of the cortical bone. The implant made of single crystal showing Young's modulus equal to that of cortical bone is highly expected.

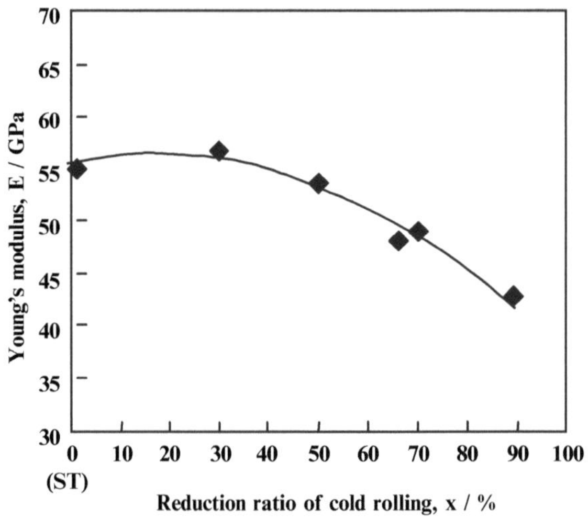

Fig. 1. Relationship between Young's modulus and reduction ratio of cold rolling in Ti–35Nb–4Sn

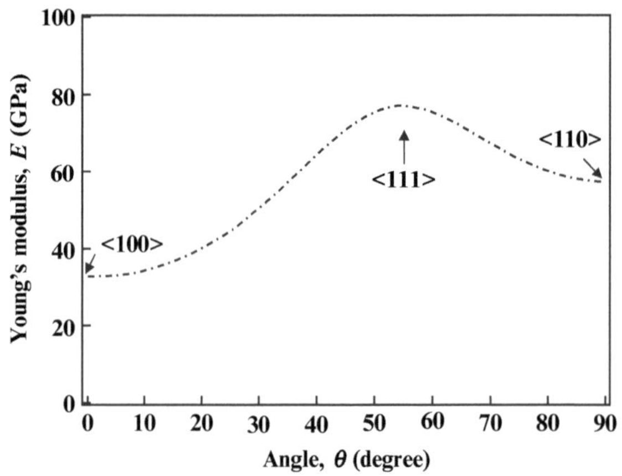

Fig. 2. Young's modulus of single-crystal Ti–29Nb–13Ta–4.6Zr with ω-phase in directions between <100> and <110>

Young's modulus of bulk β-type Ti alloy in single crystal state is still a little greater than that of the cortical bone. It has been reported elsewhere [8] that it is very effective to make Ti and its alloys porous in order to further reduce Young's moduli of Ti and its alloys; this is another way to drastically reduce Young's modulus of Ti, and in the relationship between Young's modulus and the porosity of porous Ti (pTi) samples made of Ti powders with different diameters and its comparison with Young's modulus of bulk Ti, at a porosity of approximately 30%, Young's modulus is nearly equal to that of the cortical bone, but the strength decreases drastically [9]. The decrease in the strength of pTi can be effectively inhibited by combining it with a biocompatible polymer. One of the authors (MN) has developed the method that involves the initial use of a monomer of poly(methyl methacrylate) (PMMA) [10]. pTi is first immersed into the monomer of PMMA, thereby causing the monomer to penetrate the pTi. Subsequently, the monomer in the pTi is subjected to polymerization by heating. When combined with the PMMA, the strength of the pTi is increased, but its increasing degree is not sufficient because the bonding strength between pTi and PMMA is insufficient.

The bonding strength between pTi and PMMA can be improved by silane coupling treatment. In that case, the pTi is silane coupling-treated (Si-treated) before subjecting the process mentioned above. Figures 3 and 4 [11] show tensile strengths of pTi, pTi/PMMA, and Si-treated pTi/PMMA, and Young's moduli of pTi, pTi/PMMA, and Si-treated pTi/PMMA. As a reference, the tensile strength (50–80 MPa) and Young's modulus (2–4 GPa) ranges of PMMA are also shown in these figures. The number after pTi in each material shown in the horizontal axis indicates the maximum limit of the diameter of the Ti particles, and the last number indicates the porosity. The tensile strength of each Si-treated pTi/PMMA is greater than that of each pTi/PMMA. For Si-treated pTi/PMMA, the increase in the tensile strength caused by PMMA filling can be seen in the pTi showing the relatively greater tensile strength. The Young's modulus of Si-treated pTi/PMMA is nearly equal to that of pTi or pTi/PMMA at relatively lower porosity but is relatively greater than that of pTi or pTi/PMMA at relatively greater porosity.

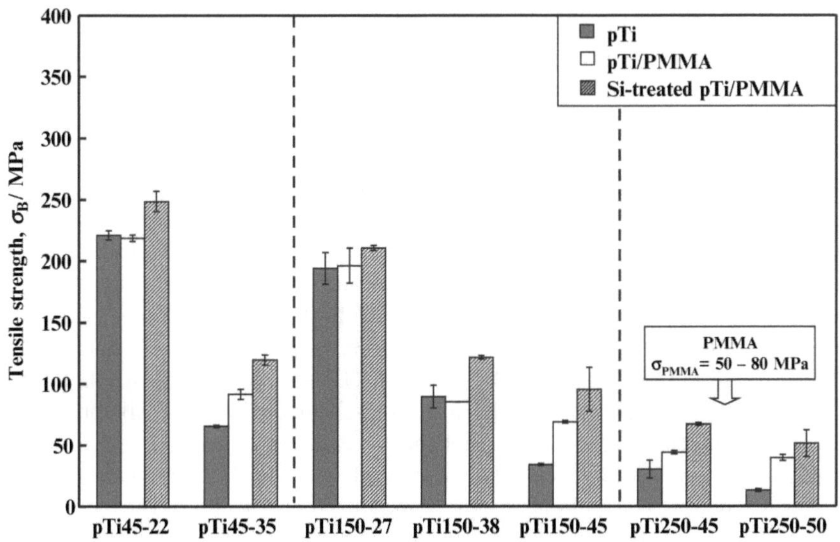

Fig. 3. Tensile strengths of pTi, pTi/PMMA, and Si-treated pTi/PMMA

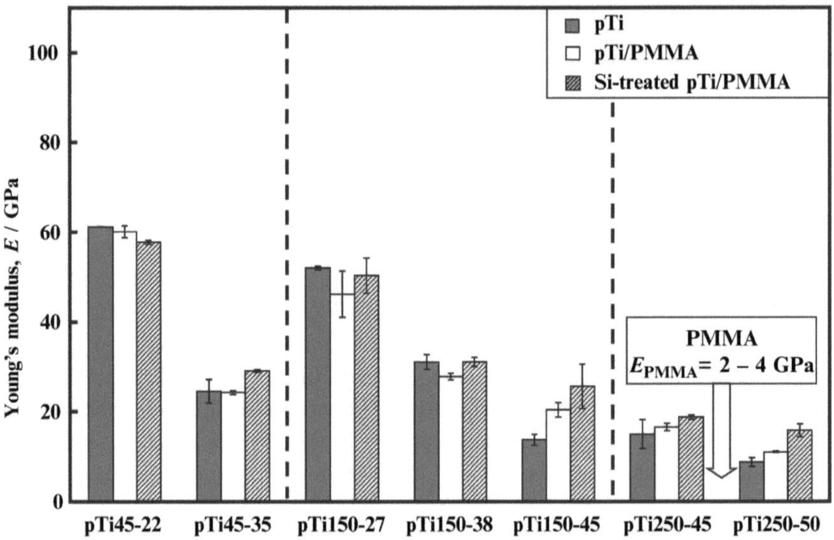

Fig. 4. Young's moduli of pTi, pTi/PMMA, and Si-treated pTi/PMMA

3 Development of Mechanical Endurance of Biomedical Titanium Alloy with Keeping Young's Modulus Low

Mechanical endurance, namely fatigue life of biomaterials, is one of the very important mechanical functionalities. Heat treatments or thermomechanical treatments that are processing composed of heat treatments and mechanical deformations are effective ways to improve fatigue life of biomedical Ti alloys, but Young's modulus increases generally because of the precipitation of the second phase when the heat treatments or thermomechanical treatments are conducted [12]. In β-type Ti alloys, such as TNTZ, precipitates are ω phase or α phase. In general, the strength and Young's modulus of the β-type Ti alloy increase more by the ω phase precipitation than by the α phase precipitation although it depends on their volume fractions, sizes, etc. Therefore, a small amount of the ω phase precipitation associated with a short-time aging is expected to improve the fatigue life of TNTZ by keeping its Young's modulus low. According to this concept, the thermomechanical treatment schematically shown in Fig. 5 [13] has been subjected to TNTZ in order to improve its fatigue life by keeping its Young's modulus low. Young's modulus and tensile properties of TNTZ subjected to the thermomechanical treatment shown in Fig. 5 is shown in Figs. 6 and 7 [13]. Balance between strength and ductility (elongation) at low Young's modulus is excellent at aging time of 3.6 and 10.8 ks. The fatigue life of TNTZ aged at 3.6 and 10.8 ks are shown in Fig. 8 [13]. The fatigue life of TNTZ aged at 10.8 ks is in the fatigue limit range of conventional biomedical Ti–6Al–4V ELI with a Young's modulus of 73 GPa, which is a much lower Young's modulus than that of Ti–6Al–4V ELI.

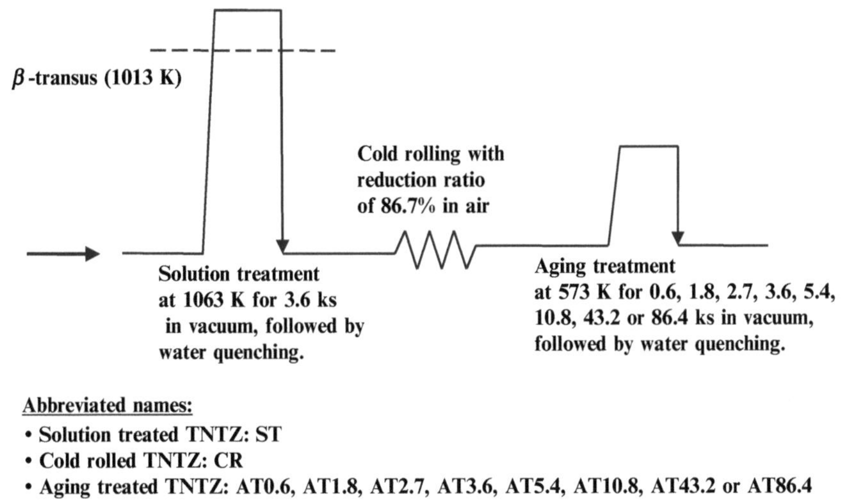

Fig. 5. Schematic drawing of thermomechanical treatment for TNTZ

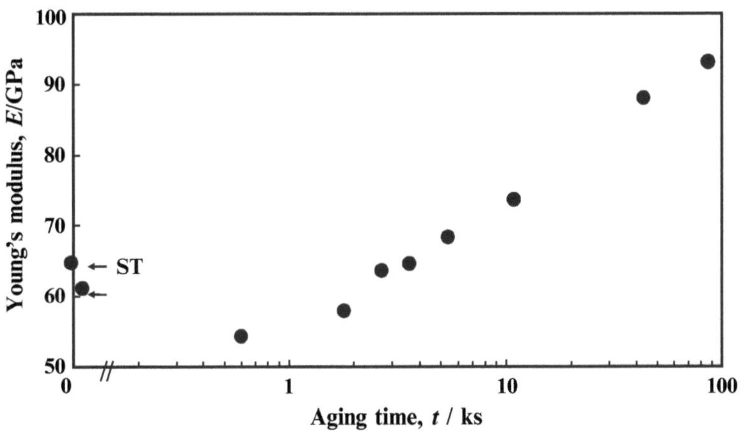

Fig. 6. Young's modulus of TNTZ subjected to thermomechanical treatments as a function of aging time

4 Young's Modulus and Bone Atrophy

Figures 9–11 [14] show images of the X-ray follow-up of the healing fractures from weeks 4 to 18 after the bone plates made of the three materials (SUS 316L stainless steel, Ti–6Al–4V ELI, and TNTZ) were implanted into the rabbit tibia fracture model; fracture healing was almost the same in each case. Initially, callus formation was observed 2 weeks after implantation, and they were still observable at 3 weeks after implantation. Bone union occurred 4 weeks after implantation, and the fracture line was barely

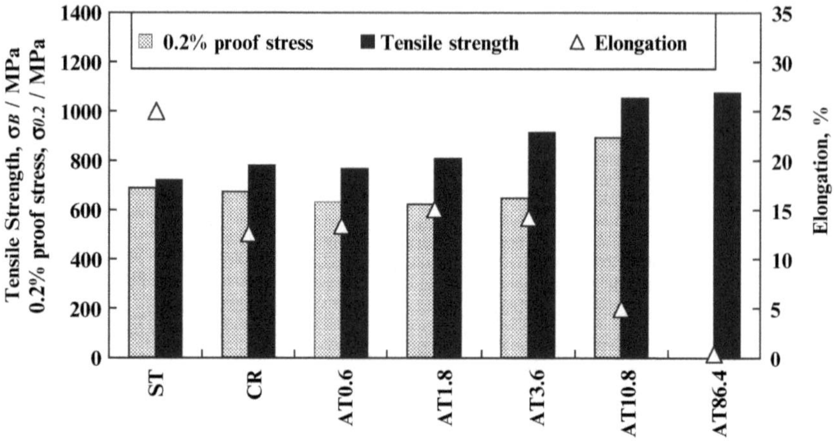

Fig. 7. Tensile properties of TNTZ subjected to thermomechanical treatments

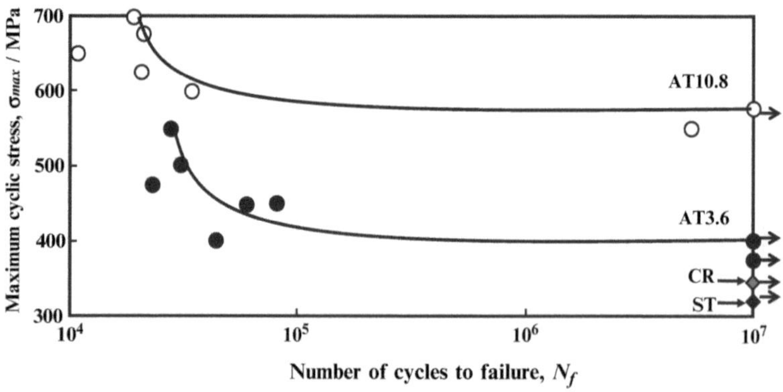

Fig. 8. S–N curves of TNTZ subjected to aging treatments for 3.6 and 10.8 ks after cold rolling

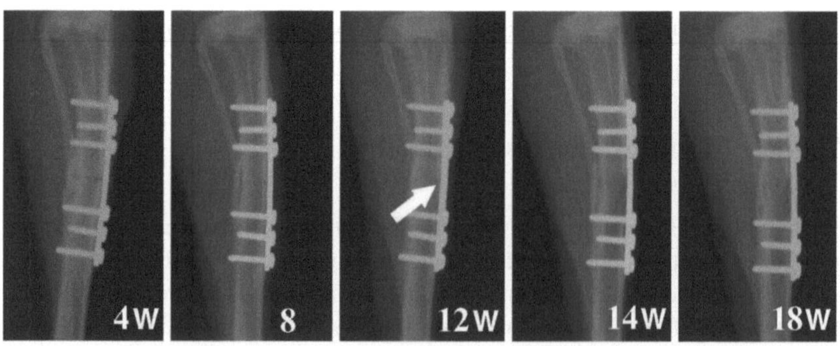

Fig. 9. X-ray follow-up 4–18 weeks after implantation for SUS 316L stainless steel

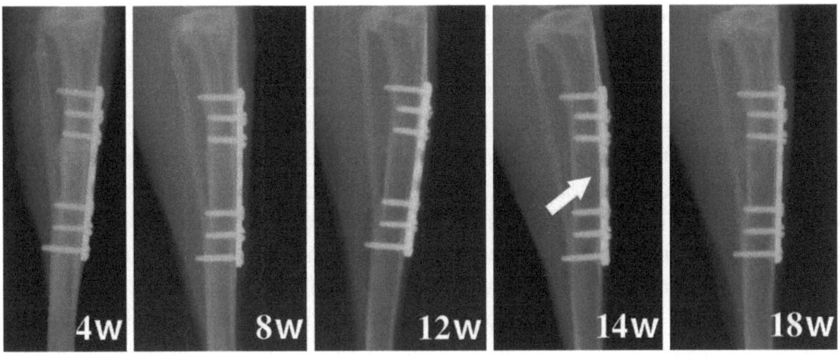

Fig. 10. X-ray follow-up 4–18 weeks after implantation for Ti–6Al–4V ELI

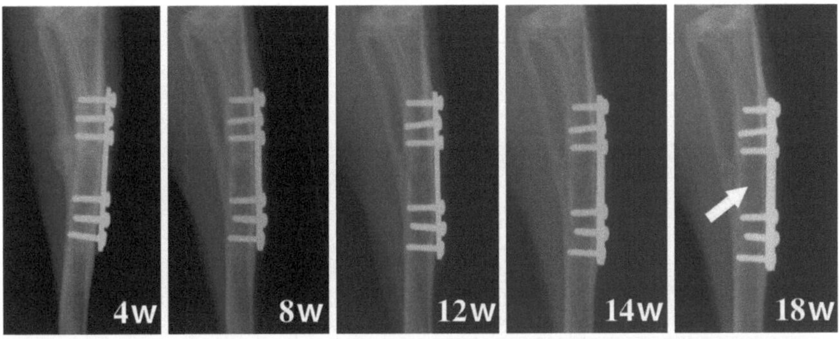

Fig. 11. X-ray follow-up 4–18 weeks after implantation for TNTZ.

visible by approximately 8 weeks after implantation. The experimental fracture trace had completely disappeared 16–20 weeks after implantation. However, under the bone plate, bone atrophy (thinning of the cortical bone) was observed; it occurred at different times for each of the materials considered. With SUS 316L stainless steel (Fig. 9), the atrophy of the cortical bone began 7 weeks after implantation, and the bone had almost disappeared by 12 weeks after implantation. With Ti–6Al–4V ELI (Fig. 10), the bone atrophy began 7 weeks after implantation, and the bone had almost disappeared by 14 weeks after implantation. And with TNTZ (Fig. 11), the atrophy of the bone began 10 weeks after implantation, and the bone had almost disappeared by 18 weeks after implantation. Therefore, since the period from the beginning of bone atrophy to the disappearance of the bone is the longest with TNTZ, it appears that a low Young's modulus is required to inhibit bone atrophy. However, the Young's modulus of TNTZ seems to be lowered more for further inhibition of bone atrophy.

5 Young's Modulus and Bone Remodeling

Figure 12 [14] shows CMR images of cross-sections of the middle and distal parts of the rabbit tibia fracture model, in which a TNTZ bone plate was implanted for 44 weeks, and of the control tibia. An increase in the tibia diameter can be observed in both the middle and distal parts. With regard to the increase in the tibia diameter in the case of TNTZ, a double-wall structure, with each wall showing different X-P densities, and a clear boundary line in the middle and distal parts are observed; the shape of the inner wall is similar to that of the original cortical bone. Therefore, it appears that the outer cortical bone is newly formed, and the intramedullar bone tissue has been formed from the remains of the old cortical bone, which is a possible result of bone remodeling with the low-modulus bone plate. This can be attributed to the fact that the increase in the tibia diameter increases the bending rigidity of the tibia, which may reduce the shear stress around the point of fixation. This phenomenon does not occur for two other materials such as SUS 316L stainless steel and Ti–6Al–4V ELI.

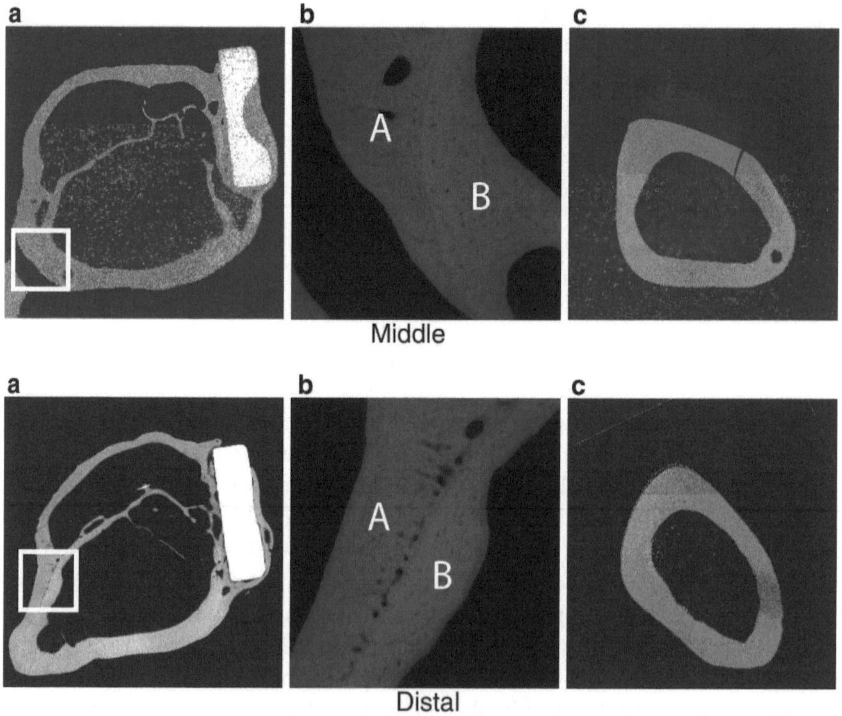

Fig. 12. CMR images of cross-sections of middle and distal parts of rabbit tibia fracture model, in which a TNTZ bone plate was implanted for 44 weeks, and the control tibia: (**a**) cross-section of fracture model, (**b**) high-magnification CMR image of branched part of outer and inner formed bone, and (**c**) cross-section of control tibia

6 Summary

The <100> direction of the single crystal of TNTZ shows 35 Gpa, which is similar to the Young's modulus of the cortical bone. The PMMA-filled pTi with silane coupling treatment shows improved strength with keeping its Young's modulus just equal to that of the cortical bone. The fatigue life of TNTZ can be improved by keeping its Young's modulus low by proper thermomechanical treatment. The bone atrophy can be inhibited by lowering the Young's modulus of the biomaterial to be similar to that of the cortical bone. Lowering the Young's modulus of the biomaterial leads to a good remodeling.

Acknowledgments This work was supported in part by the Global COE Program "Materials Integration International Center of Education and Research, Tohoku University," Ministry of Education, Culture, Sports, Science and Technology (MEXT) of Japan, and the Inter-university Cooperative Research Program "Highly-functional Interface Science: Innovation of Biomaterials with Highly-functional Interface to Host and Parasite, Tohoku University and Kyushu University," MEXT of Japan.

References

1. Niinomi M (2003) Recent research and development in titanium alloys for biomedical applications and healthcare goods. STAM 4:445–454
2. Niinomi M, Hanawa T, Narushima T (2005) Japanese research and development in metallic biomedical, dental and healthcare materials. JOM 57:18–24
3. Kuroda D, Niinomi M, Morinaga M et al (1998) Design and mechanical properties of new beta type titanium alloys for implant materials. Mater Sci Eng A 243:244–249
4. Niinomi M (2008) Mechanical biocompatibilities of titanium alloys for biomedical applications. J Mech Behav Biomed Mater 1:30–42
5. Matsumoto H, Watanabe S, Hanada S (2005) Beta TiNbSn alloys with low Young's modulus and high strength. Mater Trans 46:1070–1078
6. Sakaguchi N, Niinomi M, Akahori T et al (2005) Relationships between tensile deformation behavior and microstructure in Ti-Nb-Ta-Zr system alloys. Mater Sci Eng C 25:363–369
7. Tane M, Akita S, Nakano T et al (2008) Peculiar elastic behavior of Ti-Nb-Ta-Zr single crystals. Acta Mater 56:2856–2863
8. Niinomi M, Nakai M, Akahori T et al. (2009) Functionality of porous titanium by polymer filling. Ceram Trans 206:91–104
9. Oh IH, Nomura N, Masahashi N et al (2003) Mechanical properties of porous titanium compacts prepared by powder sintering. Scripta Mater 49:1197–1202
10. Nakail M, Niinomil M, Akahori T et al (2008) Effect of silane coupling treatment on mechanical properties of porous titanium filled with PMMA for biomedical applications. J Jpn Inst Met 72:839–845
11. Nakai M, Niinomi M, Akahori T et al. (2010) Development of biomedical porous titanium filled with medical polymer by direct polymerization of monomer solution penetrating into pores. JMMB 3:41–50
12. Akahori T, Niinomi M, Ishimizu K et al (2003) Effects of thermomechanical processings on fatigue properties of Ti-29Nb-13Ta-4.6Zr for biomedical applications. J Jpn Inst Met 67:652–660
13. Oneda T (2008) Improvement in mechanical functionality of biomedical low modulus Ti-29Nb-13Ta-4.6Zr alloy using high strength brittle phase. Master Thesis, Tohoku University
14. Sumitomo N, Noritake K, Hattori T et al (2008) Experiment study on fracture fixation with low rigidity titanium alloy – plate fixation of tibia fracture model in rabbit. J Mater Sci Mater Med 19:1581–1586

Chemical and physical factors affecting osteoconductivity of octacalcium phosphate bone substitute material

Osamu Suzuki

Abstract. The present article summarizes the factors controlling osteoconductive and biodegradable characteristics of synthetic octacalcium phosphate (OCP) when implanted in bone defects. OCP is a transient precursor, which tends to convert to hydroxyapatite (HA) in physiological environment. We recently confirmed that the subtle change of stoichiometry of OCP from Ca/P molar ratio 1.28 to 1.37, both of which are nonstoichiometric compositions compared to stoichiometric 1.33 of OCP, obtained by partial hydrolysis, makes it reduce the crystallinity and raises the bone formation rate significantly if implanted in marrow space of rat tibia more than those of original OCP and HA obtained via OCP full hydrolysis. The composite, which consists of OCP granules and collagen sponge, is vigorously resorbed by osteoclastic cells if the thick composite is implanted in subperiosteal area of murine calvaria but replaced with newly formed bone if the thin composite or OCP without collagen is used. The results suggest that the physical stress, which might be induced underneath the periosteum, controls activities of osteoblasts and osteoclasts around OCP implant. The osteoconductive characteristics of OCP appear to be controlled by its stoichiometry and the mechanical stimulation induced from surrounding tissue where OCP is implanted.

Key words. osteoconductivity, octacalcium phosphate, stoichiometry, mechanical stress, biodegradation

1 Introduction

Octacalcium phosphate [$Ca_8H_2(PO_4)_6 \cdot 5H_2O$; OCP] has been suggested to be a precursor of biological apatite crystals in bones and teeth [1]. Synthetic OCP has been investigated as a bone substitute material in various forms, such as coatings

O. Suzuki
Division of Craniofacial Function Engineering, Tohoku University Graduate School of Dentistry, 4-1 Seiryo-machi, Aoba-ku, Sendai 980-8575, Japan
e-mail: suzuki-o@m.tains.tohoku.ac.jp

T. Sasano et al. (eds.), *Interface Oral Health Science 2009*,
DOI 10.1007/978-4-431-99644-6_14, © Springer 2010

on metallic implants [2] and granules [3, 4]. These studies confirmed that synthetic OCP is an osteoconductive and biodegradable material that could be used for bone regeneration.

We have recently found by in vitro studies that synthetic OCP is capable of enhancing not only osteoblastic cell differentiation [5, 6] but also osteoclast formation in coculturing of osteoblasts and bone marrow cells [7]. Furthermore, it was apparent that the biodegradable characteristics of OCP through osteoclastic resorption are controlled by the stoichiometry in OCP [8] and the mechanical stress is caused probably under the surrounding tissues [9, 10].

The present article summarizes the osteoconductive characteristics of OCP that are significantly affected by the chemical and physical factors in association with OCP and its implantation.

2 Activation of Osteoblasts and Osteoclasts by OCP

Figure 1 shows an undecalcified histological section of OCP granules implanted in subperiosteal region of a 7-week-old BALB/c mouse calvaria for 7 days [11]. Cuboidal osteoblasts are aligned around the surface of OCP granules. An osteoclast-like multinuclear cell is also attached to the surface of OCP granule. The result suggests that OCP activates not only osteoblasts to initiate direct new

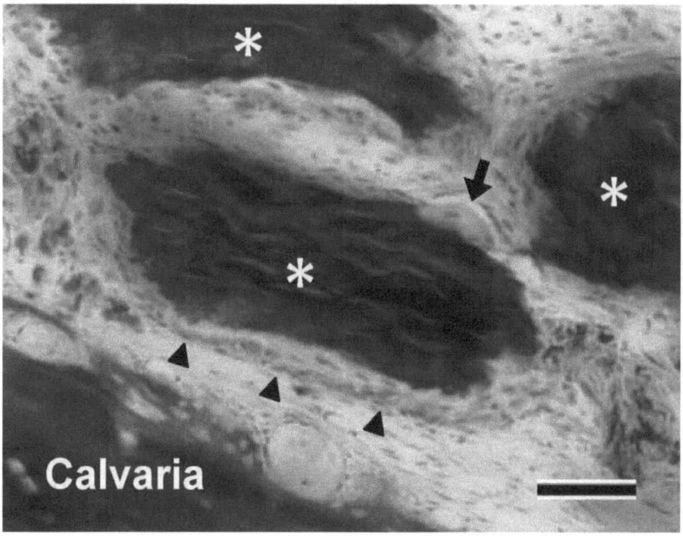

Fig. 1. Photograph of undecalcified section of OCP implantation for 7 days stained with hematoxylin and eosin. Osteoblasts with a cuboidal shape aligned on the OCP implant surface (*arrow heads*); an osteoclast-like multinuclear cell attached to the OCP implant surface (*arrow*) and *asterisks*, OCP implants. Bar = 150 μm. Reproduced from Kikawa et al. [11] with permission from Elsevier Ltd.

bone apposition onto OCP surfaces but also osteoclast-like cells to resorb OCP surfaces. In fact, it was reported that bone formation is enhanced while osteoclast-like cells are resorbing the surfaces of OCP [11, 12]. Although the resorption by osteoclastic cells having ruffled membranes has also been recognized in calcium phosphate ceramics at ultrastructural level [13], it is of interest to know whether OCP has a potential to induce osteoclast formation from the precursor cells. Coculturing of bone marrow cells and osteoblastic cells was conducted to examine the possibility that OCP is a critical mineral phase to induce osteoclast differentiation [7]. Osteoblasts expressed the receptor activator of NF-κB ligand (RANKL), an osteoclast differentiation factor, and were formed with OCP. Based on the analysis of the medium supernatant, it was hypothesized that change of calcium concentration around osteoblasts, caused by the intrinsic characteristics of OCP [tendency to convert to the thermodynamically most stable hydroxyapatite (HA)], could be a critical factor to induce osteoclast formation [7]. Other studies have confirmed that OCP stimulates osteoblastic cell differentiation with upregulating osteoblast-related genes, such as osterix [5, 6]. OCP may have a stimulatory capacity enhancing bone formation coupled with its own biodegradation through osteoclastic cellular resorption [14].

3 Effect of Stoichiometry of OCP on Osteoconductivity

It is considered that OCP exhibits a variety of compositions and structural environments differing from those expected, based on the stoichiometric structure [15, 16]. Mathew et al. [15] proposed an example of the nonstoichiometric formula of OCP, $Ca_{16}H_{4+X}(PO_4)_{12}(OH)_X{\bullet}(10{-}X)H_2O$, with excess hydrogen in the structure. In fact, the nonstoichiometric OCP has been shown to have approximately 40% HPO_4 in the structure [16]. Table 1 shows Ca/P molar ratios of synthetic nonstoichiometric OCP, the partially hydrolyzed OCP, and the fully hydrolyzed OCP having apatitic structure in comparison with the stoichiometric OCP and HA. Thus, OCP displays a variety of stoichiometry in composition and structure, most probably due to the existence of the hydrated layers, which stacks alternately with the apatitic layers [1]. The effect of subtle change in Ca/P molar ratio in OCP, caused by the partial hydrolysis in hot water, on osteoconductive property was examined by their

Table 1. Ca/P molar ratios of OCP and its partially hydrolyzed OCP

Calcium phosphate	Chemical formula	Ca/P molar ratio
OCP, nonstoichiometric	Undetermined (OCP structure by XRD)	1.26–1.28[a]
OCP	$Ca_8H_2(PO_4)_6{\bullet}5H_2O$	1.33[b]
OCP hydrolyzates	Formulae between OCP and HA	~1.48[a]
HA	$Ca_{10}(PO_4)_6(OH)_2$	1.67[b]

[a]Analytical; reproduced from Miyatake et al. [8] with permission from Elsevier Ltd.
[b]Theoretical

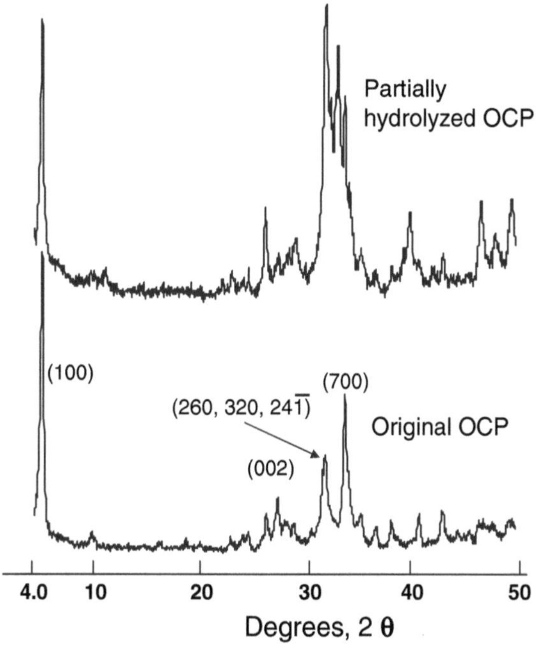

Fig. 2. X-ray diffraction patterns of synthetic OCP and its partially hydrolyzed OCP. Reproduced from Miyatake et al. [8] with permission from Elsevier Ltd.

intramedullar canal implantation in the granule form in rat tibia for 56 days [8]. The structural change of implants and tissue responses were analyzed by X-ray diffraction, histomorphometry, and expression of mRNA around the implants. The results obtained were that: (1) the partial hydrolysis lowered the crystallinity of OCP (Fig. 2); (2) the implantation caused the conversion from the OCP crystalline phase into apatitic structure; (3) the highest bone formation rate was obtained for the partially hydrolyzed OCP with Ca/P molar ratio 1.37 until 56 days; (4) the early expression of osteoclast markers TRAP and cathepsin-K was suppressed with the partially hydrolyzed OCP. The results confirmed that the partially hydrolyzed OCP with Ca/P molar ratio 1.37 enhances bone formation most in comparison with the OCP (Ca/P molar ratio 1.28) or HA obtained via full hydrolysis of OCP (Ca/P molar ratio 1.48) [8].

4 OCP–HA Conversion

It is accepted that OCP is a metastable calcium phosphate salt in physiological environment [17]. The transition of OCP to HA (Ca-deficient HA with lower Ca/P molar ratio) is thermodynamically favored and, once initiated, it advances spontaneously and irreversibly [18]. The conversion is accompanied by calcium consumption and phosphate release [17, 19]. In fact, OCP tends to convert to HA if implanted in murine tissues including bone defects [3, 5, 20]. It has been suggested that OCP can be transformed to HA via two mechanisms: (1) dissolution-reprecipitation [18] and (2) topotaxial conversion without changing its original plate-like morphology [1, 21]. The topotaxial conversion took place in vivo implantation with maintaining plate-like morphology even 3 weeks after the implantation [22]. However, the formation of many nanocrystals was simultaneously observed around the OCP crystal surfaces, suggesting that the conversion is accompanied by the dissolution-reprecipitation process in addition to the topotaxial conversion. Another possible mechanism to regulate OCP crystal morphology is selective protein adsorption induced by the implantation [20]. Protein adsorption was used to explain apatite crystal nucleation in vitro [23]. Figure 3 shows a transmission electron micrograph of a portion of an OCP granule revealing that the circulating serum proteins are accumulated around the loci corresponding to the plate-like OCP crystals (OCP crystals were already absent because of the decalcification). It is also considered that the morphology of OCP can be controlled by the protein adsorption [24]. These

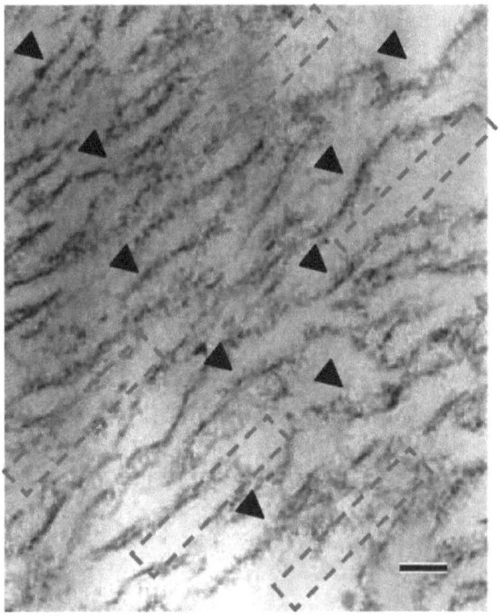

Fig. 3. Ultrastructure of a portion of an OCP granule implanted in subperiosteal region of a 7-week-old BALB/c mouse calvaria for 13 days, examined using decalcified sections. The proteins are accumulated around the loci corresponding to the plate-like OCP crystals (OCP crystals were already absent because of the decalcification). Rectangles of dotted line show the possible loci where OCP crystals were present before the decalcification. Arrows, proteins accumulated. Bar = 0.2 μm. Reproduced from Suzuki et al. [22] with permission from Bentham Science Publishers Ltd.

results suggest that OCP crystals may work as a scaffold that osteoblastic cells can attach, proliferate, and be differentiated during the conversion.

5 Effect of Mechanical Stress on Osteoconductivity of OCP

Nonresorbable and resorbable materials may differently respond to mechanical stress suffered by the surrounding tissue where they are implanted. Our previous study showed that an OCP–collagen composite (OCP/Col) is vigorously resorbed if implanted in the subperiosteal pocket of the rat calvaria while it enhances bone augmentation if the dimension (thickness) is reduced [9]. The thicker composite was ascertained to be resorbed considerably by many osteoclast-like cells, which appeared around OCP granules within collagen matrix. This study suggests that some tensions underneath the periosteum induce certain mechanical stress to those implanted materials and may stimulate osteoclast cellular activity [9]. The assumption was confirmed by a subsequent study that the alleviation of the assumed mechanical stress enhances bone augmentation coupled with moderate osteoclastic cellular biodegradation [10]. Furthermore, in vitro load-bearing test on the OCP/Col composite seeded by mouse bone marrow stromal ST-2 cells verified that the introduction of the mechanical stress induces upregulation of RANKL expression in the cells [10]. Taken together, the overall results support the proposition

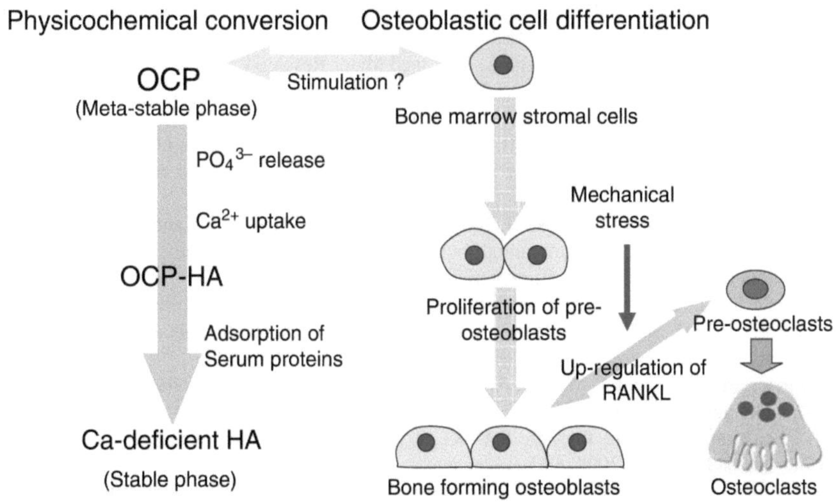

Fig. 4. Schematic view of osteoblastic differentiation activated by the process of OCP conversion into HA and osteoclast formation stimulated by both the HA–OCP conversion and the mechanical stress from the surrounding tissues where OCP was implanted, hypothesized by the experimental results obtained [5–10]

that the enhancement of bone formation by OCP is coupled with the enhancement of the biodegradation of this material by osteoclasts, induced by the presence of OCP itself, under the control of mechanical stress [14]. Figure 4 shows schematic view of the potential role of OCP in relation to OCP–HA conversion regarding the osteoblastic cell differentiation and the osteoclast formation under the control of the mechanical stress.

6 Summary

The osteoconductivity and biodegradability of OCP in vivo can be controlled by the subtle change in Ca/P molar ratio of OCP and the mechanical stress around this material caused by the surrounding tissue. The experimental evidence suggests that osteoblastic cells are activated to upregulate osteoclast differentiation factor RANKL both by transitory nature of OCP and the mechanical stress applied.

Acknowledgments This study was supported in part by Grants-in-Aid (17076001, 19390490, 20659304) from the Ministry of Education, Science, Sports, and Culture of Japan.

References

1. Brown WE, Smith JP, Lehr JR, Frazier AW (1962) Crystallographic and chemical relations between octacalcium phosphate and hydroxyapatite. Nature 196:1050–1055
2. Barrere F, van der Valk CM, Dalmeijer RA, van Blitterswijk CA, de Groot K, Layrolle P (2003) In vitro and in vivo degradation of biomimetic octacalcium phosphate and carbonate apatite coatings on titanium implants. J Biomed Mater Res A 64:378–387
3. Suzuki O, Nakamura M, Miyasaka Y, Kagayama M, Sakurai M (1991) Bone formation on synthetic precursors of hydroxyapatite. Tohoku J Exp Med 164:37–50
4. Kamakura S, Sasano Y, Homma H, Suzuki O, Kagayama M, Motegi K (1999) Implantation of octacalcium phosphate (OCP) in rat skull defects enhances bone repair. J Dent Res 78:1682–1687
5. Suzuki O, Kamakura S, Katagiri T, Nakamura M, Zhao B, Honda Y, Kamijo R (2006) Bone formation enhanced by implanted octacalcium phosphate involving conversion into Ca-deficient hydroxyapatite. Biomaterials 27:2671–2681
6. Anada T, Kumagai T, Honda Y, Masuda T, Kamijo R, Kamakura S, Yoshihara N, Kuriyagawa T, Shimauchi H, Suzuki O (2008) Dose-dependent osteogenic effect of octacalcium phosphate on mouse bone marrow stromal cells. Tissue Eng Part A 14:965–978
7. Takami M, Mochizuki A, Yamada A, Tachi K, Zhao B, Miyamoto Y, Anada T, Honda Y, Inoue T, Nakamura M, Suzuki O, Kamijo R (2009) Osteoclast differentiation induced by synthetic octacalcium phosphate through RANKL expression in osteoblasts. Tissue Eng Part A. doi:10.1089/ten.TEA.2009.0065 (in press)
8. Miyatake N, Kishimoto KN, Anada T, Imaizumi H, Itoi E, Suzuki O (2009) Effect of partial hydrolysis of octacalcium phosphate on its osteoconductive characteristics. Biomaterials 30:1005–1014
9. Suzuki Y, Kamakura S, Honda Y, Anada T, Hatori K, Sasaki K, Suzuki O (2009) Appositional bone formation by OCP-collagen composite. J Dent Res 88:1107–1112

10. Matsui A, Anada T, Masuda T, Honda Y, Miyatake N, Kawai T, Kamakura S, Echigo S, Suzuki O (2009) Mechanical stress-related calvaria bone augmentation by onlayed octacalcium phosphate-collagen implant. Tissue Eng Part A. doi:10.1089/ten.TEA.2009.0284 (in press)
11. Kikawa T, Kashimoto O, Imaizumi H, Kokubun S, Suzuki O (2009) Intramembranous bone tissue response to biodegradable octacalcium phosphate implant. Acta Biomater 5:1756–1766
12. Imaizumi H, Sakurai M, Kashimoto O, Kikawa T, Suzuki O (2006) Comparative study on osteoconductivity by synthetic octacalcium phosphate and sintered hydroxyapatite in rabbit bone marrow. Calcif Tissue Int 78:45–54
13. Takeshita N, Akagi T, Yamasaki M, Ozeki T, Nojima T, Hiramatsu Y, Nagai N (1992) Osteoclastic features of multinucleated giant cells responding to synthetic hydroxyapatite implanted in rat jaw bone. J Electron Microsc (Tokyo) 41:141–146
14. Suzuki O (2009) Biological role of synthetic octacalcium phosphate in bone formation and mineralization. J Oral Biosci (in press)
15. Mathew M, Brown W, Schroeder L, Dickens B (1988) Crystal structure of octacalcium bis(hydrogenphosphate) tetrakis(phosphate)pentahydrate, $Ca_8(HPO_4)_2(PO_4)_4•5H_2O$. J Chem Crystallogr 18:235–250
16. Suzuki O, Yagishita H, Amano T, Aoba T (1995) Reversible structural changes of octacalcium phosphate and labile acid phosphate. J Dent Res 74:1764–1769
17. Brown WE, Mathew M, Tung MS (1981) Crystal chemistry of octacalcium phosphate. Prog Crystal Growth Charact 4:59–87
18. LeGeros RZ, Daculsi G, Orly I, Abergas T, Torres W (1989) Solution-mediated transformation of octacalcium phosphate (OCP) to apatite. Scan Electron Microsc 3:129–137 discussion 137–138
19. Suzuki O, Kamakura S, Katagiri T (2006) Surface chemistry and biological responses to synthetic octacalcium phosphate. J Biomed Mater Res B Appl Biomater 77:201–212
20. Suzuki O, Nakamura M, Miyasaka Y, Kagayama M, Sakurai M (1993) Maclura pomifera agglutinin-binding glycoconjugates on converted apatite from synthetic octacalcium phosphate implanted into subperiosteal region of mouse calvaria. Bone Miner 20:151–166
21. Tseng YH, Mou CY, Chan JC (2006) Solid-state NMR study of the transformation of octacalcium phosphate to hydroxyapatite: a mechanistic model for central dark line formation. J Am Chem Soc 128:6909–6918
22. Suzuki O, Imaizumi H, Kamakura S, Katagiri T (2008) Bone regeneration by synthetic octacalcium phosphate and its role in biological mineralization. Curr Med Chem 15:305–313
23. He G, Dahl T, Veis A, George A (2003) Nucleation of apatite crystals in vitro by self-assembled dentin matrix protein 1. Nat Mater 2:552–558
24. Moradian-Oldak J, Iijima M, Bouropoulos N, Wen HB (2003) Assembly of amelogenin proteolytic products and control of octacalcium phosphate crystal morphology. Connect Tissue Res 44(Suppl 1):58–64

Session I
Biomechanical–Biological Interface

Effects of zebularine on the apoptosis of 5-fluorouracil via cAMP/PKA/CREB pathway in HSC-3 cells

Maiko Suzuki, Fumiaki Shinohara, Manabu Endo, Masaki Sugazaki, Seishi Echigo, and Hidemi Rikiishi

Abstract. During tumorigenesis, tumor suppressor and tumor-related genes are commonly silenced by aberrant DNA methylation in their promoter regions, which is one of the important determinants of susceptibility to 5-fluorouracil (5-FU) in oral squamous cell carcinoma cells. We investigated the effect of a DNA methyltransferase (DNMT) inhibitor, zebularine (Zeb), on the chemosensitivity of 5-FU and cisplatin (CDDP), and compared the molecular mechanism of action with those of a GSK3β inhibitor, LiCl, and an Hsp90 inhibitor, 17-AAG. A significant apoptotic effect by a combination of Zeb or 17-AAG was found in CDDP treatment; however, considerable suppression of 5-FU-induced apoptosis was observed after incubation with Zeb, 17-AAG, or LiCl. Zeb's suppressive effects were associated with activation of the cAMP/PKA/CREB pathway, differing from mechanisms of 17-AAG and LiCl.

Key words. zebularine, methylation, 5-fluorouracil, apoptosis, OSCC

1 Introduction

5-Fluorouracil (5-FU) and cisplatin (CDDP) are frequently used in combination therapy for the treatment of oral squamous cell carcinoma (OSCC). Altered expression based on gene mutations, gene amplifications, or epigenetic changes that influence apoptotic proteins can provide OSCC cells with resistance to chemotherapeutic drugs. Previously, we had showed the epigenetic influence on the sensitivity of oral carcinoma cell lines to 5-FU or CDDP by evaluating apoptotic inducibility [1]. Zebularine (Zeb) had chemosensitive efficacy with CDDP, whereas Zeb showed

M. Suzuki and H. Rikiishi (✉)
Department of Microbiology and Immunology, Tohoku University Graduate School of Dentistry, 4-1 Seiryo-machi, Aoba-ku, Sendai 980-8575, Japan
e-mail: riki@m.tains.tohoku.ac.jp

F. Shinohara, M. Endo, M. Sugazaki, and S. Echigo
Department of Oral Surgery, Tohoku University Graduate School of Dentistry, 4-1 Seiryo-machi, Aoba-ku, Sendai 980-8575, Japan

T. Sasano et al. (eds.), *Interface Oral Health Science 2009*,
DOI 10.1007/978-4-431-99644-6_15, © Springer 2010

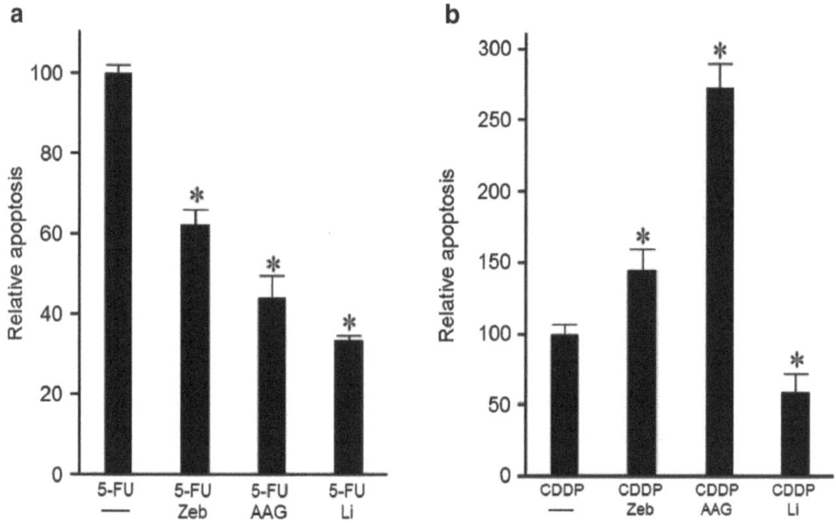

Fig. 1. Effects of each compound on 5-FU- or CDDP-induced apoptosis

inhibitory effect with 5-FU, but the actual mechanisms of combination treatment of 5-FU and Zeb in OSCC have not yet been elucidated in detail. Here, we examine the anti-apoptotic effects of Zeb on 5-FU cytotoxicity, which is involved in apoptosis, apoptosis-related proteins, and the cAMP/PKA/CREB pathway in the well-established OSCC cell line HSC-3.

2 Results and Conclusions

A significant apoptotic effect by a combination of Zeb or 17-AAG was found in CDDP treatment (Fig. 1b); however, considerable suppression of 5-FU-induced apoptosis was observed after incubation with Zeb, 17-AAG, or LiCl (Fig. 1a). Zeb's suppressive effects were associated with activation of the cAMP/PKA/CREB pathway, differing from mechanisms of 17-AAG and LiCl. Suppression of 5-FU-induced apoptosis by Zeb was not associated with increased Bcl-2 and Bcl-xL expressions dependent on transcription factor CREB and with the expression level of thymidylate synthase. In the present study, we identified a more detailed mechanism of action by which Zeb suppresses 5-FU-induced apoptosis. These results indicate that combination therapies have to be carefully investigated due to potential harmful effects in the clinical application of DNMT inhibitors.

Reference

1. Suzuki M, Shinohara F, Nishimura K et al (2007) Epigenetic regulation of chemosensitivity to 5-fluorouracil and cisplatin by zebularine in oral squamous cell carcinoma. Int J Oncol 31:1449–1456

Wnt signaling inhibits cementoblast differentiation

Eiji Nemoto, Yohei Koshikawa, Sousuke Kanaya, Masahiro Tsuchiya, Masato Tamura, Martha J. Somerman, and Hidetoshi Shimauchi

Abstract. Wnt signaling has been implicated in increased bone formation by controlling mesenchymal stem cell or osteoblastic cell functions; however the role of Wnt signaling on cementogenesis has not been examined. Exposure to Wnt3a inhibited the expression of the osteocalcin (*OCN*) gene. This effect was accompanied by decreased gene expression of *Runx2*. Pretreatment with Dickkopf-1 attenuated the suppressive effects of Wnt3a on mRNA expression of *Runx2* and *OCN* on cementoblasts. These findings suggest that canonical Wnt signaling inhibits cementoblast differentiation via regulation of expression of Runx2. Elucidating the role of Wnt in controlling cementoblast function will provide new tools needed to improve on existing periodontal regeneration therapies.

Key words. Wnt signaling, cementoblast, differentiation

1 Introduction

Cementum is an important component of the periodontal attachment apparatus and a key to establishing and regenerating a functional periodontal tissue [1]. Although cementum shares many properties with bone, most notably a remarkable similarity

E. Nemoto (✉), Y. Koshikawa, S. Kanaya, and H. Shimauchi
Department of Periodontology and Endodontology, Tohoku University Graduate School of Dentistry, 4-1 Seiryo-machi, Aoba, Sendai 980-8575, Japan
e-mail: e-nemoto@umin.ac.jp

M. Tsuchiya
Department of Aging and Geriatric Dentistry, Tohoku University Graduate School of Dentistry, Sendai, Japan

M. Tamura
Department of Biochemistry and Molecular Biology, Hokkaido University Graduate School of Dentistry, Sapporo, Japan

M.J. Somerman
Departments of Periodontics, School of Dentistry, University of Washington, Seattle, WA, USA

T. Sasano et al. (eds.), *Interface Oral Health Science 2009*,
DOI 10.1007/978-4-431-99644-6_16, © Springer 2010

in biochemical composition [1], it differs from bone in its histological profile by lacking innervation and vascularization, and has limited remodeling potential [1]. The canonical Wnt/β-catenin signaling pathway has been implicated in promotion of bone formation. Although the precise mechanism of Wnt signaling in bone

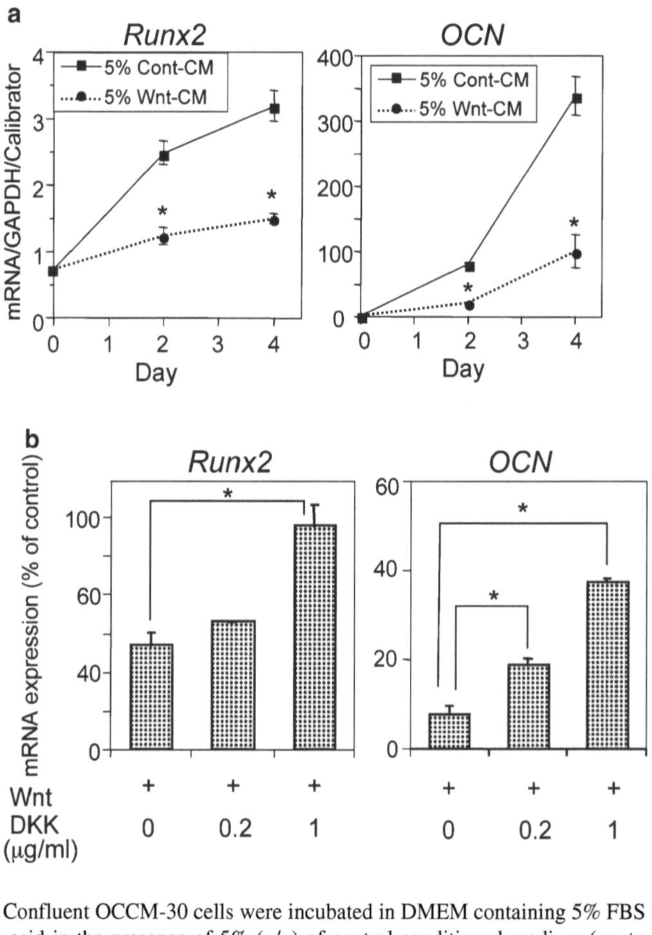

Fig. 1. (**a**) Confluent OCCM-30 cells were incubated in DMEM containing 5% FBS with 50 μg/ml ascorbic acid in the presence of 5% (v/v) of control-conditioned medium (control-CM from ATCC) or Wnt3a-conditioned medium (Wnt3a-CM from ATCC) for the indicated times. Medium was changed every 2 days. Total cellular RNA was extracted (Trizol®, Gibco), and transcripts were analyzed by real-time quantitative RT-PCR (iCycler, Bio-Rad). Representative data of three separate experiments are shown as means ± SD of triplicate assays. Statistical significances are shown (*$p<0.05$ vs. control-CM). (**b**) Confluent OCCM-30 cells were pretreated with the indicated concentration of DKK1 (R&D systems, Minneapolis, MN) for 30 min in DMEM containing 5% FBS with 50 μg/ml ascorbic acid, and further incubated by addition of 5% (v/v) of control-CM or Wnt3a-CM in the continuous presence of DKK1 for 2 days. Total cellular RNA was extracted and transcripts were analyzed by real-time quantitative RT-PCR. Representative data of two separate experiments are shown as means ± SD of mRNA expression (% of respective control) of triplicate assays. Statistical significances are shown (*$p<0.05$ vs. control)

biology remains unclear, it appears that its effects are related to both cell type and differentiation state. Here, new evidence is provided showing that Wnt signaling inhibits cementoblast differentiation.

2 Results and Discussion

An immortalized murine cementoblast cell line (OCCM-30) is established as described previously [2]. We examined the effect of Wnt3a on cementoblast differentiation by quantitative RT-PCR. Cells were incubated w/wo Wnt3a-CM or control-CM over a 4-day period. The gene expression of *OCN* and *Runx2* was increased with time in control-CM, however, significant inhibition was observed in Wnt3a-CM (Fig. 1a). Considering that Runx2 transactivates *OCN* [3] promoters in osteoblastic cells, Wnt3a signaling pathway may decrease cementoblast differentiation by inhibiting *Runx2* expression. Pretreatment of cells with DKK1, a canonical Wnt antagonist, attenuated the suppressive effects of Wnt3a on mRNA expression of *Runx2* and *OCN* (Fig. 1b). These data suggested that inhibition of cementoblast differentiation by Wnt3a was mediated by the canonical Wnt signaling pathway.

References

1. Bosshardt DD (2005) Are cementoblasts a subpopulation of osteoblasts or a unique phenotype? J Dent Res 84:390–406
2. D'Errico JA, Berry JE, Ouyang H et al (2000) Employing a transgenic animal model to obtain cementoblasts in vitro. J Periodontol 71:63–72
3. Ducy P, Zhang R, Geoffroy V et al (1997) Osf2/Cbfa1: a transcriptional activator of osteoblast differentiation. Cell 89:747–754

Prevention of necrotic actions of nitrogen-containing bisphosphonates (NBPs) in mice by non-NBPs (clodronate and etidronate)

Takefumi Oizumi, Kouji Yamaguchi, Hiromi Funayama, Hiroshi Kawamura, Shunji Sugawara, and Yasuo Endo

Abstract. Nitrogen-containing bisphosphonates (NBPs) have powerful antibone-resorptive effects (ABREs), but they induce unexpected side effect, osteonecrosis of jaw-bones (ONJ). The mechanism underlying NBP-associated ONJ is unclear. Zoledronate, the strongest NBP, exhibits the highest incidence of ONJ. We previously found that clodronate and etidronate (non-NBPs) inhibit NBP-induced inflammation. Here, we found the following. (a) Subcutaneous injection of zoledronate into ear-pinnas induced inflammation and then necrosis at these sites. These effects of zoledronate were strongest among NBPs tested, while non-NBPs lacked these effects. (b) Coinjection of clodronate or etidronate reduced the amount of zoledronate retained within the ear tissue and reduced the inflammatory and necrotic effects of zoledronate. (c) When zoledronate and clodronate were intraperitoneally injected, clodronate little affected the ABRE of zoledronate, as well as those of other NBPs. In contrast, etidronate markedly reduced the ABRE of zoledronate. Notably, etidronate reduced the ABRE of zoledronate even when it was injected 16 h after the injection of zoledronate. These results suggest that (a) clodronate and etidronate may inhibit the entry of NBPs into cells related to inflammation and/or necrosis and prevent NBPs' side effects, (b) clodronate could be useful as a combination drug with NBPs for preventing their side effects while retaining their ABREs, (c) etidronate (but not clodronate) may competitively inhibit the binding of NBPs to bone hydroxyapatite (BHA), and this reagent may reduce NBPs that have

T. Oizumi (✉), K. Yamaguchi, and H. Kawamura
Department of Oral and Maxillofacial Surgery, Graduate School of Dentistry, Tohoku University, 4-1 Seiryo-machi, Aoba-ku, Sendai 980-8575, Japan
e-mail: takefumi-oizumi@mail.tains.tohoku.ac.jp

T. Oizumi, S. Sugawara, and Y. Endo
Department of Molecular Regulation, Graduate School of Dentistry, Tohoku University, 4-1 Seiryo-machi, Aoba-ku, Sendai 980-8575, Japan

H. Funayama
Department of Pediatric Dentistry, Tsurumi University School of Dental Medicine, 2-1-3 Tsurumi, Tsurumi-ku, Yokohama 230-8501, Japan

T. Sasano et al. (eds.), *Interface Oral Health Science 2009*,
DOI 10.1007/978-4-431-99644-6_17, © Springer 2010

already accumulated within bones, and (d) etidronate, if used as a substitution drug for NBPs, may be effective at treating or preventing NBP-associated ONJ.

Key words. bisphosphonate, osteonecrosis, zoledronate, clodronate, etidronate

1 Introduction

Nitrogen-containing bisphosphonates (NBPs) have powerful antibone-resorptive effects (ABREs). They bind strongly to bone hydroxyapatite (BHA). However, in addition to their known inflammatory side effects, recent clinical applications have disclosed an unexpected side effect, osteonecrosis of jaw-bones (ONJ). We previously found in mice that clodronate and etidronate (non-NBPs), when coadministered with alendronate (an NBP), inhibited the latter's inflammatory effects. However, etidronate also reduced the ABRE of alendronate. Zoledronate is the strongest antibone-resorptive NBP, but it is also associated with the highest incidence of ONJ. The mechanism underlying NBP-associated ONJ is unclear.

2 Materials and Method

We used the following NBPs, which have been confirmed to induce ONJ, to test their antibone-resorptive activities, uptake into tissues, and necrotic actions in mice. Moreover, we examined the effects of non-NBPs (Clo and Eti) on the actions of NBPs.

3 Summary of Results

(a) Subcutaneous injection of Zol into ear-pinnas induced inflammation and then necrosis of the ear skin (relative potencies of NBPs: Zol >>Pam ≥ Ale > Ris), while non-NBPs lacked this effect.
(b) Coinjection of Clo or Eti reduced these reactions and also reduced the amount of Zol retained within the ear tissue.
(c) When Zol and Clo were intraperitoneally injected, Clo little affected the ABRE of Zol as well as those of other NBPs.
(d) In contrast, Eti, when combined with Zol or another NBP, markedly reduced the ABRE of the NBP. Notably, Eti reduced the ABRE of Zol even when it was injected 16 h after the injection of Zol.

4 Discussion

(a) Clo and Eti may inhibit the entry of NBPs into cells related to inflammation and/or necrosis and prevent NBPs' side effects.
(b) Clo could be useful as a combination drug with NBPs for preventing their side effects while retaining their ABREs.
(c) Eti (but not Clo) may competitively inhibit the binding of NBPs to BHA, and this reagent may at least partly eliminate (or substitute for) NBPs that have already accumulated within bones.
(d) Eti, if used as a substitution drug for NBPs, may be effective at treating or preventing NBP-associated ONJ.

Interface, implant, regenerated bone and recipient alveolar bone

Masahiro Nishimura, Yuuhiro Sakai, Fumio Suehiro, Masahiro Tsuboi, Koichi Kamada, Tomoharu Hori, Masanori Sakai, Mika Takeda, Koichiro Tsuji, and Taizo Hamada

Abstract. This section shows a summary of our model on how to augment alveolar ridge by minimum intervention using autologous alveolar bone-derived mesenchymal stem cells (MSCs). We collected MSC from alveolar bone using newly developed puncture needle and expanded them in vitro. We combined MSC with calcium phosphate scaffold and packed them in capsules. Two capsules were implanted underneath the ablated periosteum of edentulous site on canine maxillary bone. We successfully observed osseointegration between augmented bone and implant.

Key words. mesenchymal stem cells, alveolar ridge augmentation, implant, regenerative medicine, alveolar bone

---

M. Nishimura (✉)
Department of Prosthetic Dentistry, Graduate School of Biomedical Sciences,
Nagasaki University, 1-7-1 Sakamoto, Nagasaki 852-8588, Japan
e-mail: mnishi@nagasaki-u.ac.jp

Y. Sakai
GC Corporation, 76-1 Hasunuma-Cho, Itabashi-ku, Tokyo 174-8585, Japan

M. Sakai, M. Takeda, and K. Tsuji
Two Cells Co. Ltd., 4-5-17-501 Danbara, Minami-ku, Hiroshima 732-0811, Japan

F. Suehiro
Graduate Program for Bio-Dental Education, Graduate School of Biomedical Sciences,
Hiroshima University, 1-2-3 Kasumi, Minami-ku, Hiroshima 734-8553, Japan

M. Tsuboi, K. Kamada, and T. Hori
Department of Prosthetic Dentistry, Graduate School of Biomedical Sciences,
Hiroshima University, 1-2-3 Kasumi, Minami-ku, Hiroshima 734-8553, Japan

T. Hamada
Department of Oral Health Care Promotion, Tohoku University Graduate School of Dentistry,
4-1 Seiryo-machi, Aoba-ku, Sendai 980-8575, Japan

T. Sasano et al. (eds.), *Interface Oral Health Science 2009*,
DOI 10.1007/978-4-431-99644-6_18, © Springer 2010

1 Introduction

For long prognosis of implants, the presence of sufficient bone volume is an important prerequisite; however, implant surgeons are frequently faced with severe alveolar bone defects caused by periodontal disease or various trauma and aging. Autologous bone transplantation is used as a golden standard for bone regeneration. However, there are so many side effects caused by bone collection, so we have developed new method using mesenchymal stem cell (MSC) as a cell source. MSC are an attractive cell source for bone regenerative medicine because they can be easily expanded and have multipotentiality that includes osteogenesis. Alveolar bone-derived MSC also have multipotency [1]; however, suitable device to collect the cells from alveolar bone has not been developed. So, we have developed a new puncture needle optimized for alveolar bone marrow aspiration that dentists can approach easily. Also, we have developed a new method to pack MSC and scaffold during transplantation surgery. We have established minimum intervention method to get MSC and to augment alveolar ridge suitable for following implant insertion.

2 Material and Methods

We used an 11-year-old dog, which is equivalent to a 70-year-old human being. All premolars on the upper jaw of the dog were extracted. Bone marrow (0.5 ml) was aspirated from the lower jaw of the same dog using puncture needle (Fig. 1a) and MSCs were cultured with Dulbecco's Modified Eagle's Medium (SIGMA, St. Louis, MO) supplemented with 10% fetal bovine serum (GIBCO BRL, Gaithersburg, MD) and antibiotic–antimycotic (100 units/ml penicillin G, 100 μg/ml streptomycin, and 0.25 mg/ml amphotericin B, GIBCO BRL) at 37°C in 5% CO_2/95% air. Four months after teeth extraction, incision (1.5 cm) was made down to the bone on the distal region of the canine teeth; then, tissues were dissected and

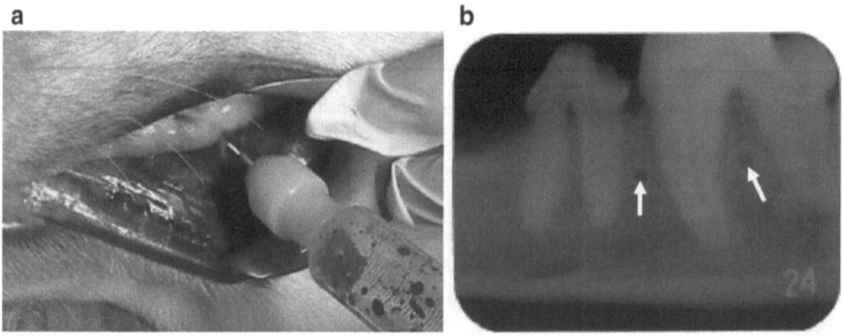

Fig. 1. (**a**) Bone marrow (0.5 ml) was aspirated from the lower jaw of the same dog using developed puncture needle. (**b**) Dental X-ray photograph after aspiration

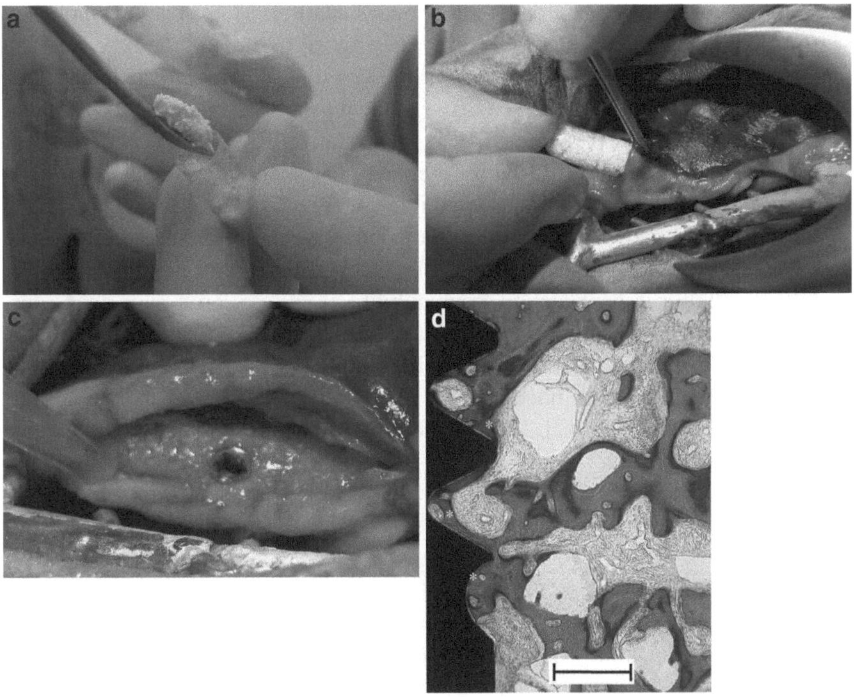

Fig. 2. (**a**) MSC and calcium phosphate scaffold (obtained from GC corporation) were packed in two capsules as transplants. (**b**) Two capsules were explanted underneath the periosteum. (**c**) Titanium implants (GENESiO, GC corporation, Φ3.8×8 mm) were inserted on the regenerated bone. (**d**) Villanueva's Goldner stain of regenerated bone around implant. *Left black area* shows implant, and white sterisks indicate the integrated bone to the inserted implant. Black bar = 0.5 mm

subperiosteal pocket was formed. MSC (5×10^7)/scaffold complexes (1 g calcium phosphate scaffold obtained from GC corporation) were packed in two capsules (Fig. 2a, patent pending) and then explanted underneath the ablated periosteum (Fig. 2b). Same explants without MSC were explanted to the opposite side as a control. After 3 months, titanium implant was inserted (Fig. 2c). Further 3 months after, dissected bone around implant was embedded into resin and sectioned for histological evaluation.

3 Results and Discussion

We could aspirate bone marrow from jaw with minimum intervention using a new puncture needle. Just a small hole on alveolar bone was observed by the puncture needle (Fig. 1b). Operativity of the capsules to transplant the complexes underneath the ablated periosteum was very simple. At control side, only a small amount of scaffolds were observed 3 months after transplantation. Major part of implanted scaffold might diffuse during the healing of gingival tissue. On the other hand, a maximum of 3 mm bone regeneration was observed at the MSC transplanted side. No obvious boundary was observed between the regenerated bone and the recipient bone. Hardness of the regenerated bone was relatively softer than the recipient bone; however, we could drill the regenerated bone smoothly. Regenerated bone was successfully attached to the inserted implant (Fig. 2d).

This review shows a successful case of bone regeneration; however, we have also got several failure cases as well. Differences of initial cell number, character of culture cells for transplantation, compatibility of culture serum with each cells, and differences of niche at recipient site are now under investigation as they may possibly be the main causes. For example, a guideline of the culture conditions required to culture alveolar bone marrow MSC derived from older individuals has not been well-established [2]. So, we should clarify these points for predictable bone augmentation.

4 Conclusions

We could augment canine alveolar ridge using alveolar bone marrow derived-MSC and scaffold. Dental titanium implant successfully integrated to the regenerated bone. So, the alveolar MSC might be useful for bone augmentation following implant treatment.

References

1. Matsubara T, Suardita K, Ishii M et al (2005) Alveolar bone marrow as a cell source for regenerative medicine: differences between alveolar and iliac bone marrow stromal cells. J Bone Miner Res 20:399–409
2. Han J, Okada H, Takai H et al (2009) Collection and culture of alveolar bone marrow multipotent mesenchymal stromal cells from older individuals. J Cell Biochem 107:1198–1204

Activation of matrix metalloproteinase-2 at the interface between epithelial cells and fibroblasts from human periodontal ligament

Mitsuru Shimonishi, Ichiro Takahashi, Masashi Komatsu, and Masahiko Kikuchi

Abstract. Matrix metalloproteinase (MMP)-2 can degrade type IV collagen, and MMP-14 can activate pro-MMP-2. Bone sialoprotein (BSP) specifically binds pro-MMP-2 and active MMP-2. The expression of MMP-2, MMP-14, and BSP were analyzed by immunohistochemistry, in situ hybridization, and RT-PCR at the interface between cells of the epithelial rests of Malassez (ERM) and fibroblasts from human periodontal ligament (HPDL). ERM cells at the interface strongly expressed MMP-2 and MMP-14 proteins. In situ hybridization analysis showed that HPDL fibroblasts expressed MMP-2 mRNA, and ERM cells expressed MMP-14 mRNA at the interface strongly. BSP and its mRNA were expressed strongly in HPDL fibroblasts at the interface. RT-PCR analysis demonstrated that the expressions of MMP-2 mRNA and BSP mRNA were significantly high. These findings indicate that upregulated MMP-2 activated by MMP-14 in ERM cells and BSP in HPDL fibroblasts could degrade matrix molecules.

Key words. MMP-2, MMP-14, bone sialoprotein, epithelial rests of Malassez

1 Introduction

Epithelial-mesenchymal interactions are responsible for morphogenesis and cell differentiation during periodontal regeneration. We demonstrated that the synthesis of type IV collagen and laminin was induced by direct interaction at the interface

M. Shimonishi (✉) and M. Kikuchi
Division of Comprehensive Dentistry, Tohoku University Dental Hospital, Tohoku University Graduate School of Dentistry, 4-1 Seiryo-machi, Aoba-ku, Sendai 980-8575, Japan
e-mail: shimo@ddh.tohoku.ac.jp

I. Takahashi
Division of Orthodontics and Dentofacial Orthopedics, Tohoku University Graduate School of Dentistry, 4-1 Seiryo-machi, Aoba-ku, Sendai 980-8575, Japan

M. Komatsu
Division of Operative Dentistry, Tohoku University Graduate School of Dentistry, 4-1 Seiryo-machi, Aoba-ku, Sendai 980-8575, Japan

T. Sasano et al. (eds.), *Interface Oral Health Science 2009*,
DOI 10.1007/978-4-431-99644-6_19, © Springer 2010

between cells of the epithelial rests of Malassez (ERM) and fibroblasts from human periodontal ligament (HPDL) [1]. Matrix metalloproteinase (MMP)-2 can degrade type IV collagen. The activated MMP-14 (MT1-MMP) binds the tissue inhibitor of metalloproteinases (TIMP)-2 by its N-terminal inhibitory domain. The C-terminal domain of the bound TIMP-2 acts as a receptor for binding the C-terminal hemopexin domain of pro-MMP-2, and MMP-14 can activate pro-MMP-2 [2]. Moreover, bone sialoprotein (BSP), a member of the SIBLING (Small, Integrin-Binding LIigand, N-linked Glycoprotein) family, specifically binds pro-MMP-2 and active MMP-2. The current study was undertaken to examine the expression of MMP-2, MMP-14, and BSP at the interface between ERM cells and HPDL fibroblasts.

2 Materials and Methods

HPDL tissues, which were sampled from the root of extracted teeth, produced outgrowths containing both ERM cells and HPDL fibroblasts in a modified serum-free medium. ERM cells were stained positively for broad-spectrum antibodies to cytokeratins, indicating their epithelial origin, while HPDL fibroblasts did not show cytokeratin expression at the interface in the same dishes. ERM cells showed higher positive signals for amelogenin mRNA. However, amelogenin mRNA signal was not detectable in HPDL fibroblasts. These results supported that ERM cells were different from HPDL fibroblasts and derived from the odontogenic epithelial origin. The expression of MMP-2, MMP-14, and BSP were analyzed by immunohistochemistry, in situ hybridization, and RT-PCR.

3 Results

ERM cells at the interface expressed MMP-2 and MMP-14 proteins strongly. In situ hybridization analysis showed that HPDL fibroblasts expressed MMP-2 mRNA, and ERM cells expressed MMP-14 mRNA at the interface strongly. BSP and its mRNA were expressed strongly in HPDL fibroblasts at the interface. RT-PCR analysis demonstrated that the expressions of MMP-2 mRNA and BSP mRNA were significantly higher, when ERM cells and HPDL fibroblasts were cocultured, than when each of them was cultured alone. However, the interaction between them did not affect the expression of MMP-14 mRNA.

4 Conclusion

These findings indicate that the ERM cells stimulate the production of MMP-2 in HPDL fibroblasts. Upregulated MMP-2 activated by MMP-14 in ERM cells and BSP in HPDL fibroblasts could degrade matrix molecules, such as Type IV collagen, in the basal membrane between ERM cells and HPDL fibroblasts.

Acknowledgment This work was supported by a Grant-in-Aid for Scientific Research (C) (No. 19592167) from Japan Society for the Promotion of Science, Japan.

References

1. Shimonishi M, Sato J, Takahashi N et al (2005) Expression of type IV collagen and laminin at the interface between epithelial cells and fibroblasts from human periodontal ligament. Eur J Oral Sci 113:34–40
2. Werb Z (1997) ECM and cell surface proteolysis: regulating cellular ecology. Cell 91:439–442

Histomorphometric study of alveolar bone-implant (miniscrew) interface used as an orthodontic anchorage

Toru Deguchi, Masakazu Hasegawa, Masahiro Seiryu, Takayoshi Daimaruya, and Teruko Takano-Yamamoto

Abstract. The use of miniscrews as an orthodontic anchorage has become widely accepted among orthodontists throughout the world. However, only few histological studies have been reported with regard to the healing process at the bone-implant interface in the past. Therefore, we have (1) analyzed the healing process of alveolar bone surrounding miniscrew by dynamic and static histomorphometric indices and (2) histomorphometrically assessed the change in the cortical bone thickness. Results indicated that small miniscrews were able to function as rigid osseous anchorage against orthodontic load with minimal (under 3 weeks) healing period. We suggest that this sufficient amount of cortical (woven) bone at the initial stage of the healing enables the immediate loading in miniscrews to resist against orthodontic force. Furthermore, less amount of cortical bone formed at the head of the miniscrew may be one reason for the higher failure rate in the mandible compared to the maxilla.

Key words. histomorphometric, miniscrew, orthodontic, dog

1 Introduction

The control of anchorage during tooth movement was one of the major concerns in practical orthodontics. Recently, innovative approach to control the anchorage was reported by the use of miniscrews. With the use of miniscrews as an anchorage, effective tooth movement such as molar intrusion and retraction of the incisors were possible without the loss of anchorage. Furthermore, compared to conventional orthodontic approach, no patient cooperation was required. In recent clinical reports,

T. Deguchi, M. Hasegawa, M. Seiryu, T. Daimaruya, and T. Takano-Yamamoto (✉)
Division of Orthodontics and Dentofacial Orthopedics, Tohoku University Graduate School of Dentistry, Sendai, Japan
e-mail: t-yamamo@mail.tains.tohoku.ac.jp

T. Sasano et al. (eds.), *Interface Oral Health Science 2009*,
DOI 10.1007/978-4-431-99644-6_20, © Springer 2010

miniscrew failure was reported as one of the problems during the use of these miniscrews. In order to assess the reasons for the failure for these miniscrews, the healing process of surrounding bone has to be analyzed to prevent miniscrews to fail. A total of 96 miniature implants (1.0×5.0 mm; 48 loaded and 48 unloaded) were placed in the mandible and maxilla of eight male dogs. The implants were allowed to heal for three different periods (3, 6, and 12 weeks) followed by 12 weeks of 2 N orthodontic force application. Bone specimens containing implants were collected for histomorphometric analysis. Analyzed histomorphometric indices were bone-implant contact (%), bone volume/total volume, woven bone volume/total volume, and bone formation rate. Cortical bone thickness was analyzed in three different locations (within 1, 1–2, and 3–4 mm away from the miniscrew).

2 Histomorphometric Indices of Alveolar Bone Surrounding Miniscrew

The results indicate that clinical rigidity was achieved by 97% of the miniscrew. After 3 weeks of healing in non-loaded miniscrews, significant amount of bone was observed (increased bone-implant contact) compared to later healing stages with increased woven bone volume in both jaws. Furthermore, bone formation rate significantly increased after 3 weeks compared to other healing stages or after the orthodontic loading. After 6 weeks of healing, no significant difference was observed between the 12-week group and loaded groups.

3 Cortical Bone Thickness Surrounding Miniscrew

The change in the cortical bone thickness resulted that in non-forced groups, significant amount of cortical bone was formed at the head of the implant at the initial stage of the healing process in the maxilla. However, less cortical bone formation was observed in the mandible compared to the control. After the force application, increased bone formation was observed within 1 mm of the miniscrew compared to other regions in both jaws. In the mandible, significantly less cortical bone was observed 3 and 6 weeks after the force application compared to the control. Bone-implant contact revealed that the osseous tissue surrounding the miniscrew matured from the apex toward the head of the implant.

4 Conclusion

Therefore, in the case of dental implants, healing duration of 2–3 months known as "osseointegration" is necessary since it has to resist heavy occlusal force and requiring long-term maintenance. However, in the case of miniscrew, immediate loading is possible since only 100–200 gm of orthodontic force is applied, and has to be taken out after the use as an anchorage as a temporary anchor. Thus, we suggest that in the case of miniscrew, "mechanical interdigitation" by the cortical bone may be required rather than "osseointegration."

Mechanical stress modulates bone remodeling signals

Hiroyuki Matsui, Naoto Fukuno, Osamu Suzuki, Kohsuke Takeda, Hidenori Ichijo, Takayasu Kobayashi, Shinri Tamura, and Keiichi Sasaki

Abstract. Mechanical stress plays an essential role in bone homeostasis. Although mechanotransduction-induced de novo gene expression is required for bone remodeling, the molecular mechanism of intracellular signaling, which leads to regulation of gene expression, is not fully understood. Here, we show that JNK and p38 [two stress-responsible mitogen-activated protein kinases (MAPKs)] are activated via ASK1 (a stress-responsible MAPK kinase) in mechanical stretch loaded MC3T3-E1 preosteoblasts. Using pharmaceutical and RNAi approaches, we demonstrated that ASK1 is activated via Ca^{2+} influx-induced reactive oxygen species generation. Furthermore, we observed that ASK1-activated JNK and p38 induced the expression of two bone remodeling related genes, *Fn14* and *MCP-3*, respectively. These findings suggest that mechanical stress-activated JNK and p38 induce cytokine cross-talks between osteoblasts and bone marrow-derived monocytes and macrophages, which may play key roles in bone remodeling.

Key words. mechanical stress, JNK/p38 MAP kinase, bone remodeling, Fn14, MCP-3

H. Matsui (✉), N. Fukuno, and K. Sasaki
Division of advanced Prosthetic Dentistry, Tohoku University Graduate School of Dentistry, Sendai, Japan
e-mail: daveevans@idac.tohoku.ac.jp

H. Matsui, N. Fukuno, T. Kobayashi, and S. Tamura
Department of Biochemistry, IDAC, Tohoku University, Sendai, Japan

O. Suzuki
Division of Craniofacial Function Engineering, Tohoku University Graduate School of Dentistry, Sendai, Japan

K. Takeda and H. Ichijo
Tokyo University Graduate School of Pharmaceutical Sciences, Tokyo, Japan

T. Sasano et al. (eds.), *Interface Oral Health Science 2009*,
DOI 10.1007/978-4-431-99644-6_21, © Springer 2010

1 Introduction

Mechanical stress is implicated in regulation of bone remodeling. Mechanical stress stretches the surface of osteoblastic cells and generates biochemical signals, which is required for the regulation of expression of bone remodeling related genes [1]. However, the detailed mechanism of intracellular signaling induced by stretch loading is not fully understood.

c-Jun N-terminal kinase (JNK) and p38 mitogen-activated protein kinase (MAPK) play important roles generally in stress responses. They are activated by biological or physicochemical stressors and control wide-variety of cellular functions such as differentiation, proliferation, apoptosis, and inflammation through regulation of de novo gene expression [2]. In the present study, we investigated the possible roles of JNK and p38 signaling pathways in mechanical stress-induced bone remodeling.

2 Materials and Methods

2.1 Cell Culture

MC3T3-E1 cells were maintained in α-MEM supplemented with 10% (v/v) fatal bovine serum at 37°C in 5% CO_2.

2.2 Mechanical Stretch Loading

Mechanical stretch experiments were performed using a ST-140 cell cyclic stretcher system (Strex, Osaka, Japan). JNK inhibitor, SP600125 (20 μM) or p38 inhibitor, SB203580 (10 μM) was added to the medium 1 h before stretch. After stretch loading, cells were harvested and subjected to western blotting [3], DNA microarray, and RT-PCR.

2.3 RNA Isolation and DNA Microarray

Total RNA extraction was performed using RNeasy kit (QIAGEN) as in manufactures' protocol. The expression level of over 39,000 genes was analyzed by DNA microarray (Genechip Mouse Genome 430 2.0; Affimetrix). Data analysis was performed using ArrayAssist (Stratagene).

2.4 Semiquantitative RT-PCR Amplification

Single-stranded cDNA was synthesized using Revtra Ace reverse transcriptase (Takara). PCR amplifications were performed with specific primers, using KOD plus DNA polymerase (Takara). After the agarose gel electrophoresis, the band fluorescent intensity was measured with ImageJ software (NIH, public domain). Relative expression level was corrected with that of GAPDH.

3 Results and Discussion

Western blot analysis showed that JNK and p38 were activated by mechanical stretch loading. These activities were substantially inhibited by the addition of EGTA (an extracellular Ca^{2+} chelator), indicating that Ca^{2+} influx is required for the activation of JNK and p38. We also observed that ASK1 (a MAP3K) was activated via Ca^{2+} influx in mechanical stretch loaded cells. In addition, transfection of siRNA for ASK1 abrogated the activation of both JNK and p38. These results indicate that ASK1 mediates the cyclic stretch-induced phosphorylation of JNK and p38. Previous studies have shown that generation of reactive oxygen species (ROS) is required for the activation of ASK1. Therefore, we asked whether ROS generation was involved in the mechanical stretch-induced activation of ASK1, JNK, and p38, and observed that pretreatment of the cells with NAC (N-acetyl cysteine, an ROS Scavenger) readily suppressed the stretch load-induced phosphorylation of these three protein kinases.

To identify the genes regulated downstream of JNK and p38, we used DNA microarray analysis. Cells were preincubated for 1 h with SP600125 (an inhibitor of JNK) or SB203580 (an inhibitor of p38), or left untreated, and then subjected to mechanical stretch loading for 6 h. Semiquantitative RT-PCR analysis of the candidate genes obtained by DNA microarray revealed that two bone remodeling related genes, *fibroblast growth factor inducible 14* (*Fn14*) and *monocyte chemoattractant protein-3* (*MCP-3*), were upregulated by activation of JNK and p38, respectively. Moreover, expression of these two genes were suppressed by either NAC application or knockdown of ASK1 by siRNA indicating that ASK1-activated JNK and p38 induced the expression of Fn14 and MCP-3, respectively. Fn14 ligand TWEAK is a macrophage-producing cytokine that is implicated in regulation of mesenchymal progenitor cell differentiation and proliferation [4]. MCP-3 induces the migration of bone marrow cells, and has an ability to enhance RANKL mediated osteoclast formation [5]. These data suggest that mechanical stress-induced activation of JNK and p38 in osteoblasts lead to cytokine cross-talks between osteoblasts and bone marrow-derived cells, which may play a key role in bone remodeling.

References

1. Zaidi M (2007) Nat Med 13:791–801
2. Takeda K, Noguchi T, Naguro I et al (2008) Annu Rev Pharmacol Toxicol 48:199–225
3. Kishimoto K, Matsumoto K, Ninomiya-Tsuji J (2000) J Biol Chem 275:7359–7364
4. Winkles JA (2008) Nat Rev Drug Discov 7:411–425
5. Yu X, Huang Y, Collin-Osdoby P et al (2004) J Bone Miner Res 19:2065–2077

Expression analysis of p51/p63 in enamel organ epithelial cells

Takashi Matsuura, Hirokazu Nagoshi, Yasuhiro Tomooka, Shuntaro Ikawa, and Keiichi Sasaki

Abstract. p51/p63, one of the tumor suppressor p53 family members, is known to maintain proliferative potential and immaturity of epithelial stem cells. p51/p63 is known to be expressed in enamel organ epithelium during tooth development. Nevertheless, its functions in tooth development have remained unclear. Thus, functions of p51/p63 in tooth development were investigated. First, expression patterns of p51/p63 in 5 cell lines established from a lower mandibular molar tooth germ of a fetal mouse were examined by western blot analysis and RT-PCR analysis. ΔNp51B/ΔNp63α expression (one of p51/p63 isoforms), which is known to maintain immaturity of epithelial stem cells, was detected in immature cell lines at high levels and in mature cell lines at low levels. These results suggest that ΔNp51B/ΔNp63α may play significant roles in amelogenesis.

Key words. p51, p63, ameloblast, tooth development, differentiation

1 Introduction

p51, also known as p63, is a homologue of the tumor suppressor and transcription factor p53 (hereafter referred to as p51) [1]. p51 isoforms are designated as TAp51A, TAp51B, ΔNp51A, and ΔNp51B. p51 appears to function mainly in embryonic development, in contrast to p53 in tumor suppression. The p51–/– mice, which suffer severe defects in epidermis and limbs, die within hours after birth

T. Matsuura and K. Sasaki (✉)
Division of Advanced Prosthetic Dentistry, Tohoku University Graduate School of Dentistry, Sendai 980-8575, Japan
e-mail: keii@mail.tains.tohoku.ac.jp

H. Nagoshi and S. Ikawa
Ikawa Group, Center for Interdisciplinary Research, Tohoku University, Aoba-ku, Sendai, Japan

Y. Tomooka
Department of Biological Science and Technology, Tissue Engineering Research Center, Tokyo University of Science, Yamazaki 2641, Noda, Chiba 278-8510, Japan

DOI 10.1007/978-4-431-99644-6_22, © Springer 2010

because of severe dehydration and also lack ectodermal appendages including teeth and mammary glands. Nevertheless, the detailed functions of p51 in tooth development remain unclear. This chapter is aimed to survey the functions of p51 in tooth development.

2 Expression Pattern of p51 in emtg-1 to -5 Cells

Five cell lines named emtg (epithelium of molar tooth germ)-1 to -5, which were established from a lower mandibular molar tooth germ of a fetal mouse, were used [2]. Each of these cell lines represents distinct differentiation stages of amelogenesis; emtg-4 cells were derived from inner enamel epithelial cells, emtg-2 and -3 cells were from preameloblasts, and emtg-1 and -5 cells were from ameloblasts. Expression pattern of p51 in emtg-1 to -5 cells was examined by western blot analysis and RT-PCR analysis.

Western blot analysis and RT-PCR analysis using emtg-1 to -5 cells revealed that the expression of TAp51 isoforms were undetectable in each of these cell lines. ΔNp51B was expressed in emtg-2, -3, and -4 cells at high levels and emtg-1 and -5 cells at low levels. ΔNp51A was expressed in each of these cell lines at very low levels.

3 Effects of p51 Knockdown on emtg-2 Cells

The effects of p51 knockdown on emtg-2 cells, which is likely to be the most immature cell line of the five, were examined using small hairpin RNAs (shRNAs) containing retroviral vector. shRNAs were designed so as to specifically target ΔNp51A, ΔNp51B, or EGFP. shEGFP was used as a control.

Western blot analysis and RT-PCR analysis in emtg-2 cells, with reduced expression of ΔNp51A and ΔNp51B revealed that the reduction of ΔNp51B expression led to the reduction of Msx2 expression, which is expressed in undifferentiated ameloblasts and downregulated in secretory stage ameloblasts.

4 Functions of p51 in Tooth Development

ΔNp51 isoforms are expressed in enamel organ epithelial cells and ΔNp51B expression may be required for maintaining the immaturity of enamel organ epithelial cells. To further examine the role of p51 expression in enamel organ epithelial cells, stable cell lines expressing various p51 isoforms or shRNA targeting various p51 isoforms using inducible retroviral systems are currently being established.

References

1. Osada M, Ohba M, Kawahara C et al (1998) Cloning and functional analysis of human p51, which structurally and functionally resembles p53. Nat Med 4:839–843
2. Komine A, Suenaga M, Nakao K et al (2007) Tooth regeneration from newly established cell lines from a molar tooth germ epithelium. BBRC 353(3):758–763

Osteogenesis by gradually expanding the interface between bone surface and periosteum: preliminary analysis of the use of novel plate and bone marrow stem cell administration in rabbits

Koichiro Sato, Naoto Haruyama, Yoshinaka Shimizu, Junichi Hara, and Hiroshi Kawamura

Abstract. The periosteum consists of the cells that are capable of differentiating into osteoblasts. Recently, gradually expanding the interface between bone surface and periosteum using the titanium-meshed plate has been suggested for osteogenesis. Meanwhile, the administration of mesenchymal stem cells (MSCs) into callus has been postulated for facilitating osteogenesis. We tested a novel mesh plate consisting of a mixture of poly-L-lactide (PLLA) and particulate resorbable uncalcined hydroxyapatite (u-HA) as an alternative material for the periosteal distraction in rabbits. In addition, we also performed preliminary analysis on the effect of rabbit MSCs administrated into the gap created by periosteal distraction. Histological analysis by hematoxylin and eosin staining revealed that the bone formation was successfully induced at the gap of periosteal distraction by the novel plate. The MSCs appeared to have a positive effect on the bone formation. Here, we showed that the novel PLLA/u-HA plate could be a replacement of the titanium-meshed plate in the periosteal distraction.

Key words. periosteal distraction, periosteum

Recently, gradually expanding the interface between bone surface and periosteum using the titanium-meshed plate has been suggested for osteogenesis [1]. However, Sencimen et al. reported that bone tissue newly formed by periosteal distraction

K. Sato (✉), J. Hara, and H. Kawamura
Division of Maxillofacial Surgery, Department of Oral Medicine and Surgery, Tohoku University Graduate School of Dentistry, Sendai 980-8575, Japan
e-mail: koichiro@mail.tains.tohoku.ac.jp

N. Haruyama
Division of Oral Dysfunction Science, Department of Oral Health and Development Sciences, Tohoku University Graduate School of Dentistry, Sendai 980-8575, Japan
Global Center of Excellence (GCOE) Program, International Research Center for Molecular Science in Tooth and Bone Diseases, Tokyo Medical and Dental University, Tokyo 113-8549, Japan

Y. Shimizu
Division of Oral and Craniofacial Anatomy, Department of Oral Biology, Tohoku University Graduate School of Dentistry, Sendai 980-8575, Japan

T. Sasano et al. (eds.), *Interface Oral Health Science 2009*,
DOI 10.1007/978-4-431-99644-6_23,

was not suitable for occlusal forces and it would be impossible to insert an endosteal implant into the area supported by newly formed bone obtained by periosteal distraction in humans [2]. In addition, further surgical procedure of removing the subperiosteal titanium mesh, which is most commonly used for the periosteal distraction, is required in periosteal distraction. We attempted to utilize the biocompatible and biodegradable meshed plate to avoid removing subperiosteal titanium mesh plate through periosteal distraction. We also performed the preliminary experiment to improve bone regeneration by administration of mesenchymal stem cells (MSCs) into the gap created by periosteal distraction.

The mesh plate (Super FIXSORB-MX; Takiron Co., Ltd., Osaka, Japan) consisting of a mixture of poly-L-lactide (PLLA) and particulate resorbable uncalcined hydroxyapatite (u-HA) was placed subperiosteally at the head of rabbits. After a latency period of 7 days, the plate was elevated by 0.5 mm/day for 20 days. On the last day of the elevation, the experimental group received rabbit MSCs, which were prepared from iliac bone, into the gap, whereas the control group received phosphate buffered saline. Histological analysis by hematoxylin and eosin staining was performed for the evaluation of bone formation.

The bone formation was successfully induced on the cortical bone at the gap of periosteal distraction by the novel plate. Compared to the previous reports [3], the new mesh plate could be an alternative to titanium mesh plate. The MSCs appeared to have a positive effect on osteogenesis. Further analysis may be required to prove whether the MSC administration is useful to induce osteogenesis at the periosteal distraction site.

In conclusion, we showed the possibility to utilize the biocompatible and biodegradable meshed plate made from the mixture of PLLA and particulate resorbable u-HA to omit the removal procedure of subperiosteal titanium-meshed plate that is most commonly used for the periosteal distraction.

References

1. Schmidt BL, Kung L, Jones C, Casap N (2002) Induced osteogenesis by periosteal distraction. J Oral Maxillofac Surg 60:1170–1175
2. Sencimen M, Aydintug YS, Ortakoglu K, Karslioglu Y, Gunhan O, Gunaydin Y (2007) Histomorphometrical analysis of new bone obtained by distraction osteogenesis and osteogenesis by periosteal distraction in rabbits. Int J Oral Maxillofac Surg 36:235–242
3. Hara J, Nei H, Kawamura H (2008) The possibility to form new bone by using osteogenesis devices placed between bone and periosteum in rabbits. J Jpn Stomatol Soc 57:38–46

Possible role of CCN family members during osteoblast differentiation

Harumi Kawaki, Makoto Suzuki, Toshiya Fujii, Masaharu Takigawa, and Teruko Takano-Yamamoto

Abstract. CCN family members share common structural characteristics, and it has been suggested that they might have similar or redundant functions. Actually, different CCN proteins were reported to share some biological functions under certain biological conditions. In this study, we investigated all CCN members during osteoblast differentiation. As a result, this study demonstrated that CCN2 promotes osteoblast proliferation and differentiation at the all steps, whereas CCN3 strongly inhibited them. Other CCN members also play their roles during osteoblast differentiation.

Key words. CCN family, osteoblast, differentiation, bone formation

CCN family consists of multifunctional proteins containing six members designated CCN1 to CCN6 [1]. The CCN members are key signaling and regulatory molecules involved in many vital biological functions, including cell proliferation, angiogenesis, tumourigenesis, and wound healing. And these members share high degree of structural homology, so it suggests that they may have similar or redundant functions [2]. An important target for CCN proteins could be bone since the expression of CCN members in osteoblasts is known from animal models and human tissues. CCN1 and CCN6 are involved in osteogenesis [3]. CCN2 is also expressed in bone, and the role in skeleletal homeostasis is strengthened by a mouse model [4]. CCN5 has been implicated in osteoblast function [5]. CCN3 and CC4 are also expressed in osteoblasts and could play a role in osteoblast differentiation [6,7]. Therefore, CCN proteins are relevant for skeletal growth and development, thus suggesting the contribution of the entire CCN family to the process of osteoblast differentiation.

H. Kawaki, M. Suzuki, T. Fujii, and T.T.-Yamamoto (✉)
Division of Orthodontics and Dentofacial Orthopedics, Tohoku University Graduate School of Dentistry, 4-1 Seiryou-machi, Aoba-ku, Sendai 980-8575, Japan
e-mail: t-yamamo@mail.tains.tohoku.ac.jp

M. Takigawa
Department of Biochemistry and Molecular Dentistry, Okayama University Graduate School of Medicine, Dentistry, and Pharmaceutial Sciences, 2-5-1 Shikata-cho, Okayama 700-8525, Japan

T. Sasano et al. (eds.), *Interface Oral Health Science 2009*,
DOI 10.1007/978-4-431-99644-6_24, © Springer 2010

In this study, to further characterize the comprehensive roles of CCN family members, we comparatively analyzed the gene expression and protein production patterns of them. The CCN proteins were differentially produced depending upon osteoblast differentiation stages in comparable patterns. Next, we isolated osteoblasts from embryonic calvariae and induced their differentiation. We established the expression pattern of CCN members. CCN1 and CCN2 mRNA levels reached their peak on day 14. CCN3 mRNA level reached its peak on day 7 and decreased. CCN4 and CCN5 mRNA levels peaked at day 21. Along with differentiation, CCN6 mRNA did not show the increase that was observed in the other CCN members. Furthermore, we evaluated the effect of exogenously-added recombinant CCN proteins (rCCNs) on the osteoblast proliferation, maturation, and calcification. The rCCN2 strongly promoted osteoblastic activities. On the contrary, all of process was remarkably inhibited by rCCN3. In addition, rCCN1 and rCCN4 slightly induced osteoblast proliferation. The rCCN1, rCCN4, and rCCN5 induced osteoblast maturation, although these effects were weaker than rCCN2. The rCCN5 enhanced osteoblast calcification.

In summary, we investigated the expression of CCN family members and their functions during osteoblast differentiation. Results showed that CCN members were differentially expressed and therefore could participate during osteoblast lineage progression.

References

1. Bork P (1993) The modular architecture of a new family of growth regulators related to connective tissue growth factor. FEBS Lett 327:125–130
2. Perbal B, Takigawa M (2005) CCN proteins: a new family of cell growth and differentiation regulators. Imperial College Press, London
3. Schütze N, Schenk R, Fiedler J et al (2007) CYR61/CCN1 and WISP3/CCN6 are chemoattractive ligands for human multipotent mesenchymal stroma cells. BMC Cell Biol 8:45
4. Kawaki H, Kubota S, Suzuki A et al (2008) Functional requirement of CCN2 for intramembranous bone formation in embryonic mice. Biochem Biophys Res Commun 366:450–456
5. Kumar S, Hand AT, Connor JR et al (1999) Identification and clonong of a connective tissue growth factor-like cDNA from human osteoblasts encoding a novel regulator of osteoblast functions. J Biol Chem 274:17123–17131
6. Canalis E (2007) Nephroblastoma overexpressed (Nov) is a novel bone morphogenetic protein antagonist. Ann N Y Acad Sci 1116:50–58
7. French DM, Kaul RJ, D'Souza AL et al (2004) WISP-1 is an osteoblastic regulator expressed during skeletal development and fracture repair. Am J Pathol 165:855–867

Inhibition of oral fibroblast growth and function by N-acetyl cysteine

Naoko Sato, Takeshi Ueno, Katsutoshi Kubo, Takeo Suzuki, Naoki Tsukimura, Keiichi Sasaki, and Takahiro Ogawa

Abstract. The effects of *N*-acetyl-L-cysteine on growth and function of oral fibroblasts were reviewed in this chapter.

Key words. antioxidant, oxidative stress, fibroblasts

1 Introduction

The management of hyperplastic gingival tissue and denture fibromatosis is of importance for successful dental treatments. Such hyper-fibrogenesis is closely linked to the production of oxidative stress. N-acetyl-L-cysteine (NAC) is a cysteinederivative and known as an antioxidant molecule that provides antioxidant precursor and directly serves as an oxidant scavenger. In this chapter, the effects of NAC on the growth and function of oral fibroblasts were reviewed.

N. Sato (✉)
The Weintraub Center for Reconstructive Biotechnology, UCLA School of Dentistry, Los Angeles, CA, USA; Maxillofacial Prosthetics Clinic, Tohoku University Hospital, Tohoku University Graduate School of Dentistry, Sendai, Japan
e-mail: kotani-n@umin.ac.jp

T. Ueno, K. Kubo, T. Suzuki, N. Tsukimura, and T. Ogawa
The Weintraub Center for Reconstructive Biotechnology, UCLA School of Dentistry, Los Angeles, CA, USA

K. Sasaki
Division of Advanced Prosthetic Dentistry, Tohoku University Graduate School of Dentistry, Sendai, Japan

T. Sasano et al. (eds.), *Interface Oral Health Science 2009*,
DOI 10.1007/978-4-431-99644-6_25, © Springer 2010

2 Effects of NAC on Growth and Function of Oral Fibroblasts

Sato et al. [2] recently revealed the effects of NAC on the growth and function of oral fibroblasts. In the study, fibroblasts harvested from the rat palatal tissue were cultured with NAC in various concentration; 2.5, 5, and 10 mM. The viability and proliferation of the cells were evaluated by annexin V-based flow cytometry and BrdU incorporation, respectively. Proliferation activity of the oral fibroblasts was significantly reduced by the addition of NAC into the culture NAC-concentration dependently. Flow cytometric analysis revealed remarkably higher percentages of viable cells treated with NAC. They also evaluated the function of the fibroblasts by Sirius Red staining for collagen production and by RT-PCR for gene expression. The NAC addition downregulated the fibroblastic gene expression, such as collagen I and II, and reduced the collagen production. To simulate an inflammatory condition, 10 μM hydrogen peroxide (H_2O_2) was added into some cultures. The H_2O_2-induced inflammatory reaction, as represented by increased fibroblastic proliferation and collagen production, was abrogated by the co-treatment with NAC. Moreover, NAC addition increased the amount of intracellular glutathione, whereas co-treatment with H_2O_2 and NAC enhanced the NAC-dependent increase of glutathione compared to that with NAC treatment alone (Fig. 1).

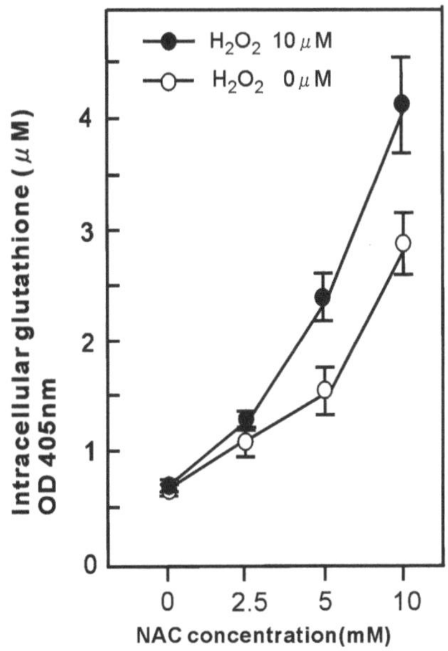

Fig. 1. Intracellular glutathione level under the treatment with NAC alone and co-treatment with NAC and H_2O_2

3 Discussion

NAC inhibited the proliferation and function of oral firoblasts with any cytotoxic effect. The controlling effects of NAC were also demonstrated under mimicking inflammatory condition and might be associated with the increase in intracellular glutathione level. These results suggested the potential therapeutic value of NAC in controlling unfavorable oral soft tissue growth.

References

1. Tsukimura N, Yamada M, Ogawa T (2009) N-acetyl cysteine (NAC)-mediated detoxification and functionalization of poly (metyl methacrylate) bone cement. Biomaterials 30(20):3378–3389
2. Sato N, Ueno T, Ogawa T et al (2009) N-acetyl cysteine(NAC) inhibits proliferation, collagen gene transcription, and redox stress in rat palatal mucosal cells. Dent Mater 25(12): 1532–1540

Computer simulation of orthodontic tooth movement using FE analysis

Masakazu Hasegawa, Taiji Adachi, Masaki Hojo, and Teruko Takano-Yamamoto

Abstract. In the present study, we propose a new simulation method of tooth movement based on the pressure-tension theory, using finite element (FE) model based on CT images of the human mandible. And with the boundary conditions, we simulate the tipping and inclination-controlled tooth movement. As a result, inclination of the tooth axis increases on the tipping model compared to the inclination-controlled model, and it reflects better the clinical tooth movement with moment. The capability of the proposed method to simulate tooth retraction with moments is suggested.

Key words. simulation, tooth movement

The purpose of the orthodontic treatment is to correct the position of the malpositioned teeth by alveolar bone remodeling by applying orthodontic force to them. It is believed that the tooth movement is highly associated with periodontal ligament (PDL) hyalinization and alveolar bone remodeling.

The recent development of the vital visualization techniques such as CT and MR imaging, and the improvement of computer performance, the large and precise finite element (FE) model, such as the one that is used on the large-scale simulation of the trabecular remodeling of the femur bone, can become analyzable. In the orthodontic field, the construction of the tooth movement simulation based upon the individual patient data is expected for order-made medicine.

In the present study, we constructed mathematical model of the tooth movement based on the pressure-tension theory. When the strain at the point in PDL exceeded

---

M. Hasegawa and T. T.-Yamamoto (✉)
Division of Orthodontics and Dentofacial Orthopedics, Tohoku University Graduate School of Dentistry, 4-1 Seiryo-machi, Aoba-ku, Sendai 980-0872, Japan
e-mail: t-yamamo@mail.tains.tohoku.ac.jp

T. Adachi and M. Hojo
Department of Mechanical Engineering and Science, Kyoto University, Yoshida-honcho, Sakyo-ku, Kyoto 606-8501, Japan

T. Sasano et al. (eds.), *Interface Oral Health Science 2009*,
DOI 10.1007/978-4-431-99644-6_26, © Springer 2010

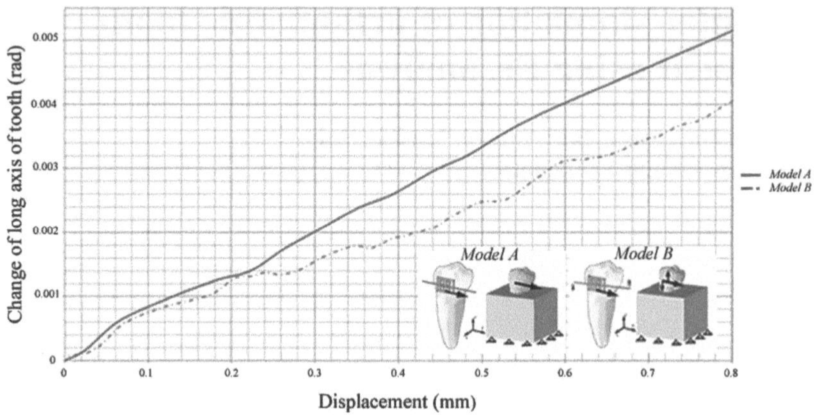

Fig. 1. Tooth inclination changes

the threshold, we assumed that the hyalinization has occurred and the driving force of tooth movement is generated at the hyalinized PDL. Then, the simulation algorithm is constructed using FE analysis based on the above mentioned mathematical model. For the FE analysis, CT image-based model of human mandibular premolar is constructed and PDL and bone element is added to that model. With the two types of boundary conditions, we evaluate the effect of the anti-tipping bend, which is used in clinical treatment. The load of 1.2 N to distal direction is applied to Model A. On Model B, in addition to the distal direction load, the load of 2.3×10^{-3} Nm was given.

As the result of simulation, the change of tooth inclination relative to displacement is shown in the Fig. 1.

It shows that:

1. On Model A, as the displacement increases, the inclination of the long axis is also increased remarkably.
2. On Model B, the inclination along with tooth movement is lower than Model A, showing that the moment applied to Model B controls the tipping effectively.

These results indicate that the method proposed at present study has the potential to simulate the clinical orthodontic tooth movement. With the other boundary conditions, it is necessary to simulate different types of tooth movement to confirm the validity of this simulation.

Mechanical-stress-induced apoptosis and angiogenesis in periodontal tissue

Mirei Chiba, Aya Miyagawa, Kaoru Igarashi, and Haruhide Hayashi

Abstract. Periodontal remodeling takes place in response to various mechanical forces. The application of excessive orthodontic force induces circulatory failure, local ischemia, tissue hyalinization, and cell death in the periodontal ligament (PDL) on the compressive side. However, the nature of compressive-force-induced tissue remodeling is not clear. We recently demonstrated that the in vitro application of a continuous compressive-force-induced apoptosis in cultured human osteoblasts enhances vascular endothelial growth factor (VEGF) production and angiogenic activity in PDL cells, which may contribute to periodontal remodeling during orthodontic tooth movement.

Key words. mechanical stress, apoptosis, angiogenesis, periodontal tissue, orthodontic tooth movement

1 Introduction

During orthodontic tooth movement, periodontal remodeling takes place in response to various mechanical stresses such as compressive and tension forces. On the compressive side, force induces circulatory failure, local ischemia, tissue hyalinization, and cell death in the periodontal ligament (PDL) [1]. Hyalinized tissue is then eliminated by scavenger cells such as multinucleated giant cells and macrophages [2]. Alveolar bone adjacent to the hyalinized tissue is also eliminated via undermining resorption by the adjacent bone marrow [3]. Finally, the compressed PDL returns to its original width, and connective tissue cells invade the degenerated tissues [4]. Newly formed blood vessels in the degenerated tissue serve

M. Chiba (✉) and H. Hayashi
Division of Oral Physiology, Department of Oral Function and Morphology, Graduate School of Dentistry, Tohoku University, 4-1 Seiryo-machi, Aoba-ku, Sendai 980-8575, Japan
e-mail: mirei@mail.tains.tohoku.ac.jp

A. Miyagawa and K. Igarashi
Division of Oral Dysfunction Science, Department of Oral Health and Development Sciences, Graduate School of Dentistry, Tohoku University, 4-1 Seiryo-machi, Aoba-ku, Sendai 980-8575, Japan

T. Sasano et al. (eds.), *Interface Oral Health Science 2009*,
DOI 10.1007/978-4-431-99644-6_27, © Springer 2010

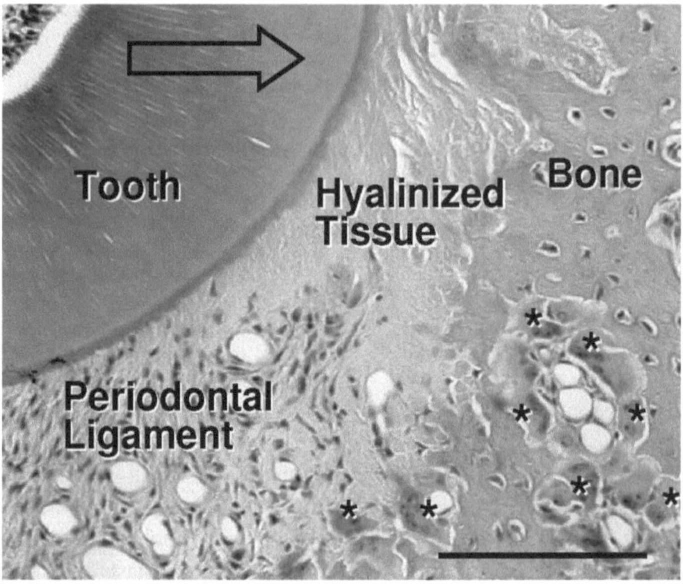

Fig. 1. Compression site of periodontal tissues of the mesiopalatal root of maxillary first molars in rats. Highly compressed areas of the periodontal ligament and "hyalinized tissues" were observed 7 days after initiating experimental tooth movement. Osteoclasts and multinucleated giant cells (*asterisk*) increased in number, when undermining resorption of the alveolar surface became apparent. *Arrow* indicated the direction of tooth movement. Bar = 100 μm

many functions such as the recruitment of hematopoietic stem cells and mesenchymal cells through blood flow and the supply of nourishment and oxygen to the re-established tissue (Fig. 1).

2 Osteoblast Apoptosis by Continuous/Compressive Force

The effects of in vitro application of continuous compressive force on the apoptosis induction in human osteoblast-like cells (MG-63 cells) and the mechanism by which apoptosis was initiated were investigated [5]. Compressive force can induce apoptosis in MG-63 cells through the activation of caspase-3 via the caspase-8 signaling cascade.

3 Continuous Compressive Force Increased the Expression of Vascular Endothelial Growth Factor in PDL Cells

The localization of endothelial growth factor (VEGF) in rat periodontal tissues during experimental tooth movement in vivo and the effects of continuous compressive force on VEGF production and angiogenic activity in human PDL cells in vitro were demonstrated [6]. PDL cells adjacent to hyalinized tissue and alveolar bone on the compressive side showed marked VEGF immunoreactivity. VEGF mRNA expression and production in PDL cells increased.

4 Conclusion

The cell-type-specific cell-death reaction to mechanical stress may regulate periodontal remodeling at sites where force is applied. Continuous compressive force enhances angiogenic activity in PDL cells, which may contribute to periodontal remodeling, during orthodontic tooth movement.

Acknowledgments This work was supported by a GrantinAid for Scientific Research (B) (16390602) from the Japanese Ministry of Education, Science, Sport, and Culture.

References

1. Rygh P (1974) Elimination of hyalinized periodontal tissues associated with orthodontic tooth movement. Scand J Dent Res 82:57–73
2. Kvam E (1972) Cellular dynamics on the pressure side of the rat periodontium following experimental tooth movement. Scand J Dent Res 80:369–383
3. Reitan K, Rygh P (1994) Biomechanical principles and reactions. In: Thomas MG, Robert LV Jr (eds) Orthodontics: current principles and techniques. Mosby-Year Book Inc, St Louis, pp 96–192
4. Proffit WR (2000) The biological basis of orthodontic therapy. In: Proffit WR, Fields HR Jr (eds) Contemporary orthodontics. Mosby-Year Book Inc., St Louis, pp 296–325
5. Goga Y, Chiba M, Shimizu Y et al (2006) Compressive force induces osteoblast apoptosis via caspase-8. J Dent Res 85:240–244
6. Miyagawa A, Chiba M, Hayashi H et al (2009) Compressive force induces VEGF production in periodontal tissues. J Dent Res 88(8):752–756

Diachronic changes of tooth wear in the deciduous dentition of the Japanese

Toshihiko Suzuki and Masayoshi Kikuchi

Abstract. Occlusal attrition in deciduous dentition was examined in the Japanese subadult skeletal samples of the Neolithic Jomon (ca. 12000–300 BC), immigrant Yayoi (ca. 300 BC–300 AD), and medieval Kamakura (1300–1600 AD) periods, compared to modern Japanese children. The dentitions of all skeletal samples showed relatively heavier attrition than in modern children. The attrition of Jomon children was not heavier than other skeletal samples.

Key words. chronological change in attrition, deciduous dentition, Japanese, skeletal remains

1 Introduction

Diachronic changes in the dental wear of the Japanese have been studied by Hanihara et al. [1] and Kaifu [2]. However, the materials of these studies were limited to permanent dentition. Regarding deciduous dentition, Saito et al. [3] investigated the progression rate of attrition by age, but the target samples were modern Japanese children. In this study, we provide information about diachronic changes in attrition of deciduous dentition in the Japanese.

2 Materials and Methods

The materials used in this study comprise deciduous teeth from the Jomon ($N=50$), Yayoi ($N=67$), and Medieval ($N=88$) series, which were unearthed from various archaeological sites in Japan. The scoring method proposed by Saito et al. [3] was

T. Suzuki (✉) and M. Kikuchi
Division of Dental and Craniofacial Anatomy, Tohoku University Graduate School of Dentistry, 4-1 Seiryo-machi, Aoba-ku, Sendai 980-8575, Japan
e-mail: suzk@anat.dent.tohoku.ac.jp

T. Sasano et al. (eds.), *Interface Oral Health Science 2009*,
DOI 10.1007/978-4-431-99644-6_28,

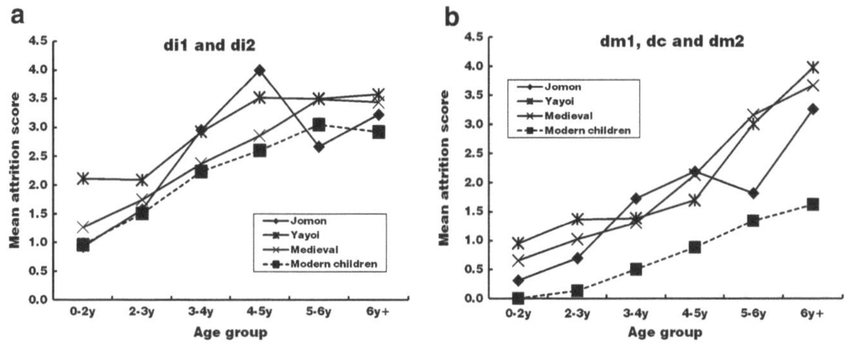

Fig. 1. Relationship between age and mean attrition scores for each population. (**a**) Deciduous first and second incisor, which erupt relatively early. (**b**) Deciduous canine, and first and second molar, which erupt relatively later

used to quantify wear severity for the deciduous tooth. As a control, data from the published source by Saito et al. [3] are cited to obtain a mean attrition score of modern children. To assess the progression of attrition rate by age, individuals in the sample were divided into six subgroups: 0–2, 2–3, 3–4, 4–5, 5–6, and, 6+.

3 Results and Discussion

Figure 1 shows the relationship between age and the mean attrition score in each population. We combined di1 and di2 as "early erupting teeth," and dm1, dc, and dm2 as "late erupting teeth." As for late erupting teeth, the skeletal samples showed heavier attrition than modern children, and the difference is more remarkable than for early erupting teeth. However, among the skeletal samples there was no clear difference. This is in contrast to previous work on attrition in permanent dentition, in which it is shown that the severity of attrition decreased from the Jomon period to the present time [1]. One likely factor for this discrepancy is that the dietary habit in Jomon children may have been different from that of Yayoi and later populations. In addition, the influence of oral habits, such as using a pacifier, should also be considered.

References

1. Hanihara K, Mizoguchi Y, Wakebe T et al (1988) Comparisons of tooth wear in the Japanese populations from the prehistoric to modern age. In: The Second Department of Anatomy, Kyushu University (ed.) The genesis of the Japanese population and culture, Rokko-shuppan, Tokyo, pp 47–53
2. Kaifu Y (1999) Changes in the pattern of tooth wear from prehistoric to recent periods in Japan. Am J Phys Anthropol 109:485–499
3. Saito K, Taura K, Shimada Y (1990) Attrition of deciduous teeth in nursery school children (in Japanese). J Dent Health 40:24–36

Dental occlusal deformation analysis of porcine mandibular periodontium using digital image correlation method

Yasuyuki Morita, Masakazu Uchino, Mitsugu Todo, Lihe Qian, Yasuyuki Matsushita, Kazuo Arakawa, and Kiyoshi Koyano

Abstract. A porcine mandible was separated to prepare thin periodontium specimens consisting of a molar, periodontal ligament (PDL), and alveolar bone. Occlusion was simulated by applying a forced compressive displacement using a table-top material tester. We photographed images of the displacing periodontium specimen and simultaneously obtained the load–displacement curve during the test. The displacement and deformation distributions were examined using digital image correlation analysis. Then, we correlated the distribution with the load–displacement curve, which was characterized by biphasic behavior, as noted in many previous studies. We found that the displacement and deformation distributions of actual periodontium correlated with the load–displacement curve during dental occlusion. Regarding the biphasic characteristics of the load–displacement curve, we showed experimentally that the first phase indicated deformation of the PDL and the second indicated deformation of the alveolar bone and tooth.

Key words. dental occlusion, load–displacement curve, deformation distribution, periodontium, digital image correlation

1 Introduction

Dental occlusal analysis is an important research theme in dentistry from the perspective of investigating problems, such as occlusion, orthodontics, and periodontal disease. Therefore, we examined the correlation of the load–displacement curve with the displacement and deformation distributions of actual periodontium under simulated

Y. Morita (✉), M. Todo, L. Qian, and K. Arakawa
Research Institute for Applied Mechanics, Kyushu University, Fukuoka 816-8580, Japan
e-mail: morita@riam.kyushu-u.ac.jp

M. Uchino
Fukuoka Industrial Technology Center, Kitakyushu 807-0831, Japan

Y. Matsushita and K. Koyano
Graduate School of Dental Science, Kyushu University, Fukuoka 812-8582, Japan

T. Sasano et al. (eds.), *Interface Oral Health Science 2009*,
DOI 10.1007/978-4-431-99644-6_29, © Springer 2010

dental occlusion. Porcine skulls, obtained from animals slaughtered a few days earlier for human consumption, were used to simulate the actual periodontium as closely as possible. The mandible was separated from the skull and specimens including a molar, periodontal ligament (PDL), and alveolar bone were prepared. The specimens were displaced forcibly in a table-top material tester. Then, the periodontium specimens were photographed during the occlusion test to obtain the load–displacement curves. Finally, the displacement and deformation distributions obtained from these images were analyzed using a digital image correlation method, and we correlated the distributions with the load–displacement curves. To our knowledge, no such research using actual periodontium has been reported. This study clarified the displacement and deformation distributions of the periodontium under dental occlusion. In addition, it suggests the ideal displacement distribution in dental implants, where titanium alloy osseointegrates into alveolar bone directly, without using the PDL as a buffer.

2 Summary

The first phase of the load–displacement curve of the periodontium shown in Fig. 1 indicated that the rate of increase of the load was relatively small, showing prominent deformation of the PDL (Fig. 2a, b), which consists of fibrous connective tissue. This is largely due to an appropriate amount of sway in the tooth. This was clarified by visualizing the displacement distributions experimentally, which showed that during the second phase, when the rate of increase of the load was relatively large, there was deformation of the alveolar bone first, and then, finally, deformation of the tooth (Fig. 2c–f).

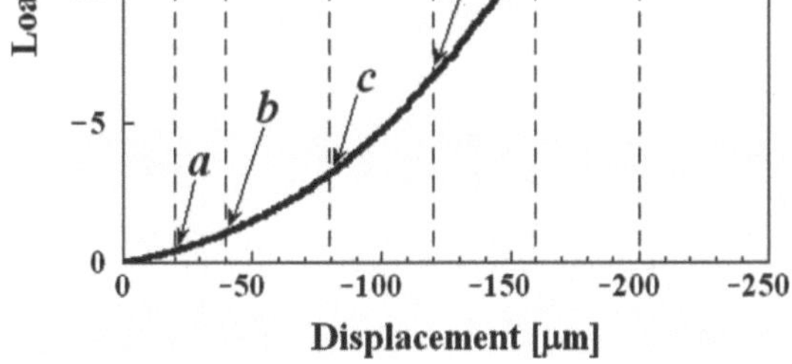

Fig. 1. Load–displacement curve of the periodontium during simulated dental occlusion

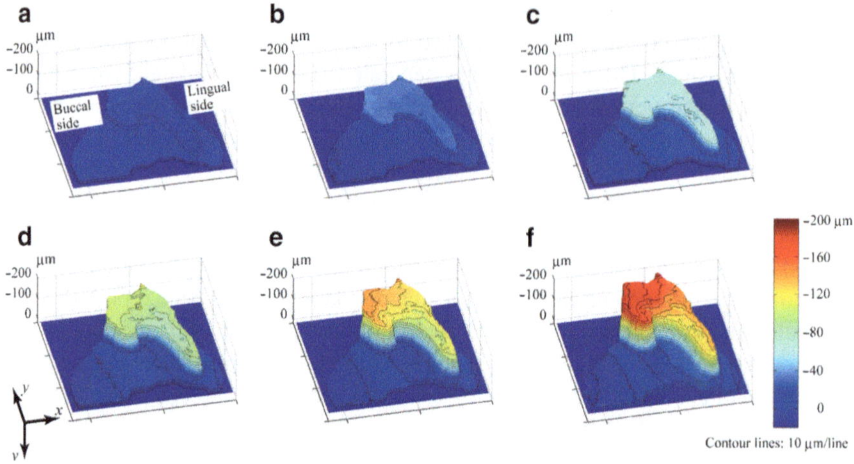

Fig. 2. Vertical (v) displacement distributions with contour lines of the fresh periodontium specimen under dental occlusion. (**a**) The distribution at Point a in Fig. 1 (δ=−20 μm). (**b**) The distribution at Point b in Fig. 1 (δ=−40 μm). (**c**) The distribution at Point c in Fig. 1(δ=−80 μm). (**d**) The distribution at Point d in Fig. 1 (δ=−120 μm). (**e**) The distribution at Point e in Fig. 1 (δ=−160 μm). (**f**) The distribution at Point f in Fig. 1 (δ=−200 μm)

Measurement of the transmitted-light through human upper incisors

Motohide Ikawa

Abstract. The light transmission through the human upper incisors was examined using infrared and green laser light in vivo. Both light simultaneously illuminated the labial surface of the tooth crown and the transmitted-light through the tooth crown was collected from the palatal surface. The intensities of transmitted infrared light and the transmitted green light (TGL) were smaller when the lights were illuminated and collected from cervical area than incisal area. TGL with nonvital teeth was almost nil. The intensities of transmitted-light through tooth crown were considered to indicate the condition of the pulp and to be applicable to pulp diagnostic testing.

Key words. human, tooth, pulp, transmitted-light, vitality

1 Introduction

The dental pulp is encapsulated by enamel and dentin, and only limited information of the condition of the pulp is available. One of the noninvasive techniques to monitor pulpal blood flow is transmitted-light photoplethysmography. In this recording technique, the use of green light has been reported to be efficient in detecting pulse waves in human central upper incisors [1–3]. Transmitted green light is considered to indicate the volume of the hemoglobin in the pulp, while transmitted red light reflected the oxygenation of the hemoglobin. The author measured the intensity of the transmitted-light through human extracted tooth crown with different contents in the artificial pulp chamber and reported the colorimetric analysis of the transmitted-light through human extracted tooth crown to the contents of the root canal [4]. To examine the applicability of colorimetric analysis to the human teeth in vivo, the intensities of the transmitted-light through the upper central and lateral incisors in young subjects were measured.

M. Ikawa
Division of Periodontology and Endodontology, Department of Oral Biology,
Tohoku University Graduate School of Dentistry, 4-1 Seiryo-machi, Aoba-ku,
Sendai 980-8575, Japan
e-mail: moto-i@mail.tains.tohoku.ac.jp

T. Sasano et al. (eds.), *Interface Oral Health Science 2009*,
DOI 10.1007/978-4-431-99644-6_30,

2 Materials and Methods

The study was approved by the Tohoku University Graduate School of Dentistry Research Ethical Committee. The purpose and method of the study were explained to the subjects and written informed consent was obtained from all of them. Upper central (vital and nonvital) and lateral (vital and nonvital) incisors of the subjects were examined. Infrared and green laser simultaneously illuminated the labial surface of the tooth crown via two optical fibers (outer diameter, 0.5 mm). The transmitted-light through the tooth crown was collected from the palatal surface and guided to photodetectors via other optical fibers. The measurement was made at three different areas of each examined tooth crown, and the intensities of transmitted infrared light (TIL) and the transmitted green light (TGL) were simultaneously measured.

3 Results

Overall, the tendency of transmitted-light intensities obtained with upper incisors was similar to that of previous results with extracted incisors. The intensities of TIL and TGL were smaller when the lights were illuminated and collected from cervical area than incisal area. TGL with nonvital teeth was almost nil.

4 Discussion and Conclusion

The intensities of transmitted light through tooth crown indicated the condition of the pulp. The prominent advantages of this analysis is that it is pain-free and harmless, which will be beneficial to the patients. Further study with different conditions of the pulp will be needed. It is very premature to draw a conclusion; however, this analysis is considered to be applicable to pulp diagnostic testing.

References

1. Ikawa M, Ikawa K, Horiuchi H (1994) Optical characteristics of human extracted teeth and the possible application of photoplethysmograpy to the human pulp. Arch Oral Biol 39:821–827
2. Ikawa M, Itagaki Y, Horiuchi H (1996) Human pulp photoplethysmography using LEDs with different power spectrum. In: Shimono M, Maeda T, Suda H, Takahashi K (eds) Dentin/pulp complex. Quintessence Publishing, Tokyo, pp 265–267
3. Kakino S, Takagi Y, Takatani S (2008) Absolute transmitted light plethysmography for assessment of dental pulp vitality through quantification of pulp chamber hematocrit by a three-layer model. J Biomed Opt 13:54023
4. Ikawa M, Uzuka R (2007) Colorimetric analysis of the transmitted-light through human teeth. Program and abstracts of papers. Jap Assoc Dent Res, 114

Three-dimensional finite element analysis of overload-induced alveolar bone resorption around dental implants

Lihe Qian, Mitsugu Todo, Yasuyuki Matsushita, and Kiyoshi Koyano

Abstract. In this study, the stress, strain, and strain energy density (SED) criteria were applied to tentatively simulate overload-induced bone resorption in implant/jawbone systems. By a comparative analysis, the SED criterion was found to be most suitable, based on which the resorption process of alveolar bone was investigated, and the effects of implant diameter and loading angle were examined. The simulations demonstrated the patterns of bone resorption that agreed well with the clinical observations published in the literature, and showed that implant diameter and loading angle influenced significantly the amount of resorbed bone and the micromotion of implant.

Key words. bone resorption, remodeling, implant, finite element analysis

1 Introduction

One of the major reasons for marginal bone loss around dental implant has been associated with unfavorable loading conditions acting on implants. Therefore, much effort has been devoted to analyzing bone's stress and strain distributions, attempting to optimize the prosthetic design and to improve the mechanical environment in bone. However, there are scarce reports on the simulation of mechanically-induced bone resorption in implant/jawbone systems and thus, how the mechanical fields influence dental bone resorption is unclear. The purpose of this study was to investigate the process of mechanical-induced bone resorption and to examine several important influencing factors.

L. Qian (✉) and M. Todo
Research Institute for Applied Mechanics, Kyushu University, 6-1 Kasuga-koen, Kasuga 816-8580, Japan
e-mail: dlhqian@yahoo.com; qian@riam.kyushu-u.ac.jp

Y. Matsushita and K. Koyano
Faculty of Dental Sciences, Kyushu University, 3-1-1 Maidashi, Higashi-ku, Fukuoka 812-0041, Japan

T. Sasano et al. (eds.), *Interface Oral Health Science 2009*,
DOI 10.1007/978-4-431-99644-6_31, © Springer 2010

2 Materials and Methods

Finite element models of dental implant/jawbone were created (Fig. 1a). Two implant diameters (3.7 and 5.2 mm) and two loading directions (axial and 15° buccolingually) were investigated. Three criteria, i.e., the equal strain, equal stress, and equal strain energy density (SED) criteria were respectively tested to choose a more suitable one for the formal simulation. Details of the simulations are given elsewhere [1].

3 Results and Conclusions

By comparison, it is found that the equal SED criterion reproduced bone resorption patterns that are most realistic to actual clinical situations published in the literature, and thus is considered to be the most suitable one for simulating bone resorption.

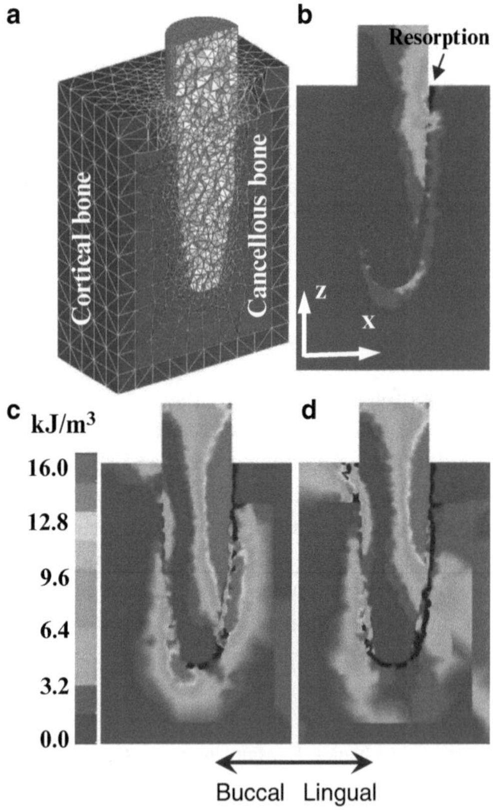

Fig. 1. (**a**) Half of the finite element model. (**b**)–(**d**) Buccolingual cross-sectional illustration of the bone resorption patterns at a buccolingual loading angle of 15°, simulated according to the equal SED criterion, at a normalized simulation time of (**b**) 0.56, (**c**) 0.84, and (**d**) 0.92. The SED contours are also shown

The bone resorption was initiated from the upper edge of the cortical bone; after penetrating the cortical bone, the bone resorption passed through the interface of the cortical/cancellous bone and subsequently extended in the cancellous bone [1] (Fig. 1). The amount of bone resorption and the movement of implant were obviously larger for the large implant than for the small implant. For oblique loading, bone resorption started earlier, and the amount of resorbed bone and the movement of implant were larger.

In conclusion, in spite of some simplifications made, the present simulations do provide a qualitative understanding of bone resorption phenomena caused by occlusive overload. This enables the prediction of bone morphology, mechanical fields, and movement of implant at various levels of bone resorption and may help optimize the design of dental implant.

Reference

1. Qian L, Todo M, Matsushita Y (2009) Finite element analysis of bone resorption around dental implants. J Biomech Sci Eng 4:365–376

Regulation of microRNA expression by bone morphogenetic protein-2

Mari M. Sato, Yasutaka Yawaka, and Masato Tamura

Abstract. MicroRNAs (miRNAs) are small noncoding RNAs that are emerging as important posttranscriptional gene regulators. Many miRNAs are expressed in a tissue-specific manner, which suggests that they have specific biological roles in the specification of tissues. In this chapter, we summarize the currently available data on the regulation of miRNA expression by bone morphogenetic protein (BMP)-2. These studies open new avenues for the study of BMP signaling and miRNA biogenesis.

Key words. miRNA, BMP-2, C2C12 cells, cell differentiation

1 Regulation of miRNA Expression by Bone Morphogenetic Protein (BMP)-2

MicroRNAs (miRNAs) are a class of noncoding regulatory RNAs approximately 22 nucleotides in length. miRNAs negatively regulate target mRNA through degradation or suppression of protein translation. More than 800 miRNAs have been discovered in mammals, and some of them are expressed in a tissue-specific manner, which suggests that they play an important role in the control of many biological processes, such as development, differentiation, proliferation, and apoptosis [1]. A small number of striated muscle-specific miRNAs, such as miR-1, miR-133a, and miR-206, have been identified [2]. Upon initiation of differentiation in a multipotent mouse myoblastic C2C12 cell line, there is steady induction of miR-1,

M.M. Sato and M. Tamura (✉)
Department of Biochemistry and Molecular Biology, Graduate School of Dental Medicine, Hokkaido University, North 13, West 7, Sapporo 060-8586, Japan
e-mail: mtamura@den.hokudai.ac.jp

M.M. Sato and Y. Yawaka
Dentistry for Children and Disabled Person, Graduate School of Dental Medicine, Hokkaido University, North 13, West 7, Sapporo 060-8586, Japan

T. Sasano et al. (eds.), *Interface Oral Health Science 2009*,
DOI 10.1007/978-4-431-99644-6_32, © Springer 2010

miR-133a, and miR-206, indicating that these miRNAs might play an important role in myogenic differentiation and cell identity [2].

Bone morphogenetic protein (BMP)-2 is known to trigger osteoblastic differentiation and to upregulate the expression of most genes encoding osteoblastic phenotype-related proteins in vitro. It has been reported that BMP-2 not only converts the differentiation pathway of C2C12 cells into that of osteoblasts, but also inhibits myogenic differentiation [3]. Our examination of BMP-2-treated C2C12 cells showed that expression of both miR-1 and miR-206 was completely suppressed [4]. Consistent with our findings, other reports have shown that treatment with BMP-2 decreases miR-206 expression in C2C12 cells for a period of 2–6 days [5].

2 Regulation of the miRNA Processing Pathway by BMP-2

miRNA expression can be controlled at either the transcriptional or posttranscriptional level. Davis et al. recently reported that BMP promotes the processing of pri-miR-21 into pre-miR-21 [6]. The transcription factor MyoD1 directly regulates transcription of the primary miR-206 transcript [5]. Although BMP-2 completely suppresses myogenin expression in C2C12 cells, we observed previously that BMP-2 does not affect the expression of MyoD1 [3]. Therefore, regulation of miR-206 expression by BMP-2 could potentially be controlled at the posttranscriptional level. BMP-2 reduced miR-206 expression both in the presence and absence of α-amanitin, a specific inhibitor of pol II-dependent transcription [4]. The pri-miR-206 level increased after BMP-2 treatment for 6 h when compared with untreated cells [4]. These results indicate that BMP-2 downregulates miR-206 expression at the posttranscriptional level by inhibiting the processing of pri-miR-206 into mature miR-206. Although our understanding is limited by the small number of miR-206 target genes that have been experimentally verified, we conclude that BMP-2 may regulate miRNA biogenesis by a novel mechanism during regulation of cell differentiation. It is also possible that BMP-2 could regulate expression of a specific gene, which is partly mediated by miRNA biogenesis. The exact nature of this regulatory mechanism awaits further investigation.

References

1. Williams A (2008) Functional aspects of animal microRNAs. Cell Mol Life Sci 65:545–562
2. McCarthy J (2008) MicroRNA-206: the skeletal muscle-specific myomiR. Biochim Biophys Acta 1779:682–691
3. Nakashima A, Katagiri T, Tamura M (2005) Cross-talk between Wnt and bone morphogenetic protein 2 (BMP-2) signaling in differentiation pathway of C2C12 myoblasts. J Biol Chem 280:37660–37668

4. Sato MM, Nashimoto M, Katagiri T et al (2009) Bone morphogenetic protein-2 down-regulates miR-206 expression by blocking its maturation process. Biochem Biophys Res Commun 383:125–129
5. Rao P, Kumar R, Farkhondeh M et al (2006) Myogenic factors that regulate expression of muscle-specific microRNAs. Proc Natl Acad Sci U S A 103:8721–8726
6. Davis BN, Hilyard AC, Lagna G et al (2008) SMAD proteins control DROSHA-mediated microRNA maturation. Nature 454:56–61

Influence of early progressive loading on implants placed into extraction sockets

Yu Ban, Ning Geng, and Ping Gong

Abstract. It has been known that the bone tissue around dental implant adapts to functional load by changes in structure and mass. However, the effects of immediate loading of implants placed into extraction sockets are uncertain. We studied the differences in early bone–biomaterials interaction and reactions between the progressive vertical loading and the nonloading of implants placed into extraction sockets. The progressive loading promoted the bone-implant osseointegration by accelerated mineralization speed many times as faster as the control groups and stimulated preosteoblast attached on the implant surface and differentiated to osteoblast. Osteoblast reacted to immediate loading with advance release of the related protein of osteogenesis. These results show that progressive loading accelerates the new bone formed around the implant and promotes osseointegration.

Key words. immediately implanted, progressive loading, osseointegration, bone-to-implant contact ratio

The aim was to detect differences in early bone–biomaterials interaction and reactions between the progressive vertical loading and the nonloading of implants placed into extraction sockets. Bilateral third, fourth and second premolar were extracted from male beagle dogs and implants were inserted on the 1st day, 14th day, and 21th day. The vertical occlusion loading instrument were used, progressive loading procedures were taken 24 h after insertion. On the 28th day, each animal was sacrificed and samples were obtained. Undecalcified sections were evaluated by scanning electron micrographic (SEM) and the bone-implant contact (BIC) ratio was measured.

Y. Ban and P. Gong (✉)
Dental Implant Center, West China College of Stomatology, Sichuan University, Block 3, No. 14, Renminnan Road, Chengdu, Sichuan, China
e-mail: wcmusunny@yahoo.com.cn

N. Geng
State Key Laboratory of Oral Diseases (Sichuan University), Block 3, No. 14, Renminnan Road, Chengdu, Sichuan, China

T. Sasano et al. (eds.), *Interface Oral Health Science 2009*,
DOI 10.1007/978-4-431-99644-6_33, © Springer 2010

Decalcified sections were used by immunohistochemistry to detect the osteopontin (OPN) and osteocalcin expression in the bone around the implants.

SEM observation of the interface matrix revealed a time-dependant mineralization process in both groups and the mineralization speed of experimental groups is many times as faster as the control groups. The BIC in experimental groups surpass than that in the control groups. The differences were statistically significant ($p<0.05$). The progressive vertical loading could stimulate preosteoblast attached on the implant surface and differentiated to osteoblast. On 7th day and 14th day, both loading and nonloading groups had OPN express in osteoblast-like cells, osteoblast, new forming bone, and bone matrix in the bone-implant interface, but OPN expression under progressive loading was significantly stronger than that in the nonloading groups. The results suggest that the stress produced by progressive vertical loading from the beginning of the experiment promoted osteoblast proliferation, differentiation, and secretary function, accelerated the early matrix mineralization, and shortened the time of mineralization. Progressive vertical loading of implants placed into extraction sockets can be performed without disturbing the osseointegration process.

In vitro gene transfection of human stromal cell derived factor-1α and its expression in rat myoblasts

Xiu-fa Tang, Deng-qi He, Yang Feng, and Cheng-ge Hua

Abstract. *Objective:* To construct a kind of genetically modified myoblast secreting human stromal cell derived factor (hSDF-1α) and study its possible effects on tissue engineering of skeletal muscle.

Methods: The recombinant plasmid pEGFP-hSDF1α was constructed by cloning hSDF1α cDNA into eukaryotic expression vector pEGFP-N1. Myoblasts from new born Sprague-Dawley rats were isolated, purified, identified, and cultured in vitro. Recombinant plasmid was transfected into passage 3 myoblasts using lipofectin method. The expression of hSDF-1α was detected at 2, 4, 6, 8, and 10 days after transfection by reverse transcription polymerase chain reaction (RT-PCR) and enzyme-linked immunosorbent assay (ELISA).

Results: The blight green fluorescence could be observed with fluorescence microscope in pEGFP-hSDF1α transfected myoblast. The expression of hSDF-1 mRNA and protein, detected by RT-PCR and ELISA, respectively, were strongly positive in engineered myoblast and negative in control group.

Conclusion: Myoblasts can act directly as a gene target cell while being infected by recombinant plasmid pEGFP-hSDF1α in vitro, and the consequent secreting of hSDF1α might lead to an angiogenesis effect in skeletal muscle tissue engineering in vivo.

Key words. SDF-1α, myoblast, gene transfection, skeletal muscle tissue engineering

---

X. Tang (✉) and C. Hua
Department of Oral and Maxillofacial Surgery (Head and Neck Tumor), West China College of Stomatology, Sichuan University, No. 14, Section 3, Renmin South Road, Chengdu, Sichuan, 610041, People's Republic of China
e-mail: tangxf1963@163.com

D. He and Y. Feng
State Key Laboratory of Oral Disease, West China College of Stomatology, Sichuan University, Chengdu, Sichuan, 610041, People's Republic of China

T. Sasano et al. (eds.), *Interface Oral Health Science 2009*,
DOI 10.1007/978-4-431-99644-6_34, © Springer 2010

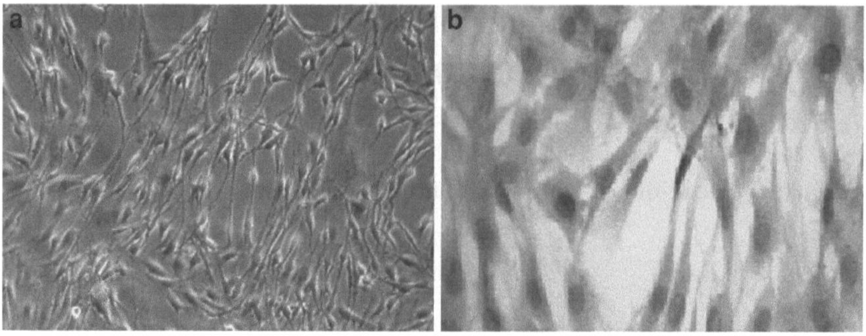

Fig. 1. The primary myoblasts (**a**) and positive stain with anti-desmin antibody (**b**)

Primary rat myoblasts were obtained from the hind limbs of 3-day-old SD rats. Satellite cells were dissociated from the minced muscles by digested in dishes which contained 0.1% collagenase type I and 0.25% trypsin for 40 min at 37°C. Isolated myoblasts could be observed 3 days after being grown in F-12 medium containing 1% penicillin/streptomycin solution and 20% new born calf serum (Fig. 1a ×200). More than 90% cells during passage 2–4 were positive for immunohistochemical stain with anti-desmin antibody (Fig. 1b ×400), which indicated a homogeneous myoblast culture.

Fluorescence microscopic examination of rat myoblasts transfected with EGFP human stromal cell derived factor (hSDF-1α) plasmid (Fig. 2a ×100, b ×200). The efficiency of hSDF-1α transfection into myoblasts was markedly influenced by vector/plasmid ratio and optimal efficiency was obtained with Lipofectamine™2000/hSDF-1α (v/w 2.5:1). Two days after transfection, more than 15% cells could be observed expressing green fluorescent protein under fluorescence microscopy.

RNA was isolated from nontransfected and transfected myoblasts 2, 4, 6, and 8 days after transfection, respectively. Nontransfected myoblasts were used as a control. A single band of approximately 294 bp containing a SDF-GFP chimeric sequence was detected only in the transfected myoblasts (Fig. 3).

Western blot analysis of nontransfected myoblasts and transfected myoblasts after 2, 4, 6, and 8 days after transfection were carried out, respectively, using anti-SDF anti-body. The result showed immunoreactive bands with a molecular mass of approximately 34 kDa only in the transfected myoblasts (Fig. 4).

Conditioned medium was collected from nontransfected and transfected myoblasts (2×10^4 cells per milliliter) 2, 4, 6, 8, and 10 days after transfection, respectively, and analyzed by a human-specific SDF-1α ELISA kit. The amount of hSDF-1α in three individual wells was calculated and is presented as picograms per milliliter of conditioned medium. SDF concentration approximately 183 pg/ml per 2×10^4 cells were detected 2 days after transfection and was increasingly reduced with time. Rare SDF was detected in nontransfected myoblasts with ELISA kit (Fig. 5).

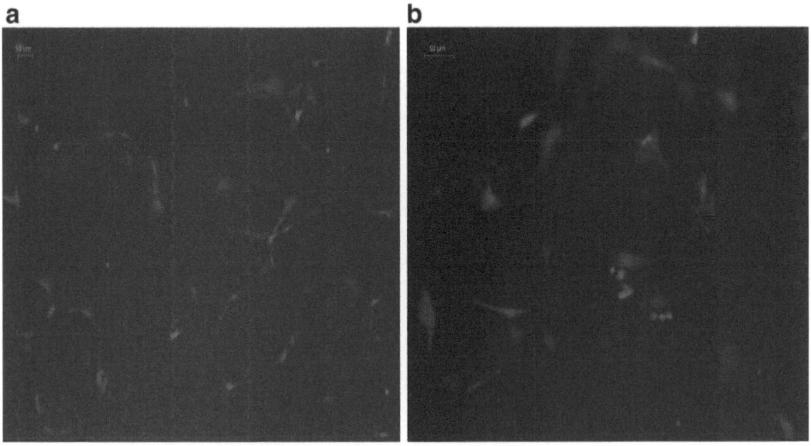

Fig. 2. Myoblasts transfected with EGFP-hSDF-1α plasmid

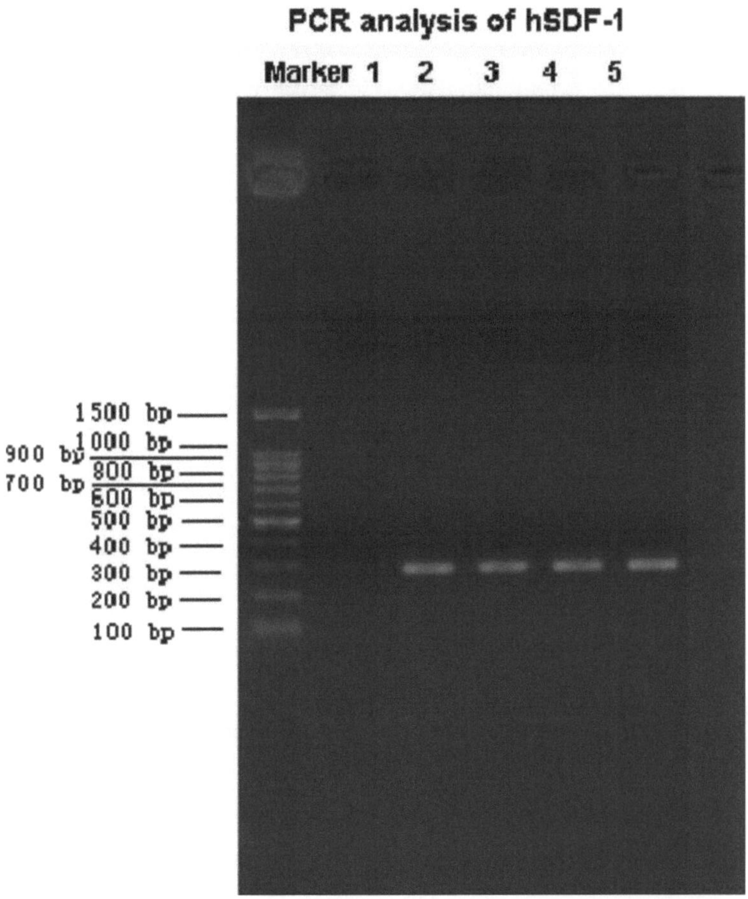

Fig. 3. RT-PCR analysis of in vitro SDF expression in transfected myoblasts

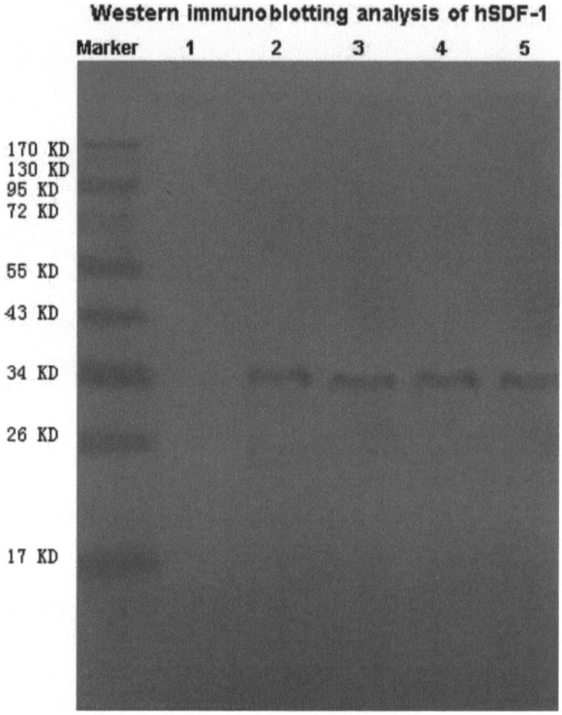

Fig. 4. Western blots analysis of in vitro SDF expression in transfected myoblasts

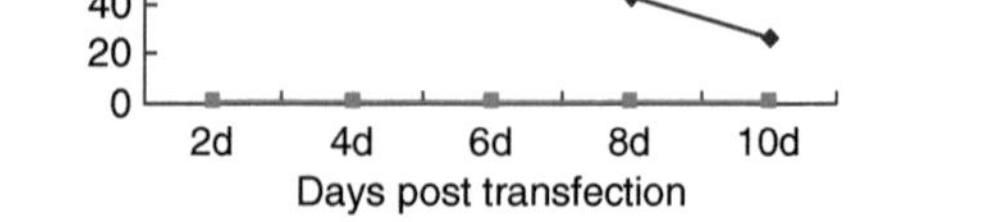

Fig. 5. ELISA assay of in vitro SDF expression in transfected myoblasts

Biomarker identification in oral cancer by using proteomics

Zhi Wang, Xiaodong Feng, Jing Li, and Ning Ji

Abstract. It is now believed that oral squamous cell carcinoma follows a similar pattern in its development, and thus is preceded by premalignant lesions such as leukoplakia, dysplasia, erythroplakia, lichen planus, and oral submucous fibrosis. It is critical to elucidate the precise nature of the genetic and protein alterations occurring at each step. Proteomics have been successfully employed in studies of other cancers with precancerous lesion like esophageal adenocarcinoma, which could be transformed from barrett's metaplasia. However, a few articles have reported the proteomics application in oral cancer cells or tissues. The article reviewed the studies which concentrate on the identification and validation of biomarkers in oral cancers using proteomics, and thus could provide some typical and applicable platform to apply proteomics technologies in a well-defined molecular and pathological framework in biomarker searching.

Key words. oral cancer, biomarker, proteomics

1 Introduction

It is now believed that oral squamous cell carcinoma (OSCC) follows a similar pattern in its development, and thus is preceded by premalignant lesions such as leukoplakia, dysplasia, erythroplakia, lichen planus, and oral submucous fibrosis [1]. It is critical to elucidate the precise nature of the genetic and protein alterations occurring at each step. Genomics has been incorporated in oncology process and now, in the postgenomic era, there is a strong drive to incorporate proteomics technologies as well. Proteomics have been successfully employed in studies of other cancers

Z. Wang (✉), X. Feng, J. Li, and N. Ji
State Key Laboratory for Dental Sciences and West China College of Stomatology, Sichuan University, No. 14, Sec. 3, Renminnan Road, Chengdu, Sichuan, 610041, People's Republic of China
e-mail: wangzhi0506@126.com

T. Sasano et al. (eds.), *Interface Oral Health Science 2009*,
DOI 10.1007/978-4-431-99644-6_35,

with precancerous lesion like esophageal adenocarcinoma, which could be transformed from barrett's metaplasia [2]. However, a few articles have reported the proteomics application in oral cancer tissues. In 2004, He et al firstly examined the expression profiles of proteins between normal and OSCC samples by using two-dimensional gel electrophoresis and matrix-assisted laser desorption/ionization-time-of-flight mass spectroscopy [3]. A number of tumor-associated proteins including heat-shock protein (HSP) 60, HSP27, B-crystalline, ATP synthase, calgranulin B, myosin, tropomyosin, and galectin 1 were consistently found to be significantly altered in their expression levels in tongue carcinoma tissues, compared with their paired normal. Similar patterns have been used in the study of buccal squamous cell carcinoma in 2004 by Chen [4], subcellular fractions from OSCC and control samples, enriched in mitochondrial and cytosolic proteins in 2006 by Turhani [1]. Due to their relationship with apoptosis, response to stimulus, metabolic regulation, etc., we may thus conclude that these proteins may play an important role in the malignant transformation process. However few of these proteins were found to vary in concert, thus reflecting their regional variability or tissue heterogeneity. Furthermore few of them have been functionally analyzed for their roles in oral carcinogenesis, and therefore there is a lack of the fundamental understanding required for clinical applications and a need for a better comprehension of the underlying biological processes. It has been widely accepted that a major challenge to cancer proteomics is the integration of biochemical, genetics, and proteomics data in the detection of biomarkers to provide the impetus for the next level of clinical application.

References

1. Turhani D, Krapfenbauer K, Thurnher D et al (2006) Identification of differentially expressed, tumor-associated proteins in oral squamous cell carcinoma by proteomic analysis. Electrophoresis 27:1417–1423
2. Zhao J, Chang AC, Li C et al (2007) Comparative proteomics analysis of Barrett metaplasia and esophageal adenocarcinoma using two-dimensional liquid mass mapping. Mol Cell Proteomics 6:987–999
3. He QY, Chen J, Kung HF et al (2004) Identification of tumor-associated proteins in oral tongue squamous cell carcinoma by proteomics. Proteomics 4:271–278
4. Chen J, He QY, Yuen AP et al (2004) Proteomics of buccal squamous cell carcinoma: the involvement of multiple pathways in tumorigenesis. Proteomics 4:2465–2475

In vivo analysis of the 3-D force on implants supporting fixed prostheses

Yoshinori Gunji, Nobuhiro Yoda, Takahiro Chiba, Toru Ogawa, Tsunemoto Kuriyagawa, and Keiichi Sasaki

Abstract. Three-dimensional (3-D) loads exerted on implants were measured in a patient with implant-supported fixed prosthesis using load-measuring devices including piezoelectric force transducers. The devices were mounted on implant-fixtures at the mandibular right molar region. Tasks analyzed were maximum voluntary clenching (MVC) and biting a piece of paraffin wax. Load measurements were made in two conditions; the superstructures were splinted and not splinted. Sum of load magnitudes exerted on the two implants showed no significant difference between with and without splinting regardless of task. During MVC, magnitudes were allocated more evenly to the two implants in the condition with splinting than without splinting. During wax biting, magnitudes and directions on both implants were not different significantly between with and without splinting.

Key words. biomechanics, implant, prosthodontics, splinting, superstructure

1 Introduction

Loads exerted on dental implants are essential factors in biomechanical and mechanobiological control of the case with implant-supported fixed prostheses. This study deals with measuring functional three-dimensional (3-D) loads on implants supporting fixed prostheses in vivo and investigating effects of splinting the superstructures on the loads.

Y. Gunji (✉), N. Yoda, T. Chiba, T. Ogawa, and K. Sasaki
Tohoku University, Graduate School of Dentistry, 4-1 Seiryo-machi, Aoba-ku, Sendai 980-8575, Japan
e-mail: gunji-thk@umin.ac.jp

T. Kuriyagawa
Tohoku University, Graduate School of Engineering, 6-6-01, Aramaki Aoba, Aoba-ku, Sendai 980-8579, Japan

T. Sasano et al. (eds.), *Interface Oral Health Science 2009*,
DOI 10.1007/978-4-431-99644-6_36, © Springer 2010

2 Measurement of 3-D Load on Implants

A 66 year-old woman participated. A 3-D load-measuring device comprised of a piezoelectric force transducer (Kistler Instruments AG) and the experimental superstructure. The devices were tightened with screws into her implant-fixtures (Straumann AG) subsisting at the mandibular right first molar (Imp 1) and second premolar (Imp 2) regions.

The tasks employed for analysis were maximum voluntary clenching (MVC) and biting a piece of paraffin wax (5 mm^3) (WAX). At first, loads were measured at the condition in which the superstructures were splinted. Subsequently, the superstructures were removed and separated between the first molar and second premolar superstructures. Then, each superstructure was set on each implant as the condition without splinting. Consequently, the occlusal contacts were the same in both conditions. Magnitude and direction of loads were analyzed in the spatial coordinates based on her Frankfort and sagittal planes [1].

3 Effects of Splinting the Superstructures on the Loads

The sum of load magnitudes exerted on the two implants showed no significant difference between the conditions with and without splinting regardless of the task. At MVC, the magnitudes and directions of loads exerted on each implant were significantly different between the two conditions ($p<0.01$). The loads were allocated more evenly to the two implants at the condition with splinting than without splinting. At WAX, the magnitudes and directions of loads on each implant were not significantly different between the two conditions.

4 Clinical Implication

The present study demonstrated the effect of splinting the superstructures on the loads on implants using in vivo measurement. At MVC, splinting of the superstructures brought even distribution of loads to the two implants than without splinting, which is consistent with other in vitro studies [2]. These results indicate that splinting is one of the key factors for load distribution on the implants supporting fixed prostheses.

However, at WAX, the loads were not allocated evenly despite splinting the superstructures, which indicates that the load distribution on the implants during biting food might depend on the biting point of food, i.e., loading point, hence the first molar is usually the major biting point.

5 Conclusions

The functional 3-D loads on the implants were affected by splinting the superstructures of implant-supported fixed prostheses. The effect of splinting superstructures at WAX was smaller than that at MVC.

References

1. Kawata T, Kawaguchi N, Yoda T et al (2007) Behavior of 3-dimensional compressive and tensile forces exerted on a tooth during function. J Oral Rehabil 34:259–266
2. DL G, Diane Y, AA C (2002) Effect of splinting and interproximal contact tightness on load transfer by implant restorations. J Prosthet Dent 87:528–535

Gap junctional communication regulates salivary gland morphogenesis

Hiroharu Suzuki, Aya Yamada, and Satoshi Fukumoto

Abstract. Connexin43 (Cx43) plays important roles of cell–cell communication as gap junctional molecules. *Cx43*-null mice showed abnormal branching morphogenesis in salivary gland. Gap junctional inhibitor 18α-GA inhibited branching morphogenesis in salivary glands organ culture. Further, exogenously addition of platelet-derived growth factor (PDGF) and fibroblast growth factor 10 (FGF10) took part from the partial rescue from the influence of the gap junctional inhibitor. These results indicate that gap junctional communication may regulate growth factor-induced branching morphogenesis in salivary glands.

Key words. salivary gland, gap junction, branching morphogenesis

1 Introduction

Salivary gland is regulated by sequential and reciprocal interactions between the neural crest-derived mesenchyme and oral ectoderm. Gap junctional protein regulates cell–cell communication involved in tissue organization. Oculodentodigital dysplasia (ODDD) is an autosomal dominant human disease caused by mutations in the Gja1 gene encoding the gap junction protein Connexin43 (Cx43). Gja1-null mice showed decrease branching morphogenesis in salivary gland indicating that cell–cell communication is important for salivary gland morphogenesis. Recently, we reported that platelet-derived growth factor (PDGF) signaling is a possible mechanism involved in the interaction between epithelial and neural crest-derived mesenchyme [1]. Exogenous administration of PDGF in organ culture of salivary glands enhanced branching via mesenchymal FGF expression. Here, we analyzed

H. Suzuki (✉), A. Yamada, and S. Fukumoto
Division of Pediatric Dentistry, Tohoku University Graduate School of Dentistry, Sendai 980-8575, Japan
e-mail: hiroharu@mail.tains.tohoku.ac.jp

T. Sasano et al. (eds.), *Interface Oral Health Science 2009*,
DOI 10.1007/978-4-431-99644-6_37, © Springer 2010

the interaction between PDGF signaling and gap junction during branching morphogenesis of salivary gland using in vitro organ culture system.

2 Materials and Methods

Submandibular and sublingual salivary gland rudiments (SMGs) from either E13 ICR mice were cultured on Cell Culture Insert track-etched filters (0.4 mm pore size) as previously described [2].The filters were floated on serum-free Dulbecco's modified Eagle's medium-F12. The medium contained 200 U/ml penicillin, 200 mg/ml streptomycin, 0.5 mg/ml amphotericin, 150 mg/ml vitamin C, and 50 mg/ml transferrin. SMGs were cultured. In case of addition of 18α-GA, exogenous 10 ng/ml PDGF-BB, 500 ng/ml FGF7, or 1 mg/ml fibroblast growth factor 10 (FGF10) was added to the culture medium, and photographed at 2, 24 and 48 h. The numbers of terminal buds at 24 and 48 h were compared with those at 2 h.

3 Results and Discussion

Gap junctional inhibitor 18α-GA inhibited branching morphogenesis of SMGs. In contrast, exogenous addition of PDGF had accelerated branching. Further, exogenous PDGF and FGF10 partially rescued branching inhibited by 18α-GA. Exogenous PDGF in organ culture of salivary glands enhanced branching via mesenchymal FGF expression as described above. Consequently, these results indicate that gap junctional communication is important for branching morphogenesis and may regulate PDGF signaling.

References

1. Yamamoto S, Fukumoto E, Yoshizaki K et al (2008) PDGF regulate salivary gland morphogenesis via FGF expression. JBC 283:23139–23149
2. Steinberg Z, Myers C, Heim VM et al (2005) FGFR2b signaling regulates ex vivo submandibular gland epithelial cell proliferation and branching morphogenesis. Development 132:1223–1234

Pulpal blood flow in human permanent teeth with different root formation

Hideji Komatsu, Motohide Ikawa, and Satoshi Fukumoto

Abstract. The purpose of this study was to examine the relation between root formation and the pulpal blood flow (PBF) in human young permanent teeth using laser Doppler flowmeter. Recordings were made on clinically healthy upper permanent central incisors in healthy participants (age: 7 years 7 months to 16 years 2 months). According to Nolla's method, the state of roots of the teeth was classified into four groups by radiographs. The mean PBF signals tended to decrease with the progress of the root formation. PBF measurement was considered to be applicable to assess root formation in young permanent teeth.

Key words. pulp, blood flow, laser Doppler, permanent teeth, root formation

1 Introduction

Komatsu et al. [1] reported that the pulpal blood flow (PBF) in the human primary tooth reduces with age. This age-related reduction of PBF is considered to be produced by the reduction of the blood supply due to the root formation. The purpose of this study was to examine the relationship between root formation and PBF in human young permanent teeth.

H. Komatsu (✉) and S. Fukumoto
Division of Pediatric Dentistry, Department of Oral Health and Development Sciences, Tohoku University Graduate School of Dentistry, 4-1 Seiryo-machi, Aoba-ku, Sendai 980-8575, Japan
e-mail: komatsu@pedo.dent.tohoku.ac.jp

M. Ikawa
Division of Periodontology and Endodontology, Department of Oral Biology, Tohoku University Graduate School of Dentistry, 4-1 Seiryo-machi, Aoba-ku, Sendai 980-8575, Japan

T. Sasano et al. (eds.), *Interface Oral Health Science 2009*,
DOI 10.1007/978-4-431-99644-6_38, © Springer 2010

2 Materials and Methods

The repeated measurement of the PBF recordings using laser Doppler flowmeter (LDF) were made on clinically healthy upper primary central incisors in healthy participants (age: 7 years 7 months to 16 years 2 months). The interval between the repeated measurements in each individual ranged between 4 and 22 months. Prior to the measurement, individual resin caps with a hole at approximately 1.5 mm incisal to the gingival margin of the examined tooth were prepared, and the probe of the LDF (MBF3D, Moor instruments, UK) was fitted in the hole during the recording. Measurements were made with and without opaque black rubber dam application to the examined tooth. Signals were stored in a computer (Power Macintosh G3, Apple Computer Inc.) via a laboratory interface (MacLab/8s, ADInstruments Pty Ltd., Australia) and signal processing software (Chart, ADInstruments Pty Ltd., Australia). The state of roots of the teeth examined were assessed by radiographs. According to Nolla's method [2], the state of roots of the teeth were classified.

3 Results

The mean PBF signals tended to decrease with the progress of the root formation. The decrease was, however, statistically not significant ($p=0.141>0.05$, Speaman rank correlation).

4 Discussion and Conclusion

The authors [3, 4] reported the age-related reduction of PBF in human primary teeth, and the reduction is considered to be due to the decrease of blood supply due to the root resorption. Ikawa et al. [5] has reported that the hemodynamics in the human permanent tooth is reduced with age. In general, in the human permanent teeth, the number of blood vessels is decreased and the size and volume of the pulp is also reduced due to the increase in calcified tissue with age. In the present study, PBF in young permanent teeth was decreased with the increase of the age of the subjects. This age-related reduction of PBF is considered to be produced by the reduction of the blood supply due to the root formation.

References

1. Komatsu H, Ikawa M, Mayanagi H (2003) Pulpal blood flow measurement in human young permanent teeth. J Dent Res 82:C-437 Abst #105
2. Nolla CM (1960) The development of the permanent teeth. J Dent Child 27:254–266
3. Komatsu H, Ikawa M, Mayanagi H (2007) Age-related changes of pulpal blood flow in primary teeth measured by laser Doppler blood flowmetery. Ped Dent J 17:27–31

4. Komatsu H, Ikawa M, Mayanagi H (2007) Pulpal blood flow in human primary teeth with different root resorption. In: Watanabe M, Okuno O (eds) Interface oral health science 2007. Springer, Tokyo, pp 197–198
5. Ikawa M, Komatsu H, Ikawa K et al (2003) Age-related changes in the human pulpal blood flow measured by laser Doppler flowmetry. Dent Traumatol 19:36–40

Immunohistological study on STRO-1 in developing rat dental tissues with light and electron microscopy

Ryuta Kaneko, Hirotoshi Akita, Hidetoshi Shimauchi, and Yasuyuki Sasano

Abstract. Cell population positive for the anti-STRO-1 antibody has been shown to contain mesenchymal stem cells. STRO-1-positive cells have been reported to reside in dental pulp and periodontal ligaments, while their tissue localization in developing teeth is not clear. The present study investigated localization of STRO-1 in developing rat teeth by immunohistochemistry by using light microscopy and electron microscopy.

Key words. STRO-1, tooth, development, immunohistochemistry, electron microscopy

1 Introduction

A cell population with STRO-1 antigens has been shown to contain mesenchymal stem cells in dental tissues such as pulp and periodontal ligaments [1–5]. However, the detailed localization of STRO-1 in developing dental tissues is not known. The present study was designed to investigate localization of STRO-1 in developing rat teeth and periodontal tissues by immunohistochemistry.

2 Materials and Methods

Wistar rats at 2, 3, 6, and 12 weeks postnatum and 18 days postciutum were fixed with 4% paraformaldehyde. Mandibles and maxillae were resected and decalcified in 10% EDTA. The specimens were embedded in paraffin and sections with 5 μm

R. Kaneko and H. Shimauchi
Divisions of Periodontology and Endodontology, Tohoku University Graduate School of Dentistry, Sendai 980-8575, Japan

R. Kaneko, H. Akita, and Y. Sasano (✉)
Craniofacial Development and Regeneration, Tohoku University Graduate School of Dentistry, Sendai 980-8575, Japan
e-mail: sasano@anat.dent.tohoku.ac.jp

T. Sasano et al. (eds.), *Interface Oral Health Science 2009*,
DOI 10.1007/978-4-431-99644-6_39, © Springer 2010

thickness were cut and processed for immunohistochemistry using the anti-STRO-1 antibody. Selected specimens were frozen, sectioned, and processed for immunohistochemistry, and the visualized immunoreaction was observed using electron microscopy.

3 Results

Immunoreactions for STRO-1 were identified in some bone marrow cells. Some odontoblasts and dental pulp cells showed positive immunoreactivity in developing rat molar crowns, roots, and incisors. Alveolar osteoblasts, cementoblasts, and periodontal ligament cells were also immunoreactive. The electron microscopy localized the antigen in plasma membrane, rough endoplasmic reticulum, and some vesicles in dental pulp cells and odontoblasts.

4 Conclusion

The present study suggests that the STRO-1 antigen is involved in the differentiation of mesenchymal cell lineages and in the formation of the matrix in dental tissues.

References

1. Gronthos S, Mankani M, Brahim J et al (2000) Postnatal human dental pulp stem cells (DPSCs) in vitro and in vivo. Proc Natl Acad Sci U S A 97:13625–13630
2. Seo BM, Miura M, Gronthos S et al (2004) Investigation of multipotent postnatal stem cells from human periodontal ligament. Lancet 364(9429):149–155
3. Mina M, Braut A (2004) New insight into progenitor/stem cells in dental pulp using Col1a1-GFP transgenes. Cells Tissues Organs 176:120–133
4. Iohara K, Nakashima M, Ito M et al (2005) Dentin regeneration by dental pulp stem cell therapy with recombinant human bone morphogenetic protein 2. J Dent Res 83:590–595
5. Zhang W, Walboomers XF, Wolke JG et al (2005) Differentiation ability of rat postnatal dental pulp cells in vitro. Tissue Eng 11:357–368

The physiological calcification process is replicated in a rat embryonic calvarial culture

Yasuko Kimura, Shigeshi Kikunaga, Ichiro Takahashi, Yuji Hatakeyama, Satoshi Fukumoto, and Yasuyuki Sasano

Abstract. The present study was designed to investigate the progress of physiological calcification in the bone matrix by using a rat embryonic calvarial organ culture. We analyzed the content and concentration of calcium in the cultured bone matrix for assessing the organ culture system to replicate the progress in calcification.

Key words. bone, calcification, organ culture, rat, calvaria

1 Introduction

Bone calcification is a complex biological process involving various cells and extracellular matrices [1–5]. Calcification progresses during bone development, but the mechanism of the progress of the physiological calcification has been largely unknown. The present study was designed to establish an organ culture system to replicate the physiological calcification process in rat embryonic calvaria as a device for assessing the progress of calcification.

Y. Kimura and S. Fukumoto
Divisions of Pediatric Dentistry, Tohoku University Graduate School of Dentistry, Sendai 980-8575, Japan

Y. Kimura and Y. Sasano (✉)
Division of Craniofacial Development and Regeneration, Tohoku University Graduate School of Dentistry, Sendai 980-8575, Japan
e-mail: sasano@anat.dent.tohoku.ac.jp

S. Kikunaga
Faculty of Sciences of Human Life, Notre Dame Seishin University, Okayama 700-8516, Japan

I. Takahashi
Division of Orthodontics and Dentofacial Orthopedics, Tohoku University Graduate School of Dentistry, Sendai 980-8575, Japan

Y. Hatakeyama
Functional Structure Section, Department of Morphological Biology, Fukuoka Dental College, Fukuoka 814-0193, Japan

T. Sasano et al. (eds.), *Interface Oral Health Science 2009*,
DOI 10.1007/978-4-431-99644-6_40, © Springer 2010

2 Materials and Methods

The standardized calvarial explants were dissected from embryonic day 18 and 20 (E18 and E20) rats and cultured for 1, 3, and 5 days. The total calcium content of the cultured explants was quantified using atomic absorption spectrophotometry. Alternatively, the explants were fixed, embedded in paraffin, sectioned, and stained with von Kossa combined with hematoxylin-eosin or processed for energy dispersive X-ray spectroscopy to analyze the concentration of calcium, phosphorus, and carbon in the tissue.

3 Results

The total calcium content after 5 days of culture increased 2.9 and 2.0 times in E18 and E20 cultured calvaria (E18cc and E20cc), respectively. All cultured calvarias were von Kossa positive, although the staining was intensified only in E18cc over 5 days. Sound ostoblasts and osteocytes were observed in the bone matrix in E18cc, while the cells often degenerated in E20cc. Concentrations of calcium and carbon increased 2.1 and 2.7 times in E18cc for 5 days, respectively, while those in E20cc showed little increase. The phosphorus concentration was constant for 5 days in both E18cc and E20cc.

4 Conclusion

The physiological calcification proceeded in E18cc, but not in E20cc. These results indicate that the organ culture system using E18cc is useful for replication of the physiological calcification process in vitro.

References

1. Bonucci E (1971) The locus of initial calcification in cartilage and bone. Clin Orthop 78: 108–139
2. Hoshi K, Kenmotsu S, Takeuchi Y et al (1999) The primary calcification in bones follows removal of decorin and fusion of collagen fibrils. J Bone Miner Res 14:273–280
3. Hoshi K, Ozawa H (2000) Matrix vesicle calcification in bones of adult rats. Calcif Tissue Int 66:430–434
4. Sasano Y, Li HC, Zhu JX et al (2000) Immunohistochemical localization of type I collagen, fibronectin and tenascin C during embryonic osteogenesis in the dentary of mandibles and tibias in rats. Histochem J 32:591–598
5. Marks SC, Odgren PR (2002) Structure and development of the skeleton. In: Bilezikian JP, Raisz LG, Rodan GA (eds) Principles of bone biology, vol 1, 2nd edn. Academic, San Diego, pp 3–15

Tonometric measurement of the gingiva in young and elder humans

Kyoko Ikawa, Motohide Ikawa, and Takeyoshi Koseki

Abstract. We have already reported the hardness of the gingiva in young human subjects measured by a tonometer. In this study, the age-related changes in the hardness of the human gingival tissue were examined. The measurement was made on the young and the elder people who had clinically healthy gingival tissue using a tonometer. The labial surfaces of attached gingival site were selected as examination sites. The gingiva was significantly harder with elder subjects than the young subjects. The results indicated the age-related changes of gingival hardness.

Key words. human, gingiva, pressure, tonometer, aging

1 Introduction

It is generally accepted that the gingival tissue tends to soften with the progress of gingival inflammation. One of the techniques to measure tissue hardness is tonometer, which is widely used for the detection of eye diseases. We have already measured the hardness of the gingival tissue using a commercially available tonometer and reported the difference of the hardness between sound and inflamed gingiva in the young human subjects. In the previous study, the information was obtained from limited ages of the subject, and age-related changes in the hardness of the human gingival tissue have not been reported, yet.

K. Ikawa (✉) and T. Koseki
Division of Preventive Dentistry, Department of Oral Health and Development Sciences, Tohoku University Graduate School of Dentistry, 4-1 Seiryo-machi, Aoba-ku, Sendai 980-8575, Japan
e-mail: kkikw@mail.tains.tohoku.ac.jp

M. Ikawa
Division of Periodontology and Endodontology, Department of Oral Biology, Tohoku University Graduate School of Dentistry, 4-1 Seiryo-machi, Aoba-ku, Sendai 980-8575, Japan

T. Sasano et al. (eds.), *Interface Oral Health Science 2009*,
DOI 10.1007/978-4-431-99644-6_41, © Springer 2010

2 Materials and Methods

The measurement was made on the young (males average 24.7 year) and elder people (both males and females, average 72.2 year) who had clinically healthy gingival tissue. Prior to the measurement, informed consent was obtained from all of them. The depths of the periodontal pockets of the recording sites were measured by one of the authors who was qualified as a periodontal specialist. The labial surfaces of proximal area of attached gingival sites of teeth (13–23, 33–43) were selected as examination sites.

Induction-based impact method for measuring intraocular pressure was applied; a very light probe was used to make a momentary contact with the gingiva. A ball of 1.0 mm diameter was attached to one end of a thin steel bar of 40 mm long. The other end of the bar was inserted into the tonometer where a 20 mm-long permanent magnet solenoid was mounted. The ball-like end of the probe hits the gingival surface and bounces back, and the solenoid detected the movement and impact of the probe. The device displays the digital value of the gingival pressure (mmHg) [1]. No participants claimed any pain sensation during measurement.

3 Results

Pocket depth and bleeding on probing were not significantly different between two subject groups. The lower gingival hardness of elderly people was significantly harder than that of young people ($p<0.001$, average Young < Elder). The average of gingival hardness of elderly people was 1.3 times harder than that of younger people. However, upper gingival hardness was not significant between young and elderly people. The individual hardness between upper and lower gingiva were not significantly different.

4 Discussion and Conclusion

The gingiva was significantly harder with elder subjects than that with the young subjects. Although the information obtained in the present study was limited, the results indicated age-related changes in the human gingival hardness [2]. The measurement of gingival hardness seems to be useful to diagnose gingivitis. In this study, we measured the gingival hardness by the direct contact of the measurement probe with the gingival surface. The development of the measurement technique without any contact of the probe and the gingiva will be required.

References

1. Kontiola AI (2000) A new induction-based impact method for measuring intraocular pressure. Acta Ophthalmol Scand 78:142–145
2. Müller HP, Könönen E (2005) Variance components of gingival thickness. J Periodontal Res 40:239–244

Isolation and comparison of mesenchymal stem cells derived from human wisdom tooth germs and periodontal ligament in vitro

Daisuke Nishihara, Yoko Iwamatsu-Kobayashi, Masatsugu Hirata, Koji Kindaichi, Junko Kindaichi, and Masashi Komatsu

Abstract. We examined the isolation of mesenchymal stem cells from human wisdom tooth germs (hWTGs) or human periodontal ligament (hPDL) and compared their characteristics. Collection rate of STRO-1 positive cells from WTGs was 6.01% but from PDL was 2.22%. After 7 days' culture, alkaline phosphatase activity of stem cells derived from WTGs was increased and it was statistically higher than those of PDL stem cells. After 14 days, protein content of stem cells derived from WTGs became statistically higher than those of PDL stem cells. It is suggested that hWTGs are useful for stem cell treatment and tissue engineering.

Key words. mesenchymal stem cell, wisdom tooth germ

1 Introduction

Mesenchymal stem cells (MSCs) have the possibility for stem cell treatment and tissue engineering. They have been isolated from the dental pulp, the deciduous tooth, and the periodontal ligament (PDL). Human wisdom tooth germs (hWTGs) are considered as one of the sources of MSCs in odontology. If the stem cells of WTGs are separated and identified, it is possible to use them for regenerative medicine. In this research, we examined the isolation of MSCs from hWTGs or hPDL and compared their characteristics.

D. Nishihara (✉), Y. Iwamatsu-Kobayashi, M. Hirata, and M. Komatsu
Division of Operative Dentistry, Tohoku University Graduate School of Dentistry, Sendai, Japan
e-mail: nishihara-th@umin.ac.jp

K. Kindaichi
Division of Oral and Craniofacial Anatomy, Tohoku University Graduate School of Dentistry, Sendai, Japan

J. Kindaichi
National Center for Child Medical Health and Development, Tokyo, Japan

T. Sasano et al. (eds.), *Interface Oral Health Science 2009*,
DOI 10.1007/978-4-431-99644-6_42, © Springer 2010

2 Materials and Methods

After informed consent, hWTGs were extracted from healthy subjects (aged 8–10 years). hPDL was obtained from clinically healthy third molars (aged 20–25 years). For isolation of MSCs, WTGs were dissociated into cells. PDL was removed from the root surface and subcultured until passage 1. MSCs were isolated by immunomagnetic bead selection (Dynabeads®, Invitrogen, Carsbad, CA) using the STRO-1 (MAB1038R&D Systems, Minneapolis, MN) antibody and plated at 1×10^4/ml in 24-well dishes. After 3, 5, 7, 14, and 21 days' culture, we measured protein content and alkaline phosphatase (ALP) activities.

3 Results

Collection rate of STRO-1 positive cells from WTGs was 6.01% but from PDL was 2.22%. After 7 days, ALP activity of stem cells derived from WTGs was increased and it was statistically higher than those of PDL stem cells. After 14 days, protein content of stem cells derived from WTGs became statistically higher than those of PDL stem cells. After 21 days, stem cells derived from WTGs became confluent but PDL-derived stem cells did not.

4 Discussion

Seo et al. reported that MSCs that had the high differentiation and proliferation potency were separated from PDL [1]. They were expressed in the MSC makers STRO-1 and CD146. In our previous research, it was confirmed that MSCs existed in hWTGs. Moreover, a lot of these cells existed especially around the blood vessels like the stem cells in the adult PDL. Nagatomo et al. reported that the percentage positivity with STRO-1 was $1.2\pm0.1\%$ in the PDL [2]. In this study, the collection rate of STRO-1 positive cells was $6.01\pm1.36\%$ from hWTGs and $2.22\pm0.46\%$ from PDL. In addition, protein content and ALP activity of stem cells derived from WTGs became higher than those from PDL. It was suggested that using immunomagnetic beads was more efficient to collect STRO-1 positive cells and hWTGs could be a good source of stem cells. Recently, Itaya et al. reported that the STRO-1 positive fraction in PDL was 33.5% at passage 0, which was reduced to 14.7% at passage 3 [3]. In this study, we used PDL-derived cells subcultured at passage 1 because it was difficult to obtain sufficient cell numbers from PDL without subculture. This might influence the collection rate. Further studies require to examine the characteristics of stem cells and the affection of different MSCs to tissue engineering in detail.

References

1. Seo BM, Miura M, Gronthos S et al (2004) Investigation of multipotent postnatal stem cells from human periodontal ligament. Lancet 364(9429):149–155
2. Nagatomo K, Komaki M, Sekiya I et al (2006) Stem cell properties of human periodontal ligament cells. J Periodontal Res 41(4):303–310
3. Itaya T, Kagami H, Okada K et al (2009) Characteristic changes of periodontal ligament-derived cells during passage. J Periodontal Res 44(4):425–433

Unitary discharges of TMJ mechanosensitive neurons during cortically induced jaw movement

Yasuo Takafuji, Akito Tsuboi, Takayoshi Tabata, Osuke Suzuki, and Makoto Watanabe

Abstract. The purpose of this study was to investigate the response properties of the temporomandibular joint (TMJ) neurons in the trigeminal ganglion during cortically induced jaw movement of the rabbit. The discharges of TMJ units were recorded from the left trigeminal ganglion with metal microelectrodes. All TMJ units recorded in this experiment were slowly adapting type. The firing frequencies of TMJ units increased with displacement of the condyle movement in the jaw opening phase and returned to the initial response level in the jaw closing one. The TMJ neurons are assumed to carry sensory information about the jaw movement to the central nervous system in the jaw opening phase.

Key words. trigeminal ganglion, mechanosensitive neurons, temporomandibular joint, jaw movement, rabbit

1 Introduction

Mechanoreceptors in the temporomandibular joint (TMJ) are considered to play an important role in controlling chewing, based on sensory information about the jaw position and movement [1–3]. However, there are no reports on physiological properties of the mechanosensitive neurons innervating the TMJ during jaw movement. The purpose of this study was to investigate the response properties of the mechanosensitive TMJ neurons in the trigeminal ganglion during cortically induced rhythmical jaw movement.

Y. Takafuji (✉), A. Tsuboi, T. Tabata, O. Suzuki, and M. Watanabe
Division of Aging and Geriatric Dentistry, Tohoku University Graduate School of Dentistry, Sendai 980-8575, Japan
e-mail: takafuji@mandible.dent.tohoku.ac.jp

T. Sasano et al. (eds.), *Interface Oral Health Science 2009*,
DOI 10.1007/978-4-431-99644-6_43, © Springer 2010

2 Methods

Adult female rabbits were initially anesthetized with a urethane–chloralose mixture (α-chloralose 40 mg/kg, urethane 500 mg/kg). The discharges of TMJ neurons sensitive to mechanical stimulation of the condyle were recorded from the left trigeminal ganglion with metal microelectrodes. The response properties of TMJ neurons were investigated during rhythmical jaw movement induced by repetitive stimulation to the caudolateral part of the right cortical masticatory area.

3 Response Properties of TMJ Units

Previous studies have found fewer rapidly adapting responses than slowly adapting responses in the primary afferents supplying the TMJ mechanoreceptors [1–3]. Our present study also demonstrated that the majority of TMJ units were of the SA type, and a few of TMJ units were of the RA type. These findings led us to surmise that RA responses may contribute less to regulation of TMJ movement.

4 Neuronal Activities of TMJ Neurons During Cortically Induced Jaw Movement

The firing frequencies of TMJ units increased with displacement of the condyle movement in the jaw opening phase and returned to the initial response level in the jaw closing one (Fig. 1). This finding suggested that TMJ neurons in the trigeminal ganglion carry sensory information about the jaw movement to the central nervous system only in the jaw opening phase.

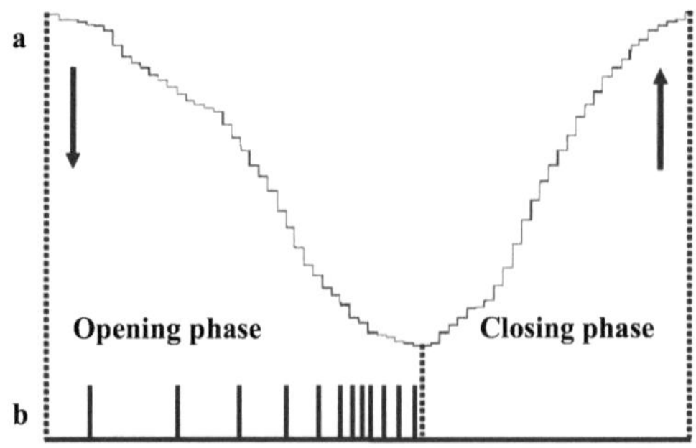

Fig. 1. Pattern diagrams of response of a TMJ unit during cortically induced jaw movement. (**a**) Jaw movement, (**b**) spike discharges of TMJ units

References

1. Kawamura Y, Abe K (1974) Role of sensory information from temporomandibular joint. Bull Tokyo Med Dent Univ 21:78–92
2. Lund JP, Matthews B (1981) Responses of temporomandibular joint afferents recorded in the Gasserian ganglion of the rabbit to passive movements of the mandible. In: Kawamura Y, Dubner R (eds) Oral-facial sensory and motor functions. Quintessence, Tokyo, pp 153–160
3. Tsuboi A, Takafuji Y, Itoh S et al (2009) Response properties of trigeminal ganglion mechanosensitive neurons innervating the temporomandibular joint of the rabbit. Exp Brain Res. doi:10.1007/s00221-009-1978-z

Evaluation of bone metabolism of temporomandibular joint by using high resolution PET scanner

Miou Yamamoto, Masayoshi Yokoyama, Shigeto Koyama, Yoshihito Funaki, Youhei Kikuchi, Kenji Nakamura, Kouichi Nakazawa, Hiromichi Yamazaki, Keizo Ishii, and Keiichi Sasaki

Abstract. Bone metabolism of temporomandibular joint (TMJ) induced by loss of unilateral occlusal support was evaluated using the high resolution positron emission tomography scanner. Highly enhanced metabolic activity of TMJ in the extracted side was observed in the early stage after tooth extraction, and it was gradually decreased.

Key words. bone metabolism, temporomandibular joint, PET scanner, occlusal support

1 Introduction

There is little information regarding longitudinal evaluation of bone remodeling at temporomandibular joint (TMJ), although it is generally believed that loss of occlusal support affects bone remodeling activity of TMJ. This chapter aims to

---

M. Yamamoto (✉), M. Yokoyama, and K. Sasaki
Division of Advanced Prosthetic Dentistry, Tohoku University Graduate School of Dentistry, 4-1 Seiryo-machi, Aoba-ku, Sendai 980-8575, Japan
e-mail: miyamamoto-thk@umin.net

S. Koyama
Maxillofacial Prosthetics Clinic, Tohoku University Hospital, Sendai, Japan

Y. Funaki
Division of Radiopharmaceutical Chemistry, Tohoku University Cyclotron Radioisotope Center, Sendai, Japan

H. Yamazaki and K. Ishii
Division of Radiation Protection and Safety Control, Tohoku University Cyclotron Radioisotope Center, Sendai, Japan

Y. Kikuchi, K. Nakamura, K. Nakazawa, and K. Ishii
Division of Quantum Science and Energy Engineering, Tohoku University Graduate School of Engineering, Sendai, Japan

T. Sasano et al. (eds.), *Interface Oral Health Science 2009*,
DOI 10.1007/978-4-431-99644-6_44, © Springer 2010

review the influence of occlusal support loss on dynamic changes of bone metabolism of TMJ using a newly developed high resolution positron emission tomography (PET) scanner (Tohoku university, Sendai, Japan) [1].

2 Quantitative Evaluation of Bone Metabolic Activity by Using PET Scanner

The right maxilla molars of Wistar rats (5 weeks) were extracted under anesthesia. The rats with ^{18}fluoride ion ($^{18}F^-$) (5mCi/rat) intravenously-injected were scanned by a practical semiconductor animal PET scanner at 4, 7, 14, 21, 28 days after teeth extraction. Controls with sham operation were scanned in the same way. Spheric region of interests were defined around the TMJ in the extracted side (r-TMJ) and in the intact side (l-TMJ). Accumulation counts (ACs) of $^{18}F^-$ at the TMJs of the scanned data were examined.

3 Bone Metabolic Activity in TMJ Induced by Tooth Support Loss

AC at the r-TMJ was significantly higher than that of l-TMJ at 7 days after extraction and then decreased to the same level as that of control at 21 days (Fig. 1a). This tendency was also observed in AC at the extracted socket of alveolar bone (Fig. 1b).

The accumulation of $^{18}F^-$, which chemically binds to the mineral crystal of bone tissue during new bone formation, indicates osteogenesis. In this study, initially elevated AC value at the r-TMJ decreased to the control level with time. These results suggested that mechanical stress induced by missing of unilateral occlusal support may initially enhance bone metabolic activity of TMJ at the extracted side, however, TMJ might adapt to the mechanical stress with time.

a b c

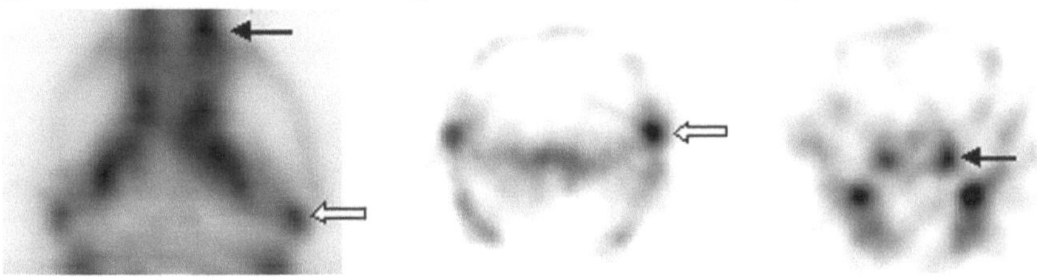

Fig. 1. (**a**) PET axial image at 7 days after teeth extraction. *White arrow* shows r-TMJ, and black arrow shows the extracted socket. (**b**) PET coronal image of the TMJ area. (**c**) PET coronal image of the extracted socket area

Reference

1. Ishii K, Kikuchi Y, Matsuyama S et al (2007) First achievement of less than 1 mm FWHM resolution in practical semiconductor animal PET scanner. Nucl Instrum Methods Phys Res A 576:435–440

Physiological role of type II bone morphogenetic protein receptor and its interacting molecules in bone morphogenetic protein signaling

Tada-aki Kudo, Akira Watanabe, Masanobu Asano, Ye Zhang, Fei Zhao, Mitsuhiro Kano, Yoshinaka Shimizu, Hiroyasu Kanetaka, Shinri Tamura, and Haruhide Hayashi

Abstract. Type II bone morphogenetic protein (BMP) receptor (BMPRII) is a membrane receptor containing a unique tail domain. This kinase receptor is involved in various biological processes, including bone formation. Recent studies have revealed that novel BMPRII-interacting proteins are involved in the Smad signaling pathway or in the non-Smad signaling pathway. This report focuses on BMPRII-mediated signal transduction and provides advanced insights into the physiological roles of BMPRII.

Key words. bone morphogenetic protein (BMP), BMP receptor

T. Kudo(✉), M. Asano, F. Zhao, and H. Hayashi
Division of Oral Physiology, Tohoku University Graduate School of Dentistry, 4-1 Seiryo-machi, Aoba-ku, Sendai 980-8575, Japan
e-mail: tada@m.tains.tohoku.ac.jp

A. Watanabe
Genome Science Division, Research Center for Advanced Science and Technology, The University of Tokyo, 4-6-1 Komaba, Meguro-ku, Tokyo 153-8904, Japan

Y. Zhang
Department of Physical Medicine and Rehabilitation, Tohoku University Graduate School of Biomedical Engineering, 4-1 Seiryo-machi, Aoba-ku, Sendai 980-8575, Japan

M. Kano and Y. Shimizu
Division of Oral and Craniofacial Anatomy, Tohoku University Graduate School of Dentistry, 4-1 Seiryo-machi, Aoba-ku, Sendai 980-8575, Japan

H. Kanetaka
Department of Physical Medicine and Rehabilitation, Tohoku University Graduate School of Biomedical Engineering, 4-1 Seiryo-machi, Aoba-ku, Sendai 980-8575, Japan
Division of Advanced Prosthetic Dentistry, Tohoku University Graduate School of Dentistry, 4-1 Seiryo-machi, Aoba-ku, Sendai 980-8575, Japan

S. Tamura
Department of Biochemistry, Institute of Development, Aging and Cancer, Tohoku University, 4-1 Seiryo-machi, Aoba-ku, Sendai 980-8575, Japan

T. Sasano et al. (eds.), *Interface Oral Health Science 2009*,
DOI 10.1007/978-4-431-99644-6_45, © Springer 2010

1 Introduction

Bone morphogenetic proteins (BMPs) elicit diverse cellular responses. They bind to heterogeneous complexes of kinase receptors known as type I and II BMP receptors (BMPRI and BMPRII, respectively). BMPRII phosphorylates BMPRI, activates Smads through phosphorylation, and they in turn upregulate the expression of target genes. The intracellular domain of BMPRII consists of a kinase domain from amino acids 200 to 500 and a unique tail domain from amino acids 500 to 1,038 [1]. The kinase domain also interacts with other targets, including the receptor for activated c-kinase 1 (Rack1) [2]. The BMPRII tail domain, which is conserved among vertebrates, has several poorly understood functions, including regulation of p38 [1] and interaction with LIM domain kinase (LIMK), Src, Tctex, and c-Jun NH_2-terminal kinase (JNK) [2, 3].

2 Regulation of Smad Signaling

Several BMPRII-interacting molecules, including caveolin-1 and tropomyosin-related kinase C (TrkC), modulate the ability of BMPRII to activate Smad via BMPRI phosphorylation [4, 5]. Caveolin-1 negatively regulates BMPRII-mediated Smad signaling by regulating the BMPRII localization and its accessibility to BMPRI [5]. TrkC also suppresses BMP signaling by binding with BMPRII to inhibit formation of the BMPRI–BMPRII complex [6].

3 Regulation of Non-Smad Signaling

Several other BMPRII-interacting proteins are involved in non-Smad signaling. The interaction between LIMK and the BMPRII tail inhibits the ability of LIMK to alter the actin cytoskeleton dynamics [2]. Moreover, the interaction of Src with the BMPRII tail inhibits Src activity and inhibits downstream cell cycle regulators that have been suggested to influence cell proliferation [2]. Interestingly, termination mutation by substitution with arginine at amino acid 899 (R899X) in the BMPRII tail domain leads to increased basal activity of p38 with unaltered Smad activity in vivo [1]; thus, we expect that the BMPRII region that was disrupted by the R899X mutation (amino acids 899–1,038) may have important site(s) associated with the regulation of the BMP-p38 pathway.

References

1. West J, Harral J, Lane K et al (2008) Mice expressing BMPR2^{R899X} transgene in smooth muscle develop pulmonary vascular lesions. Am J Physiol Lung Cell Mol Physiol 295:744–755

2. Zakrzewicz A, Hecker M, Marsh LM et al (2007) Receptor for activated C-kinase 1, a novel interaction partner of type II bone morphogenetic protein receptor, regulates smooth muscle cell proliferation in pulmonary arterial hypertension. Circulation 115:2957–2968
3. Kudo T, Kobayashi T, Tamura S (2007) Bioinformatics-based cyclopedic search for novel JNK binding molecules. Tohoku Univ Dent J 26:1–11
4. Wertz JW, Bauer PM (2008) Caveolin-1 regulates BMPRII localization and signaling in vascular smooth muscle cells. Biochem Biophys Res Commun 375:557–561
5. Jin W, Yun C, Kim HS et al (2007) TrkC binds to the bone morphogenetic protein type II receptor to suppress bone morphogenetic protein signaling. Cancer Res 67:9869–9877

Role of the protein serine/threonine phosphatase dullard in cell differentiation

Fei Zhao, Tada-aki Kudo, Ye Zhang, Mitsuhiro Kano, Shinri Tamura, Yoshinaka Shimizu, Hiroyasu Kanetaka, and Haruhide Hayashi

Abstract. Bone morphogenetic protein (BMP) signaling is essential for neurogenesis and bone formation. Dullard phosphatase functions as a negative regulator of BMP signaling by promoting the dephosphorylation or degradation of BMP receptors. In contrast, dullard regulates nuclear membrane biogenesis. This report aims to characterize dullard and determine the role played by it in cellular signaling.

Key words. bone morphogenetic proteins, phosphatase dullard

1 Introduction

Cellular protein phosphorylation is modulated by protein kinases and phosphatases to regulate a series of biological functions. A phosphatase dullard was initially isolated as a neural region-specific gene from a cDNA library of the lateral regions

F. Zhao, T. Kudo (✉), and H. Hayashi
Division of Oral Physiology, Tohoku University Graduate School of Dentistry, 4-1 Seiryo-machi, Aoba-ku, Sendai 980-8575, Japan
e-mail: tada@m.tains.tohoku.ac.jp

Y. Zhang
Department of Physical Medicine and Rehabilitation, Tohoku University Graduate School of Biomedical Engineering, 4-1 Seiryo-machi, Aoba-ku, Sendai 980-8575, Japan

H. Kanetaka
Department of Physical Medicine and Rehabilitation, Tohoku University Graduate School of Biomedical Engineering, 4-1 Seiryo-machi, Aoba-ku, Sendai 980-8575, Japan
Division of Advanced Prosthetic Dentistry, Tohoku University Graduate School of Dentistry, 4-1 Seiryo-machi, Aoba-ku, Sendai 980-8575, Japan

M. Kano and Y. Shimizu
Division of Oral and Craniofacial Anatomy, Tohoku University Graduate School of Dentistry, 4-1 Seiryo-machi, Aoba-ku, Sendai 980-8575, Japan

S. Tamura
Department of Biochemistry, Institute of Development, Aging and Cancer, Tohoku University, 4-1 Seiryo-machi, Aoba-ku, Sendai 980-8575, Japan

T. Sasano et al. (eds.), *Interface Oral Health Science 2009*, 196
DOI 10.1007/978-4-431-99644-6_46, © Springer 2010

of *Xenopus* [1]. Dullard comprises 244 amino acids and has N-terminal transmembrane regions, which make it hydrophobic [2]. Dullard also has a DXDX(T/V) catalytic motif in a C-terminal phosphatase domain [1].

2 Function of Dullard

In vitro kinetic analysis using a phosphopeptide array showed that dullard preferably dephosphorylates peptides containing phospho-Ser or phospho-Thr [2]. ALK3 was the first substrate of dullard to be identified [1]. ALK3 is a cell-surface receptor, which is a bone morphogenetic protein (BMP) receptor of type I (BMPRI). BMPRI binds BMPs with BMP receptor type II (BMPRII) and subsequently phosphorylates the C-terminal motif of BMP-Smads (Smad1/5/8). Dullard blocks the BMP-dependent phosphorylation of ALK3; this is followed by the inhibition of BMP-dependent Smad1/5/8 activation. Dullard recognizes Lipin1 as another substrate. Lipin1 also belongs to a DXDX(T/V)-containing phosphatase family and has multiple Ser/Thr-Pro phosphorylation sites, also known as mitogen-activated protein kinase (MAPK) phosphorylation sites, and is considered to participate in a signaling cascade that regulates nuclear membrane biogenesis during cell division. Dullard also promotes proteosomal degradation of BMPRII to inhibit BMP signaling, suggesting that dullard is involved in ubiquitination of BMPRII by an unknown E3 ligase [2].

3 Regulation of Dullard

Although the inactivation of dullard has a profound effect on brain development in *Xenopus* through inhibition of BMP signaling [1], further research may reveal the regulatory mechanisms of dullard, including the modulation of its subcellular localization. Interestingly, short-term mechanical stress can induce gene expression of both dullard and BMP2 in smooth muscle cells of the bladder [3]. Moreover, dullard gene expression is upregulated on day 4, similar to that of the neural stem cell marker Mash1, during neural differentiation of P19 cells (data not shown), in which multiple signaling pathways are activated, including the c-Jun N-terminal kinase (JNK) pathway [4]. Furthermore, since dullard contains putative MAPK phosphorylation sites – Ser^{49}-Pro and Thr^{89}-Pro, – which are highly conserved in its phosphatase domain among vertebrates, the activity of dullard phosphatase may be regulated posttranscriptionally by MAPK(s) at these phosphorylation sites.

References

1. Satow R, Kurisaki A, Chan TC et al (2006) Dullard promotes degradation and dephosphorylation of BMP receptors and is required for neural induction. Dev Cell 11:763–774

2. Kim Y, Gentry MS, Harris TE et al (2007) A conserved phosphatase cascade that regulates nuclear membrane biogenesis. Proc Natl Acad Sci U S A 104:6596–6601
3. Adam RM, Eaton SH, Estrada C et al (2004) Mechanical stretch is a highly selective regulator of gene expression in human bladder smooth muscle cells. Physiol Genomics 20:36–44
4. Akiyama S, Yonezawa T, Kudo T et al (2004) Activation mechanism of c-Jun amino-terminal kinase in the course of neural differentiation of P19 embryonic carcinoma cells. J Biol Chem 279:36616–36620

The role of extracellular signal-regulated kinase 5 signaling pathway in neurons

Ye Zhang, Tada-aki Kudo, Yunchia Ku, Fei Zhao, Mitsuhiro Kano, Yoshinaka Shimizu, Haruhide Hayashi, Taizo Hamada, and Hiroyasu Kanetaka

Abstract. Extracellular signal-regulated kinase 5 (ERK5) plays an important role in the regulation of cell proliferation and differentiation. Recent data suggest that ERK5 is activated by neurotrophic factors in various types of neurons and induces neurotrophin-mediated neuronal survival. This report provides advanced findings on the mechanism underlying ERK5-mediated neuroprotection.

Key words. extracellular signal-regulated kinase 5 (ERK5), neuron

1 Introduction

The mitogen-activated protein kinase (MAPK) cascade is a conserved module involved in various cellular functions. In mammals, the following four MAPK pathways have been identified: extracellular signal-regulated kinase 1 and 2 (ERK1/2), c-jun

Y. Zhang
Department of Physical Medicine and Rehabilitation, Tohoku University Graduate School of Biomedical Engineering, 4-1 Seiryo-machi, Aoba-ku, Sendai 980-8575, Japan

H. Kanetaka
Department of Physical Medicine and Rehabilitation, Tohoku University Graduate School of Biomedical Engineering, 4-1 Seiryo-machi, Aoba-ku, Sendai 980-8575, Japan
Division of Advanced Prosthetic Dentistry, Tohoku University Graduate School of Dentistry, 4-1 Seiryo-machi, Aoba-ku, Sendai 980-8575, Japan

T. Kudo (✉), F. Zhao, and H. Hayashi
Division of Oral Physiology, Tohoku University Graduate School of Dentistry, 4-1 Seiryo-machi, Aoba-ku, Sendai 980-8575, Japan
e-mail: tada@m.tains.tohoku.ac.jp

Y. Ku and T. Hamada
Department of Oral Health Care Promotion, Tohoku University Graduate School of Dentistry, 4-1 Seiryo-machi, Aoba-ku, Sendai 980-8575, Japan

M. Kano and Y. Shimizu
Division of Oral and Craniofacial Anatomy, Tohoku University Graduate School of Dentistry, 4-1 Seiryo-machi, Aoba-ku, Sendai 980-8575, Japan

T. Sasano et al. (eds.), *Interface Oral Health Science 2009*,
DOI 10.1007/978-4-431-99644-6_47, © Springer 2010

N-terminal kinase (JNK), p38, and ERK5 [1]. ERK5 contains a Thr-Glu-Tyr activation motif similar to ERK1/2. However, ERK5 has an unusually large C-terminal nonkinase domain, which makes ERK5 different from ERK1/2 and other members of the MAPK family. This domain contains (1) a nuclear localization signal, which is essential for the nuclear localization of ERK5, and (2) a proline-rich region that may serve as a site for interaction with other proteins [1, 2].

2 ERK5 Pathway

The signaling modules in the ERK5 pathway consist of kinases, MAPK kinase kinase 2 and 3 (MEKK2/3), MAPK kinase 5 (MEK5), and ERK5 [1]. The ERK5 pathway is activated by mitogens, including serum, brain-derived neurotrophic factor (BDNF), and nerve growth factor (NGF), as well as by stress stimuli, including hyperosmotic shock and oxidative stress [2]. A number of molecules have been identified as substrates of the ERK5 pathway, including p90 ribosomal S6 kinase (RSK), myocyte enhancer factor 2 (MEF2) family of transcriptional factors (MEF2A, MEF2C, and MEF2D), and BCL2-associated agonist of cell death (BAD), which is a proapoptotic B-cell leukemia/lymphoma 2 (BCL2) family member [2].

3 Neuronal Survival

ERK5 plays a neuroprotective role in various neurons. Treatment of PC12 cells with hydrogen peroxide induces ERK5 activation and increases MEF2C activity, which counteracts cellular damage [3]. Moreover, BDNF induces the ERK5–MEF2–neurotrophin-3 pathway to promote survival of rat cerebellar granule neurons [4]. ERK5 is also activated locally by NGF in the distal axons of dorsal root ganglion neurons of the peripheral nervous system. Activated ERK5 promotes retrograde neuronal survival through RSK activation in order to regulate neuronal survival [5]. Furthermore, MEF2D expression is stimulated in response to activation of the ERK5–MEF2 pathway by Trk, an NGF receptor. MEF2D regulates the expression of BCL-W, an antiapoptotic BCL2 family member, and promotes sensory neuron survival [6].

References

1. Morimoto H, Kondoh K, Nishimoto S et al (2007) Activation of a C-terminal transcriptional activation domain of ERK5 by autophosphorylation. J Biol Chem 282:35449–35456
2. Hayashi M, Lee JD (2004) Role of the BMK1/ERK5 signaling pathway: lessons from knockout mice. J Mol Med 82:800–808

3. Suzaki Y, Yoshizumi M, Kagami S et al (2002) Hydrogen peroxide stimulates c-Src-mediated big mitogen-activated protein kinase 1 (BMK1) and the MEF2C signaling pathway in PC12 cells: potential role in cell survival following oxidative insults. J Biol Chem 277:9614–9621
4. Shalizi A, Lehtinen M, Gaudilliere B et al (2003) Characterization of a neurotrophin signaling mechanism that mediates neuron survival in a temporally specific pattern. J Neurosci 23:7326–7336
5. Watson FL, Heerssen HM, Bhattacharyya A et al (2001) Neurotrophins use the Erk5 pathway to mediate a retrograde survival response. Nat Neurosci 4:981–988
6. Pazyra-Murphy MF, Hans A, Courchesne SL et al (2009) A retrograde neuronal survival response: target-derived neurotrophins regulate MEF2D and bcl-w. J Neurosci 29:6700–6709

Regulation of bone morphogenetic protein-mediated signaling by tumor necrosis factor-α

Keisuke Okayama, Tada-aki Kudo, Yoshinaka Shimizu, Ye Zhang, Fei Zhao, Mitsuhiro Kano, Hiroyasu Kanetaka, and Keiichi Sasaki

Abstract. Exposure of cells to compressive strain results in rapid induction of bone morphogenetic protein 2 (BMP2). On the other hand, BMP signaling is inhibited by tumor necrosis factor-α (TNF-α), a proinflammatory cytokine, in chronic inflammatory disorders in which the level of TNF-α increases. However, the mechanism underlying this signaling crosstalk is still unclear. This report focuses on the recently found advantages of the regulation of BMP signaling via TNF-α signaling.

Key words. bone morphogenetic protein, tumor necrosis factor

1 Introduction

Bone morphogenetic proteins (BMPs) are key signaling molecules associated with vertebrate development. However, little is known about *BMP* gene regulation. BMP signals are transduced by BMP receptors of type I and type II (BMPRI and BMPRII, respectively); signal transduction via BMPRI leads to the activation of transcription factors called BMP-Smads (Smad1/5/8). These Smads form dimers

K. Okayama, Y. Shimizu, and M. Kano
Division of Oral and Craniofacial Anatomy, Tohoku University Graduate School of Dentistry, 4-1 Seiryo-machi, Aoba-ku, Sendai 980-8575, Japan

T. Kudo (✉) and F. Zhao
Division of Oral Physiology, Tohoku University Graduate School of Dentistry, 4-1 Seiryo-machi, Aoba-ku, Sendai 980-8575, Japan
e-mail: tada@m.tains.tohoku.ac.jp

Y. Zhang and H. Kanetaka
Department of Physical Medicine and Rehabilitation, Tohoku University Graduate School of Biomedical Engineering, 4-1 Seiryo-machi, Aoba-ku, Sendai 980-8575, Japan

H. Kanetaka and K. Sasaki
Division of Advanced Prosthetic Dentistry, Tohoku University Graduate School of Dentistry, 4-1 Seiryo-machi, Aoba-ku, Sendai 980-8575, Japan

T. Sasano et al. (eds.), *Interface Oral Health Science 2009*,
DOI 10.1007/978-4-431-99644-6_48,

with Smad4 before they are translocated into the nucleus. Smad6 and Smad7 are distantly related to the other Smads and inhibit BMP signaling [1, 2]. Tumor necrosis factor-α (TNF-α) acts as a pleiotropic proinflammatory cytokine. TNF-α is synthesized by a variety of cell types. When TNF-α binds to its receptors, it results in the activation of an inflammatory response. Additionaly, intracellular signal transduction induced by TNF-α elicits a wide range of other cellular responses, including modulation of cell differentiation and proliferation. Nuclear factor-kappa B (NF-κB), a transcription factor that is activated by TNF-α, controls the expression of various genes associated with cell survival and proliferation and plays key roles in inflammatory responses [1, 3].

2 Regulation of Genes Encoding BMP2 and BMPRs

To date, several studies have shown that expression of the *BMP* gene is regulated at the transcriptional level by the TNF-signaling molecules present downstream of BMP in the signaling pathway. For example, NF-κB facilitates growth-plate chondrogenesis by inducing *BMP2* expression and activity [4]. In addition, TNF-α increases *BMP2* expression in human mesenchymal stem cells through NF-κB signaling during early osteogenic differentiation; however, *BMP2* expression is not altered in MC3T3-E1-14 osteoblastic cells and in undifferentiated ATDC5 chondrogenic cells after TNF-α stimulation [3, 5]. TNF-α also modulates the expression of *BMPR* transcripts. TNF-α downregulates *BMPRIA* and *BMPRII* mRNA expressions whereas upregulates *BMPRIB* mRNA expression in human bone cells [1].

3 Regulation of the Gene Encoding Inhibitory Smad

Smad7, an inhibitor of BMP signaling, is a transcriptional target of NF-κB. NF-κB is associated with TNF-α-mediated upregulation of Smad7 protein at the posttranscriptional level in several cell lines. For example, in human osteosarcoma Saos2 cells, NF-κB represses BMP/Smad signaling and BMP2-induced osteoblast differentiation via upregulation of Smad7 [2].

References

1. Singhatanadgit W, Salih V, Olsen I (2006) Bone morphogenetic protein receptors and bone morphogenetic protein signaling are controlled by tumor necrosis factor-alpha in human bone cells. Int J Biochem Cell Biol 38:1794–1807
2. Eliseev RA, Schwarz EM, Zuscik MJ et al (2006) Smad7 mediates inhibition of Saos2 osteosarcoma cell differentiation by NF-κB. Exp Cell Res 312:40–50
3. Hess K, Ushumorov A, Fiedler J et al (2009) TNFalpha promotes osteogenic differentiation of human mesenchymal stem cells by triggering the NF-kappaB signaling pathway. Bone 45:367–376

4. Wu S, Flint JK, Rezvani G et al (2007) Nuclear factor-kappaB p65 facilitates longitudinal bone growth by inducing growth plate chondrocyte proliferation and differentiation and by preventing apoptosis. J Biol Chem 282:33698–33706
5. Fukui N, Ikeda Y, Ohnuki T et al (2006) Pro-inflammatory cytokine tumor necrosis factor-alpha induces bone morphogenetic protein-2 in chondrocytes via mRNA stabilization and transcriptional up-regulation. J Biol Chem 281:27229–27241

Mechanosensitive TRP channels in osteoblasts

Takashi Yoshida, Yuki Miyajima, and Minoru Wakamori

Abstract. Mechanical stress plays a vital role in maintaining bone architecture. The process by which osteoblasts convert mechanical signal into biochemical responses leading to bone remodeling is not fully understood. The earliest cellular response detected in mechanically stimulated osteoblasts is an increase in intracellular calcium concentration ($[Ca^{2+}]_i$). This increase in $[Ca^{2+}]_i$ results from Ca^{2+} influx from the extracellular space via calcium channels located on the plasma membrane. In spite of the significant role of calcium channels in osteoblast, the molecular identity of mechanosensitive channels remains unknown. However, in recent study, the expression patterns of mammalian homologues of Drosophira transient receptor potential (TRP) channel superfamily in the clonal osteoblastic cell lines were investigated. RT-PCR analysis revealed that some mechanosensitive TRP channel genes are expressed in osteoblastic cells. This chapter attempts to provide an overview of the current knowledge on mechanosensitive TRP proteins and discuss their possible roles in osteoblastic cells.

Key words. mechanical stress, channel, TRP, osteoblast

1 Introduction

Mechanical loading is essential for maintaining skeletal integrity and bone mass. Osteoblasts lining the endosteal and periosteal surfaces of bone have been shown to be sensitive to mechanical loading. It has been shown that mechanical stimuli regulate various osteoblast functions, including gene expression, protein synthesis, cell proliferation, and cell differentiation. It is widely accepted that ion channels,

T. Yoshida (✉) and M. Wakamori
Division of Molecular Pharmacology and Cell Biophysics, Department of Oral Biology, Tohoku University Graduate School of Dentistry, 4-1 Seiryo-machi, Aoba-ku, Sendai 980-8575, Japan
e-mail: t-yoshida@mail.tains.tohoku.ac.jp

Y. Miyajima
Division of Orthodontics and Dentofacial Orthopedics, Tohoku University Graduate School of Dentistry, 4-1 Seiryo-machi, Aoba-ku, Sendai 980-8575, Japan

T. Sasano et al. (eds.), *Interface Oral Health Science 2009*,
DOI 10.1007/978-4-431-99644-6_49, © Springer 2010

especially those permeating Ca^{2+}, occupy a central position in the cellular responses to mechanical forces via changes in $[Ca^{2+}]_i$. However, the underlying signaling pathways and the molecular identity of mechanosensors in osteoblast are largely unknown. Recently, several members of the transient receptor potential (TRP) family of cation channels have been proposed as mechanosensitive ion channels [1].

2 Mechanosensitive TRP Channels

Mechanosensitive TRP channels are mainly investigated in the cardiovascular system. In artery myocytes, TRPM4 and TRPV2 cation channels are activated by membrane stretch [2, 3]. TRPV4 and TRPM7 have been reported that cell swelling activates these channels and induces Ca^{2+} influx [4, 5]. The flow-mediated Ca^{2+} transient requires the presence of functional TRPP2 and TRPP1 [6].

3 Expression Pattern of TRPs in the Osteoblastic Cells

Using RT-PCR, Abed et al. detected that signals for some TRP channels belong to TRPC, TRPV, and TRPM subfamilies in osteoblastic cells [7]. In members of TRPV subfamily, TRPV2 and TRPV4 mRNAs were expressed. In channels of TRPM subfamily, TRPM7 mRNA was found at highest levels. TRPM3 and TRPM4 mRNA were found at significant levels. In addition, strong signals were found for TRPP1 and TRPP2 in our survey using MC3T3-E1 cells.

Taken together, osteoblastic cells showed a strong expression of diverse mechanosensitive TRP channels.

References

1. Pedersen SF, Nilius B (2007) Transient receptor potential channels in mechanosensing and cell volume regulation. Methods Enzymol 428:183–207
2. Morita H, Honda A, Inoue R et al (2007) Membrane stretch-induced activation of a TRPM4-like nonselective cation channel in cerebral artery myocytes. J Pharmacol Sci 103:417–426
3. Muraki K, Iwata Y, Katanosaka Y et al (2003) TRPV2 is a component of osmotically sensitive cation channels in murine aortic myocytes. Circ Res 93:829–838
4. Numata T, Shimizu T, Okada Y (2007) TRPM7 is a stretch- and swelling-activated cation channel involved in volume regulation in human epithelial cells. Am J Physiol Cell Physiol 292:C460–C467
5. Vriens J, Watanabe H, Janssens A et al (2004) Cell swelling, heat, and chemical agonists use distinct pathways for the activation of the cation channel TRPV4. Proc Natl Acad Sci U S A 101:396–401
6. Nauli SM, Alenghat FJ, Luo Y et al (2003) Polycystins 1 and 2 mediate mechanosensation in the primary cilium of kidney cells. Nat Genet 33:129–137
7. Abed E, Labelle D, Martineau C et al (2009) Expression of transient receptor potential (TRP) channels in human and murine osteoblast-like cells. Mol Membr Biol 26:146–158

Immunohistochemical localization of CD134 ligand, CD137 ligand, GITR ligand, and BAFF in Sjögren's syndrome-like autoimmune sialadenitis of MRL/lpr mice

Keiichi Saito, Shiro Mori, Masao Ono, Ryoichi Hosokawa, and Takeyoshi Koseki

Abstract. Although the mechanism provoking Sjören's syndrome has been extensively studied, the pathogenesis remains obscure. We examined immunohistochemically the expression of CD134 ligand, CD137 ligand, glucocorticoid-induced TNF receptor ligand, and B cell-activating factor of the tumor necrosis factor family in autoimmune saialadenitis of submandibular glands of MRL/lpr mice compared with those of MRL/+ control mice to elucidate the pathogenesis of Sjögren's syndrome. These four molecules are costimulatory factors usually expressed in antigen presenting cells. Unexpectedly, we find that ductal epithelial cells of MRL/lpr mice strongly express the four costimulatory factors, whereas those of MRL/+ control mice show negative or very weak expression.

In autoimmune diseases, self-reactive T and B cells play a crucial role in organ dysfunction through autoimmune reactions. Our results indicate that these four molecules are involved in the pathogenesis of Sjögren's syndrome.

Key words. Sjögren's syndrome, CD134 ligand, CD137 ligand, GITR ligand, BAFF

1 Introduction

Sjögren's syndrome is an autoimmune disease characterized by xerostomia and xerophthalmia. These symptoms are related to hypofunctions of salivary and lacrimal glands caused by self-reactive T and B cells. In addition to T cell receptor stimulation, mutual actions of costimulatory factors such as CD134 ligand (CD134L), CD137

K. Saito(✉), R. Hosokawa, and T. Koseki
Division of Preventive Dentistry, Department of Oral Health and Development Sciences, Tohoku University Graduate School of Dentistry, Sendai, Japan
e-mail: upup@mail.tains.tohoku.ac.jp

S. Mori
Department of Maxillofacial Surgery, Tohoku University Hospital, Sendai, Japan

M. Ono
Division of Histopathology, Department of Pathology, Tohoku University Graduate School of Medicine, Sendai, Japan

T. Sasano et al. (eds.), *Interface Oral Health Science 2009*,
DOI 10.1007/978-4-431-99644-6_50, © Springer 2010

ligand (CD137L), and glucocorticoid-induced TNF receptor (GITR) ligand (GITRL) have been shown to contribute to potent effector T cell activation. Moreover, B cell-activating factor of the tumor necrosis factor family (BAFF) activates B cells without CD40–CD154 costimulation, that is, in a T cell-independent manner. Herein, we investigated immunohistochemically the expression of CD134L, CD137L, GITRL, and BAFF in submandibular glands of MRL/lpr mice affected with autoimmune sialadenitis and MRL/+ control mice to explore the implication of these four stimulatory molecules for the pathogenesis of Sjören's syndrome.

2 Materials and Methods

Specimens of submandibular gland tissues were obtained from 20-weeks-old MRL/lpr mice inducing autoimmune sialadenitis and age-matched MRL/+ control mice. Paraffin-embedded tissue sections (3 μm) were immunostained against CD134L, CD137L, GITRL, and BAFF, using streptoavidin-biotin procedure (Histofine SAB-PO Kit, Nichirei Co., Tokyo, Japan). The primary antibodies we utilized were four goat polyclonal antibodies (Santa Cruz Biotechnology, Inc., CA). The proportion of tissue sections expressing these four stimulatory molecules in MRL/lpr mice were individually compared with those of MRL/+ control mice, and the estimated values were statistically analyzed by Fisher's exact test.

3 Results and Discussion

The submandibular gland tissue sections prepared from autoimmune saialadenitis of MRL/lpr mice showed intense immunolocalization of CD134L, CD137L, GITRL, and BAFF in ductal epithelial cells, in which of MRL/+ control mice, however, these four molecules-positive cells were negligible mostly. Additionally, the percentages of CD134L-, CD137L-, GITRL-, and BAFF-positive specimens in MRL/lpr mice were raised significantly compared with those of MRL/+ mice specimens.

We found that some MRL/lpr specimens were immunostained intensely against all the four stimulatory molecules, and showed extensive invasion of enormous inflammatory cells. It has been verified that CD134L and CD137L demonstrate, especially, more potent capacities to stimulate effector T cells compared with GITRL, although all of them augment effector T cell proliferation. In addition, synergistic stimulation of CD134 and CD137 contributes to more robust activation of T cells than each alone. Moreover, GITRL has been noticed to render effector T cells capable of counteracting repressive revactions of regulatory T cells more dominantly than CD134L and CD137L. Meanwhile, BAFF has been ascertained to facilitate proliferation of not only B cells but also T cells. Furthermore, elevated levels of BAFF are detected in both serum and saliva of patients with Sjögren's syndrome.

In this context, our results suggest that reciprocal influences of the four stimulatory molecules induce severe inflammatory cell infiltration of autoimmune sialadenitis, and these influences are inextricably linked to the development of Sjögren's syndrome-like autoimmune sialadenitis in MRL/lpr mice.

Expression of microRNA during tooth development

Kojiro Tanaka, Aya Yamada, Hiroharu Suzuki, Makiko Arakaki, and Satoshi Fukumoto

Abstract. To understand the control of tooth-specific gene expressions, we analyzed microRNA (miRNA) expressions in tooth germ using microarray methods. miRNAs in tooth germ were isolated from embryonic day 16 (E16), postnatal day 1 (P1), and P3 mice. Most of the miRNAs were decreased in P3 molar compared with E16. Prediction of target miRNA for Gja1 gene was identifived using Sanger miRNA database. Those miRNAs were mmu-miR-1, 101a, 101b, 200a, 338-3p, 376a, and 434-3p. Among those miRNAs, mmu-miR1 binds to 3′-region of Gja1 mRNA and inhibits their expression. This result indicates that the profile of miRNA expression may be useful for understanding the regulation of tooth-specific gene expression.

Key words. microRNA, tooth development, oculodentodigital dysplasia, Gja1

1 Introduction

MicroRNAs (miRNAs) are a class of short noncoding RNA molecules that post-transcriptionally regulate gene expression in plants and animals [1]. Since their discovery as regulators of developmental timing in *Cenorhabditis elegans*, hundreds of miRNAs have been identified. miRNAs play a critical role in the regulation of gene expression and that their deregulation may underlie many human diseases such as Oculodentodigital dysplasia (ODDD), an autosomal dominant pleiotropic disorder caused by mutations in Gja1. ODDD patients showed severe enamel hypoplasia indicating that Gja1 regulates ameloblast differentiation and enamel formation. To understand the control of tooth-specific gene expression, especially Gja1, we analyzed miRNA expression during tooth development.

K. Tanaka (✉), A. Yamada, H. Suzuki, M. Arakaki, and S. Fukumoto
Division of Pediatric Dentistry, Tohoku University Graduate School of Dentistry, Sendai 980-8575, Japan
e-mail: kojiro-tanaka@mail.tains.tohoku.ac.jp

T. Sasano et al. (eds.), *Interface Oral Health Science 2009*,
DOI 10.1007/978-4-431-99644-6_51, © Springer 2010

2 Materials and Methods

miRNA microarray: miRNA was isolated from first molar of E16, P1, and P3 ICR mice using acid-phenol method. Isolated miRNA labeled with fluorescence and then performed Genopal DNA microarray analysis.

3 Results and Discussion

We analyzed the expression of total 178 miRNAs in tooth germ isolated from E16, P1, and P3 molar compared with E16. Expression of mmu-miR-18, 124a, 127, and 301 were decreased in P3 compared to E16. Expression of mmu-miR-1, 15b, 19a, 20b, 31, 93, 106b, 125a, 130b, 133a, and 296 were downregulated in P3 compared to P1. Prediction of the target miRNA for Gja1 was identified using the Sanger miRNA database. Target miRNAs for Gja1 gene were mmu-miR-1, 101a, 101b, 200a, 338-3p, 376a, and 434-3p. Among those miRNAs, expression of mmu-miR-1, 101a, 376a, and 434-3p in P3 molars were decreased compared with E16.

The expression of Gja1 was increased in secretory ameloblasts. The expression of miR-1 was decreased in secretory stage ameloblast compared with presecretory stage. The previous report [2] showed that miR-1 overexpression in cardiac myocytes showed conduction and depolarized the cytoplasmic membrane by posttranscriptionally repressing Gja1. Therefore, miR-1 may downregulate the expression of Gja1 through the same mechanism in tooth development.

References

1. Bartel DP (2004) MicroRNAs: genomics, biogenesis, mechanism, and function. Cell 116:281–297
2. Baofeng Y, Huixian L, Jiening X et al (2007) The muscle-specific microRNA miR-1 regulates cardiac arrhythmogenic potential by targeting GJA1 and KCNJ2. Nat Med 13:486–491

Session II
Host-Parasite Interface

New quantitative fluorometry for evaluating oral bacterial adhesion to biomaterials

Yoko Sakuma, Jumpei Washio, Yasuhisa Takeuchi, Keiichi Sasaki, and Nobuhiro Takahashi

Abstract. This study aimed to develop a simple, quick, and widely-applicable nonradioisotope quantitative method for evaluating adhesion of oral bacteria to dental materials. So, we tried to establish a method using Alamar Blue®. We used four representative oral bacteria, *Streptococcus mutans, Streptococcus sanguinis, Actinomyces naeslundii*, and *Viellonella atypica*. Plastic wells with acrylic resin bottoms were filled with cell suspensions for 2 h. After discarding, the wells were filled with Alamar Blue® solution for 3 h, then, the fluorescence intensity of the solution was measured by fluorophotometer. The fluorescence intensity increased with amounts of bacteria in all the species, and the relationship seemed proportional. Alamar blue method is thought to be useful to quantify oral bacterial adhesion to biomaterials.

Key words. acrylic resin, bacterial adhesion, resazurin, *Streptococcus*, *Actinomyces*

1 Introduction

It is suggested that biofilm formed on the surfaces of dental biomaterials, such as acrylic resin, causes various oral diseases. However, only a few oral bacteria such as mutans streptococci and *Candida* species have been investigated for adhesion to dental materials [1]. In addition, amounts of bacteria attached to biomaterials are usually evaluated by the method using radioisotope (RI)-labeled bacteria, which requires an exclusive-use facility [2]. Therefore, the development of a simple, quick, and widely-applicable non-RI quantitative method for evaluating the adhesion of oral bacteria is needed.

Y. Sakuma, Y. Takeuchi, and K. Sasaki
Division of Advanced Prosthetic Dentistry, Tohoku University Graduate School of Dentistry, 4-1 Seiryo-machi, Aoba-ku, Sendai 980-8575, Japan

Y. Sakuma, J. Washio, Y. Takeuchi, and N. Takahashi (✉)
Division of Oral Ecology and Biochemistry, Tohoku University Graduate School of Dentistry, 4-1 Seiryo-machi, Aoba-ku, Sendai 980-8575, Japan
e-mail: nobu-t@mail.tains.tohoku.ac.jp

T. Sasano et al. (eds.), *Interface Oral Health Science 2009*,
DOI 10.1007/978-4-431-99644-6_52, © Springer 2010

2 The Non-RI Quantitative Method using Alamar Blue®

A non-RI method using this Alamar Blue® solution was applied for evaluating bacterial adhesion to acrylic resin. Alamar Blue® fluorescence dye reflects bacterial reduction–oxidation metabolic activity.

Type starins of four oral bacteria, *Streptococcus mutans NCTC10449* (Sm), *Streptococcus sanguinis ATCC10556* (Ss), *Actinomyces naeslundii ATCC12104* (An), and *Veillonella atypica* ATCC17744 (Va) were anaerobically grown at 37ºC in tryptone–yeast extract medium containing 0.5% glucose (for *Streptococcus* and *Actinomyces*), 0.5% sodium chloride (for *Streptococcus* and *Actinomyces*), 0.1% ammonium bicarbonate (for *Actinomyces*), 1.8% sodium lactate (for *Veillonella*), washed and suspended in 50 mM potassium phosphate buffer (pH 6.0; PPB). Acrylic resin plate was prepared and washed with distilled water and dried completely. Then, columnar tubes, made of non-protein-adhesive plastic, were attached to the plate surface. This plastic wells with acrylic resin bottoms were treated overnight with PPB and filled with the cell suspension of each bacterial suspension and incubated for 2 h at 37°C. After discarding the cell suspension, the wells were rinsed with PPB and filled with Alamar Blue® solution. After 3 h at 37°C, fluorescence intensity of solution was measured by fluorophotometer. The amounts of attached bacteria were calculated using standard curves of amounts of bacteria vs. fluorescence intensity.

The fluorescence intensity increased with amounts of bacteria in all the species, and the relationship seemed proportional.

3 Conclusion

The non-RI method using fluorescence dye developed in the present study is thought to be useful to quantify oral bacterial adhesion to biomaterials.

Acknowledgments This study was supported partly by JSPS (19390539, 20791632) from the Japan Society for the Promotion of Science.

References

1. Faltermeier A, Burgers R, Rosentritt M (2007) Bacteria adhesion of *Streptococcus mutans* to orthodontic adhesives with various filler-volume fraction. Am J Orthod Dentofacial Orthop 132:728.e15–728.e19
2. Shimamotoyodome A, Koudate T (2007) Reducation of *Streptococcus mutans* adherence and dental biofilm formation by surface treatment with phosphorylated polyethylene glycol. Antimicrob Agents Chemother 51:3635–3641

Analgesic effects of NOD1 and NOD2 agonists

Tadasu Sato, Hidetoshi Shimauchi, Yasuo Endo, and Haruhiko Takada

Abstract. In this chapter, we have demonstrated analgesic activities of NOD1 as well as NOD2 agonists. Intravenous injection of NOD2-agonistic muramyldipeptide (MDP) and its derivatives, 6-*O*-stearoyl-MDP (L18-MDP) and MDP-Lys(L18), and NOD1-agonistic FK156 and FK565 exhibited analgesic activity in BALB/c mice. FK565 exhibited the analgesic activity by various administration routes: intraperitoneal, intramuscular, sublingual, subcutaneous, intragingival, intragastric, and intracerebroventricular. The analgesic effect of intravenously administered FK565 starts after 30 min and persists for 24 h, and is observed even in tumor necrosis factor (TNF)-α knock out (KO), interleukin (IL)-1α/β double KO, and triple KO mice. Naloxane, a nonselective antagonist for opioid-receptors, completely inhibited the analgesic effect of FK565. These findings suggest that NOD1 and NOD2 activation induces analgesic effect via opioid-receptor(s) in a TNF-α and IL-1α/β-independent manner.

Key words. NOD1/2, FK565, MDP, analgesic effect, opioid

In the 1970s, muramyldipeptide (MDP; MurNAc-L-Ala-D-*iso*Gln) was demonstrated to be a minimum essential structure for the immunoadjuvant activity of bacterial peptidoglycan (PGN) and has been demonstrated to exhibit various bioactivities in vivo and in vitro [1]. In 1987, Ogawa and Kotani [2] reported that MDP exerted analgesic activity to decrease the frequency of acetic acid-induced writhing

T. Sato (✉) and H. Shimauchi
Division of Periodontology and Endodontology, Tohoku University Graduate School of Dentistry, Sendai 980-8575, Japan
e-mail: tadasu@m.tains.tohoku.ac.jp

Y. Endo
Division of Oral Molecular Regulation, Tohoku University Graduate School of Dentistry, Sendai 980-8575, Japan

H. Takada
Division of Oral Microbiology, Tohoku University Graduate School of Dentistry, Sendai 980-8575, Japan

T. Sasano et al. (eds.), *Interface Oral Health Science 2009*,
DOI 10.1007/978-4-431-99644-6_53, © Springer 2010

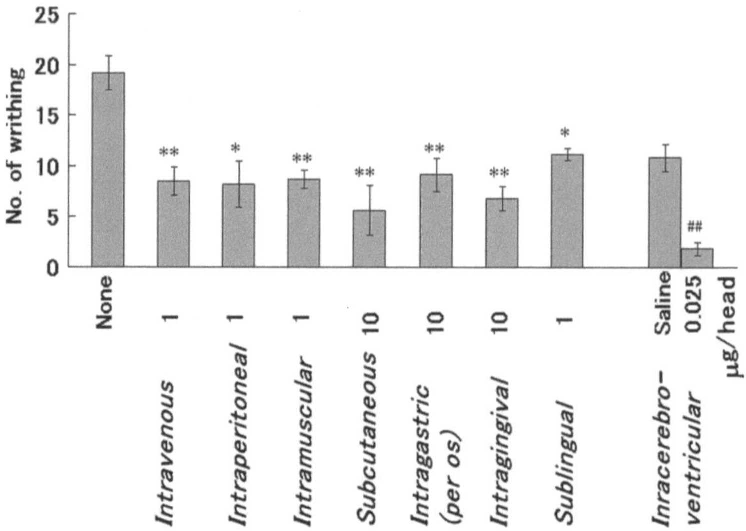

Fig. 1. FK565 exhibited analgesic effects in various administration routes. Groups (6/group) of female BALB/c mice were administered with FK565 via various routes. After 1 h, the mice received an intraperitoneal injection of 1% acetic acid (0.1 ml/10 g of body weight), then the number of writhing movements was counted during a 20 min period. *$p<0.05$ and **$p<0.01$ vs. none, ##$p<0.01$ vs. saline (Quoted from Sato et al. [3] with modification)

movements in mice. In 2003, it was demonstrated that intracellular NOD2 and NOD1 sensed the MDP moiety and the diaminopimelic acid (DAP)-containing peptide moiety of PGN, respectively [1].

Recently, we found analgesic activities of NOD1 as well as NOD2 agonists in mice. Intravenous injection of NOD1-agonistic FK156 (D-lactyl-L-Ala-D-Glu-*meso*-DAP-L-Gly) (50 μg/head) and FK565 (heptanoyl-D-Glu-*meso*-DAP-D-Ala) (1.0 μg/head), which were supplied by Astellas Pharmaceutical Co. (Tokyo, Japan), and NOD2-agonistic MDP (10 μg/head), 6-*O*-stearoyl-MDP (L18-MDP) (50 μg/head) and MDP-Lys(L18) (2.0 μg/head) exhibited significant analgesic activity. Furthermore, FK565 exhibited definite analgesic activity by various adminstration routes (Fig. 1). The analgesic activity of FK565 was observed even in tumor necrosis factor (TNF)-α knock out (KO), interleukin (IL)-1α/β double KO, and triple KO mice. Naloxane (160 μg/head), which is a nonselective antagonist for opioid-receptors, completely inhibited the analgesic effect of FK565 (10 μg/head). In conclusion, NOD1 and NOD2 activation induces analgesic effects via opioid-receptor(s) in a TNF-α and IL-1α/β-independent manner.

References

1. Takada H, Uehara A (2006) Enhancement of TLR-mediated innate immune responses by peptidoglycans through NOD signaling. Curr Pharm Design 12:4163–4172
2. Ogawa T, Kotani S (1987) Analgesic effects of *N*-acetylmuramyl-L-alanyl-D-isoglutamine in decreasing the acetic acid-induced abdominal-writhing response. Infect Immun 55:494–496
3. Sato T, Shikama Y, Shimauchi H et al (2009) Analgesic effects of chemically synthesized NOD1 and NOD2 agonists in mice. Innate Immun, doi 10.1117/1753425909351904

Porphyromonas gingivalis-induced alveolar bone loss in interleukin-18 transgenic mice

Noriaki Shoji, Kotaro Yoshinaka, Takashi Nishioka, Yumiko Sugawara, Shunji Sugawara, and Takashi Sasano

Abstract. This study was designed to assess the functional role of interleukin-18 (IL-18) in periodontal tissues. IL-18 transgenic (IL-18Tg) mice and wild-type (WT) mice were inoculated intraorally with Porphyromonas gingivalis. Alveolar bone loss, gingival cytokine levels, and gingival gene expression were assessed using morphometric analysis, enzyme-linked immunosorbent assay (ELISA), and semiquantitative reverse transcription polymerase chain reaction, respectively. P. gingivalis-infection induced periodontal bone loss in IL-18Tg mice, whereas bone loss did not occur in WT mice. Interferon gamma was downregulated only in IL-18Tg mice by P. gingivalis-infection. P. gingivalis-infection upregulated the expression of receptor activator of nuclear factor kappa B ligand (RANKL) in IL-18Tg mice vs. WT-type mice.These results suggest that IL-18Tg mice are highly susceptible to bone loss induced by the periodontal pathogen P. gingivalis, which may be mediated via RANKL-dependent pathway.

Key words. interleukin-18, transgenic mice, *Porphyromonas gingivalis*, alveolar bone loss

---

N. Shoji (✉), K. Yoshinaka, T. Nishioka, Y. Sugawara, and T. Sasano
Division of Oral Diagnosis, Department of Oral Medicine and Surgery, Tohoku University Graduate School of Dentistry, 4-1 Seiryo-machi, Aoba-ku, Sendai 980-8575, Japan
e-mail: tiger@mail.tains.tohoku.ac.jp

S. Sugawara
Division of Oral Immunology, Department of Oral Biology, Tohoku University Graduate School of Dentistry, 4-1 Seiryo-machi, Aoba-ku, Sendai 980-8575, Japan

T. Sasano et al. (eds.), *Interface Oral Health Science 2009*,
DOI 10.1007/978-4-431-99644-6_54, © Springer 2010

1 Introduction

There is an association between the severity of periodontal disease and interleukin-18 (IL-18) levels [1]. To understand the role of IL-18, we examined the periodontal bone loss of IL-18 transgenic (Tg) mice following infection with the periodontal pathogen *Porphyromonas gingivalis*.

2 Materials and Methods

Eight- to ten-week-old IL-18Tg mice were kindly provided by T. Hoshino (Kurume University, Kurume, Japan). The mice were bred in the animal facility of Tohoku University Graduate School of Dentistry. *P. gingivalis* W83 was grown and periodontal infection with *P. gingivalis* was carried out. Animals were killed by CO_2 inhalation 70 days after *P. gingivalis* infection. Mandibles and maxillae were removed, and mandibles were hemisected. To extract protein, gingival tissue removed from around the lower left molars was homogenized. The left hemisected mandible was defleshed, bleached, and mounted on a microscope slide for bone loss measurements. The right hemisected mandible was used for histological analysis. Maxillae were used as a source of RNA. Gingival tissue around the bilateral upper molars was isolated under a surgical microscope, weighed, and kept at −70°C until further analysis. Total gingival RNA was extracted using TRIzol (Invitrogen, Carlsbad, CA, USA). cDNA was reverse transcribed using SuperScript™ II RT and oligo-dT_{12-18} primer (both from Invitrogen). Gene expression levels of β-actin, mature IL-18, receptor activator of nuclear factor kappa B ligand (RANKL) and osteoprotegerin in gingival tissues were determined using semiquantitative reverse transcription polymerase chain reaction.

3 Results and Discussion

We demonstrated that infection with *P. gingivalis* induced periodontal bone loss in IL-18Tg mice but not in wild-type (WT) mice. Analyses of cell-associated molecules showed that infection with *P. gingivalis* upregulated expression of RANKL (the key stimulator of osteoclast development and activation) in IL-18Tg mice compared to WT mice. In IL-18Tg mice, however, the Interferon gamma (IFN-γ) level was reduced by infection with *P. gingivalis* but unchanged in infected WT mice. It has been reported that a balance between the levels of RANKL and IFN-γ may regulate osteoclast formation [2]. For example, during acute immune reactions, enhanced production of IFN-γ counterbalances increased RANKL expression thereby reducing aberrant osteoclast formation. In chronic synovitis from rheumatoid arthritis, however, this balance may be skewed in favor of RANKL expression, which leads to bone destruction [2]. Similarly, the results from our study suggest that, for IL-18Tg mice, lower IFN-γ levels and increased expression of RANKL after *P. gingivalis* infection cause alveolar bone loss, whereas in WT mice, increased production of IFN-γ counterbalances the increased expression of RANKL, thereby inhibiting alveolar bone loss.

References

1. Orozco A, Gemmell E, Bickel M et al (2006) Interleukin-1beta, interleukin-12 and interleukin-18 levels in gingival fluid and serum of patients with gingivitis and periodontitis. Oral Microbiol Immunol 21:256–260
2. Takayanagi H, Ogasawara K, Hida S et al (2000) T-cell-mediated regulation of osteoclastogenesis by signalling cross-talk between RANKL and IFN-gamma. Nature 408:600–605

Anaphylaxis-like shock induced by LPS plus antineutrophil monoclonal antibodies in mice

Yukinori Tanaka, Yasuhiro Nagai, Toshinobu Kuroishi, Haruhiko Takada, Yasuo Endo, and Shunji Sugawara

Abstract. During the course of experiments on neutrophil depletion, we happened to find that intravenous injection of antineutrophil monoclonal antibodies (RB6-8C5 and 1A8) induces anaphylaxis-like shock in mice pretreated with lipopolysaccharide. The shock reaction depends largely on neutrophils and Toll-like receptor 4, and significantly on macrophages. Complement C5 may be involved in the shock reaction. We expect that elucidation of behavior of neutrophils may contribute to understand the pathology of septic shock.

Key words. neutrophils, RB6-8C5, 1A8, LPS, shock

Neutrophils are essential effector cells of the innate immune system. They are normally found in the blood stream. During infection, however, they migrate to the site of inflammation where they have a crucial role in the clearance of microorganisms. Neutrophil depletion with RB6-8C5, which is an antineutrophil monoclonal antibody, has been performed to investigate their role in immune responses. During the course of experiments on neutrophil depletion, we happened to find that intravenous injection of RB6-8C5 induces anaphylaxis-like shock in mice pretreated with lipopolysaccharide (LPS). From this finding, we speculate that there might be a phase in which neutrophils play an important role as effector cells during the development of septic shock. Here, we examined the mechanism underlying the shock reaction induced by LPS and RB6-8C5. First, another antineutrophil monoclonal antibody, 1A8, showed the same result with RB6-8C5 while no signs of shock were

Y. Tanaka(✉), Y. Nagai, T. Kuroishi, Y. Endo, and S. Sugawara
Division of Oral Immunology, Tohoku University Graduate School of Dentistry, Sendai 980-8575, Japan
e-mail: y-tanaka@mail.tains.tohoku.ac.jp

H. Takada
Division of Oral Microbiology, Tohoku University Graduate School of Dentistry, Sendai 980-8575, Japan

DOI 10.1007/978-4-431-99644-6_55, © Springer 2010

seen by administration of control antibodies. These antineutrophil monoclonal antibodies may directly activate neutrophils. Second, the shock reaction depends largely on Toll-like receptor (TLR) 4 and significantly on macrophages. Finally, complement C5 (but neither histamine, interleukin-1, TNF-α, nor active oxygen) may be involved in the reaction. We expect that clarifying the detailed mechanism may contribute to understand the pathology of septic shock and to find effective strategies against infectious diseases.

Concentrations of metal ions in murine nickel allergy and its cross-reactions: effects of lipopolysaccharide

Masayuki Kinbara, Toshinobu Kuroishi, Teruko Takano-Yamamoto, Shunji Sugawara, and Yasuo Endo

Abstract. Few adequate animal models for metal allergies have existed so far. However, we found that lipopolysaccharide (LPS) promotes and augments metal allergies in mice and established a murine model close to the actual situations. Here, we investigated the effects of LPS on metal concentrations and on cross-reactions in this murine model. LPS markedly reduced the threshold concentration at both sensitization and elicitation steps in Ni allergy and at its cross-reactions. Our findings suggest that the bacterial milieu is a critically important factor leading to metal allergies.

Key words. nickel allergy, lipopolysaccharide, hypersensitivity, dermatitis, inflammation

Animal models are required for the analysis of the mechanism by which metal allergies are established, but few adequate models have existed so far. Numerous studies have been performed in vitro by using sensitized human lymphocytes, and it is believed that like contact hypersensitivity to classical haptens, T cells, the players of acquired immunity, are essentially responsible for metal allergies. However, unlike classical haptens, metal ions form geometrically highly defined, but reversible, coordination complexes with partner molecules. Thus, it is difficult to define the allergenic epitopes of such complexes, and the host recognizes metal ions in complicated ways. Allergic reactions consist of sensitization and elicitation steps.

M. Kinbara (✉) and T. Takano-Yamamoto
Department of Orthodontics and Dentofacial Orthopedics, Tohoku University,
4-1 Seiryo-machi, Aoba-ku, Sendai 980-8575, Japan
e-mail: kinbara@m.tains.tohoku.ac.jp

M. Kinbara, T. Kuroishi, S. Sugawara, and Y. Endo
Department of Molecular Regulation, Graduate School of Dentistry, Tohoku University,
4-1 Seiryo-machi, Aoba-ku, Sendai 980-8575, Japan

T. Sasano et al. (eds.), *Interface Oral Health Science 2009*,
DOI 10.1007/978-4-431-99644-6_56, © Springer 2010

We previously found that lipopolysaccharide (LPS) promotes and augments metal allergies in mice via innate immunity at either the priming or elicitation step [1]. Strangely, Ni allergy induced by the sensitization with LPS + Ni is induced even in nude (T-cell deficient) mice [1]. Here, we investigated the effects of LPS on metal concentrations and on cross-reactions. LPS markedly reduced the threshold concentrations of Ni at the steps of both sensitization and elicitation. Its threshold concentration, especially, at the elicitation step was reduced by a markedly low concentration. There was an inverse correlation between the concentration of LPS and the threshold concentration of Ni at both sensitization and elicitation steps. Either Pd, Cr, or Co induced cross-allergic reactions in mice sensitized with Ni + LPS, and their threshold concentrations were markedly reduced by LPS similarly to the true antigen Ni. Ag and Cu also induced cross-allergic reactions. There may be a similarity in the conformation of metal–protein complexes irrespective of the kind of metal ions, although we need confirmation using ultrapurified metal salts. Bacterial milieu may be critically important to promote metal allergies. Although a high Ni concentration is needed for sensitization to Ni, once sensitization has been established, a markedly low concentration of Ni (and other metals, too) induces allergic inflammation.

Reference

1. Sato N, Kinbara M, Kuroishi T et al (2007) Liopolysaccharide promotes and augments metal allergies in mice, dependent on innate immunity and histidine decarboxylase. Clin Exp Allergy 37:743–751

Muramyldipeptide augments the actions of LPS via multiple fashions in mice

Yosuke Shikama, Toshinobu Kuroishi, Yasuhiro Nagai, Hidetoshi Shimauchi, Haruhiko Takada, Shunji Sugawara, and Yasuo Endo

Abstract. Muramyldipeptide (MDP) is the minimum essential structure responsible for the immunoadjuvant activity of peptidoglycan, and it is known to be a ligand of nuclear-binding domain 2 (NOD2). MDP augments the activity of lipopolysaccharide (LPS), but the mechanism underlying this effect is unclear. Here, we used mice to examine the effects of MDP on endotoxin-shock and the LPS-induced productions of cytokines. Intravenously injected MDP augmented LPS-induced hypothermia, although mice deficient in IL-1αβ and/or tumor necrosis factor (TNF)-α did not exhibit this response. MDP also increased pro-IL-1β in tissues, downregulated the expression of suppressor of cytokine signaling (SOCS) 1, and augmented the LPS-induced productions of TNF-α, IL-12 p40, and IFN-γ. Moreover, by performing in vivo and in vitro experiments, we obtained evidence that macrophages were involved in these effects of MDP. These results suggest that two different mechanisms may underlie the above augmenting effect of MDP namely stimulation of pro-IL-1β production and downregulation of SOCS1 in tissues, which are involved in macrophages.

Key words. MDP, LPS, SOCS1, inflammatory cytokines, macrophages

---

Y. Shikama (✉) and H. Shimauchi
Division of Periodontology and Endodontology, Tohoku University, 4-1 Seiryo-machi, Aoba-ku, Sendai 980-8575, Japan
e-mail: shikama-thk@umin.ac.jp

T. Kuroishi, Y. Nagai, S. Sugawara, and Y. Endo
Division of Oral Immunology, Tohoku University, 4-1 Seiryo-machi, Aoba-ku, Sendai 980-8575, Japan

H. Takada
Division of Oral Microbiology, Graduate School of Dentistry, Tohoku University, 4-1 Seiryo-machi, Aoba-ku, Sendai 980-8575, Japan

T. Sasano et al. (eds.), *Interface Oral Health Science 2009*,
DOI 10.1007/978-4-431-99644-6_57, © Springer 2010

1 Introduction

Muramyldipeptide (MDP) is the minimum essential structure responsible for the immuno-adjuvant activity of bacterial cell wall peptidoglycan that is sensed mainly by Toll-like receptor 2 (TLR2). MDP is not recognized by TLR2 but by intracellular receptors, nuclear-binding domain 2 (NOD2). MDP enhances the activities of lipopolysaccharide (LPS), including endotoxic shock [1]. Although various inflammatory cytokines – including tumor necrosis factor (TNF)-α, interleukin (IL)-1, IL-12 p40, and interferon-γ (IFN-γ) – have been shown to be involved in endotoxic shock, the mechanism underlying such augmentations remains unclear.

Suppressor of cytokine signaling (SOCS) proteins comprise a family of intracellular proteins several of which have been shown to regulate the responses of immune cells to cytokines. SOCS1$^{-/-}$ mice are hypersensitive to LPS, leading to increased productions of TNF-α, IL-12, and IFN-γ [2]. Therefore, we examined the effects of MDP on LPS-induced productions of inflammatory cytokines and SOCS1 expression in vivo and in vitro.

2 Materials and Methods

MDP and/or *Escherichia coli* LPS was intravenously injected into BALB/c mice [wild type, IL-1α/β knock out (KO), TNF-α KO, or both IL-1α/β and TNF-α KO]. Clodronate-liposome (Clo-lip), a reagent for depleting phagocytic macrophages in vivo, was injected a day before MDP injection. Decrease in rectal temperature was used as the index of endotoxin-shock. The murine macrophage cell-line RAW264 was stimulated by MDP and/or LPS. Cytokines, other proteins, and mRNAs were measured by ELISA, Western blotting, and qRT-PCR, respectively.

3 Results and Discussion

Intravenously injected MDP augmented LPS-induced hypothermia in wild-type mice, but not in IL-1α/β and/or TNF-α KO mice. MDP also (a) increased pro-IL-1β in tissues but did not increase IL-1β in serum (since caspase-1 was not activated by the MDP), (b) downregulated the expression of SOCS1 in spleen, and (c) augmented the LPS-induced productions of TNF-α, IL-12 p40, and interferon-gamma (IFN-γ) in tissues and/or serum. Moreover, the augmenting effect of MDP on LPS-induced TNF-α production was markedly diminished in Clo-lip-injected mice, and MDP augmented LPS-induced TNF-α production in RAW264. These results suggest that two different mechanisms may underlie the above augmenting effect of MDP, namely stimulation of pro-IL-1β production and downregulation of SOCS1, which are involved in macrophages. Details will be shown in our report [3].

References

1. Takada H, Galanos C (1987) Enhancement of endotoxin lethality and generation of anaphylactoid reactions by lipopolysaccharides in muramyl-dipeptide-treated mice. Infect Immun 55:409–413
2. Yoshimura A, Naka T, Kubo M (2007) SOCS proteins, cytokine signaling and immune regulation. Nat Rev Immunol 7:454–465
3. Shikama Y, Kuroishi T, Nagai Y et al (2009) Muramyldipeptide augments the actions of LPS in mice by stimulating macrophages to produce pro-IL-1β and by downregulation of suppressor of cytokine signaling 1 (SOCS1). Innate Immun Nov 6 [Epub ahead of print]

Isolation and identification of viable bacteria within acrylic resin denture bases

Yasuhisa Takeuchi, Kazuko Nakajo, Takuichi Sato, Yoko Sakuma, Keiichi Sasaki, and Nobuhiro Takahashi

Abstract. This study aimed to isolate and identify the microbial species within acrylic resin denture bases. Bacteria were detected from within acrylic resin denture bases. *Lactobacillus* species, *Actinomyces*, *Streptococcus*, and *Propionibacterium* species were predominantly identified. The present study revealed that viable bacteria existed within acrylic resin denture bases suggesting that the bacteria and their metabolites may be related to the deterioration of the base materials and denture-specific malodor and oral infections.

Key words. acrylic resin denture, bacterial detection, bacterial identification

1 Introduction

It has been estimated that approximately 500–700 species of microorganisms inhabit the human oral microflora biofilm, e.g., on the surface of teeth, tongue, and oral mucosa, as well as on the surface of acrylic resin dentures. These microorganisms on the denture surface can invade into the acrylic resin dentures through the micropores, the interfaces to other materials, and the microcracks of acrylic resin. Moreover, these microorganisms can also colonize and form biofilm within dentures using the supplied nutrients, such as carbohydrates and proteins, in the oral cavity. Similarly, to the biofilm on the denture surfaces [1, 2], the biofilm within dentures may be associated with oral and respiratory infections, and which may be hard to remove. Besides, there are few reports on the microbial existence of denture bases,

Y. Takeuchi, Y. Sakuma, and K. Sasaki
Division of Advanced Prosthetic Dentistry, Tohoku University Graduate School of Dentistry, 4-1 Seiryo-machi, Aoba-ku, Sendai, Japan

K. Nakajo, T. Sato, Y. Sakuma, and N. Takahashi (✉)
Division of Oral Ecology and Biochemistry, Tohoku University Graduate School of Dentistry, 4-1 Seiryo-machi, Aoba-ku, Sendai, Japan
e-mail: nobu-t@mail.tains.tohoku.ac.jp

DOI 10.1007/978-4-431-99644-6_58, © Springer 2010

which was shown by scanning electron microscope [3], and no information is available on microbial viability and identification. Thus, this study aimed to detect and identify microorganisms within acrylic resin dentures.

2 Detection of Viable Bacteria within Acrylic Resin Denture Bases

Samples were obtained from the inner layers of used denture bases with a special attention to contamination of bacteria attached to denture surfaces and cultured on blood agar plates under anaerobic conditions. Bacteria were detected from five out of 15 samples, indicating that bacteria invaded within acrylic resin denture bases and continued to survive within the bases for a long time. This finding suggests that the internal environment of the denture bases could be suitable for the bacteria to live.

3 Identification of Bacteria Isolated within Denture Bases

Lactobacillus, *Actinomyces*, *Streptococcus*, and *Propionibacterium* species were predominantly found by the method using 16S rRNA gene sequence analysis. Predominance of bacteria was different among samples, suggesting that microflora in the internal layer of acrylic resin bases was diverse, probably due to its structural and environmental difference.

4 Clinical Implication of Acrylic Resin Denture

These results suggest that acrylic resin denture can act as a bacterial reservoir and that the bacteria and their metabolite within the acrylic resin dentures may cause the denture-specific malodor, oral and respiratory infections, and the deterioration with age of the base materials.

References

1. Gusberti FA, Gada TG, Lang NP et al (1985) Cultivable microflora of plaque from full denture bases and adjacent palatal mucosa. J Biol Buccale 13:227–236
2. Sumi Y, Miura H, Sunakawa M et al (2002) Colonization of denture plaque by respiratory pathogens in dependent elderly. Gerodontology 19:25–29
3. Walter B, Frank RM (1985) Ultrastructural relationship of denture surfaces, plaque and oral mucosa in denture stomatitis. J Biol Buccale 13:145–166

Bactericidal effect of photodynamic therapy

Keisuke Nakamura, Mika Tada, Taro Kanno, Hiroyo Ikai, Eisei Hayashi, Takayuki Mokudai, and Masahiro Kohno

Abstract. Photodynamic therapy (PDT) is a procedure whereby undesired tissue can be destroyed by the combined action of light, oxygen, and a photosensitizer. It is thought that a photosensitizer excited by light with a specific wavelength transfers its excitation energy to ground state oxygen molecules resulting in the generation of singlet oxygen. Singlet oxygen is a reactive oxygen species with high oxidative reactivity. Therefore, the reaction of singlet oxygen with cellular constituents can result in oxidative damage leading to cell death. Although PDT is mainly used in cancer treatment, several studies have shown that PDT also has antimicrobial properties. It has been proposed that the bactericidal effect of singlet oxygen generated by PDT could be used in clinical dentistry. Applications of PDT for the treatment of periodontitis and peri-implantitis, endodontic disease, and caries are now under study.

Key words. photodynamic therapy, singlet oxygen, bactericidal effect, reactive oxygen species, disinfection

Photodynamic therapy (PDT) is a procedure whereby undesired tissue can be destroyed by the combined action of light, oxygen, and a photosensitizer [1]. PDT is mainly used for cancer treatment. Photosensitizers used in cancer treatment are taken in selectively by the cancer cells. When the photosensitizer is exposed to light with a specific wavelength, the photosensitizer is excited and transfers its excitation

K. Nakamura (✉), M. Tada, T. Mokudai, and M. Kohno
New Industry Creation Hatchery Center, Tohoku University, 6-6-10 Aoba, Aramaki, Aoba-ku, Sendai 980-8579, Japan
e-mail: keisuke1@mail.tains.tohoku.ac.jp

K. Nakamura, T. Kanno, H. Ikai, and E. Hayashi
Division of Fixed Prosthodontics, Department of Restorative Dentistry,
Tohoku University Graduate School of Dentistry, 4-1 Seiryo-machi, Aoba-ku, Sendai 980-8575, Japan

T. Sasano et al. (eds.), *Interface Oral Health Science 2009*,
DOI 10.1007/978-4-431-99644-6_59, © Springer 2010

energy to ground state oxygen molecules resulting in the generation of singlet oxygen. Singlet oxygen is a reactive oxygen species with high oxidative reactivity. Therefore, the reaction of singlet oxygen with cellular constituents can result in oxidative damage leading to cell death.

Oxidation by singlet oxygen also has antimicrobial properties. Applications of PDT for clinical dentistry have been studied [2] because most dental diseases, such as periodontitis, peri-implantitis, endodontic disease, and caries, are caused by oral bacteria. It is thought that PDT bactericidal treatment does not cause the appearance of drug-resistant bacteria because singlet oxygen attacks several cell structures, such as organelles, proteins, nucleic acids, etc. Thus, bactericidal effect by PDT has advantages in the treatment of infectious dental diseases.

Several in vitro studies have shown that PDT can kill oral pathogens, such as *Streptococcus mutans* [3], *Porphyromonas gingivalis* [4], *Candida albicans* [5], etc. Our research group also confirmed that PDT could effectively kill *Staphylococcus aureus* even though the concentration of the photosensitizer was very low (10 μM). Several animal studies have also shown that PDT is effective in the treatment of periodontitis [6] and peri-implantitis [7]. There are, however, only a few studies that clinically investigate the effect of PDT for dentistry. It is necessary to investigate which photosensitizer is most effective for killing oral pathogens as well as to evaluate the clinical efficacy of PDT in the future.

References

1. Clo E, Snyder JW, Ogilby PR et al (2007) Control and selectivity of photosensitized singlet oxygen production: challenges in complex biological systems. Chembiochem 8:475–481
2. Konopka K, Goslinski T (2007) Photodynamic therapy in dentistry. J Dent Res 86:694–707
3. Wood S, Metcalf D, Devine D et al (2006) Erythrosine is a potential photosensitizer for the photodynamic therapy of oral plaque biofilms. J Antimicrob Chemother 57:680–684
4. Dobson J, Wilson M (1992) Sensitization of oral bacteria in biofilms to killing by light from a low-power laser. Arch Oral Biol 37:883–887
5. Donnelly RF, McCarron PA, Tunney MM et al (2007) Potential of photodynamic therapy in treatment of fungal infections of the mouth. Design and characterisation of a mucoadhesive patch containing toluidine blue O. J Photochem Photobiol B 86:59–69
6. Komerik N, Nakanishi H, MacRobert AJ et al (2003) In vivo killing of *Porphyromonas gingivalis* by toluidine blue-mediated photosensitization in an animal model. Antimicrob Agents Chemother 47:932–940
7. Shibli JA, Martins MC, Ribeiro FS et al (2006) Lethal photosensitization and guided bone regeneration in treatment of peri-implantitis: an experimental study in dogs. Clin Oral Implants Res 17:273–281

Induction of Tregs from PBMC by interacting with immunosuppressive molecule B7-H3 on oral mesenchymal stem cells

Yasuhiro Nagai, Toshinobu Kuroishi, Daisuke Shiraishi, Akiko Ohki, and Shunji Sugawara

Abstract. Mesenchymal stem cells show tolerogenic property. It is also known that periodontal tissues contain many kinds of mesenchymal stem cells, such as the dental pulp stem cells, dental follicle cells, and periodontal ligament stem cells. In this study, we attempted to establish mesenchymal stem cells from the human periodontal tissues and investigated the immunosuppressive function of the cells. A fibroblast-like cell line named H1C1 was established by limiting dilution method and the cell line had abilities to differentiate into adipocytes, osteocytes, and chondrocytes; therefore, we termed H1C1 as the human mesenchymal stem cells. When human PBMC were cocultured with H1C1, the population of regulatory T cells (Tregs) was increased compared with monoculture of PBMC, and neutralization of B7-H3 suppressed this induction. These results suggest that the human oral mesenchymal stem cells such as H1C1 have the abilities to suppress immunoreactions by costimulatory molecules and increasing regulatory T cells.

Key words. mesenchymal stem cell, costimulatory molecules, Tregs

1 Introduction

Despite high bacterial colonization, acute infections are rare in the oral mucosa. A recent study showed that mesenchymal stem cells show tolerogenic property. It is also known that periodontal tissues contain many kinds of mesenchymal stem cells such as the dental pulp stem cells, dental follicle cells, and periodontal ligament stem cells. In addition, our recent study showed that anterior pituitary progenitor cells strongly express immunosuppressive costimulatory molecule B7-H3 [1, 2]. These reports lead us to the assumption that oral mesenchymal stem cells can

Y. Nagai (✉), T. Kuroishi, D. Shiraishi, A. Ohki, and S. Sugawara
Divisions of Oral Immunology, Tohoku University, Sendai 980-8575, Japan
e-mail: nagai13@m.tains.tohoku.ac.jp

D. Shiraishi
Divisions of Periodontology and Endodontology, Tohoku University, Sendai 980-8575, Japan

T. Sasano et al. (eds.), *Interface Oral Health Science 2009*,
DOI 10.1007/978-4-431-99644-6_60, © Springer 2010

contribute to the oral tolerogenic property by immunosuppressive costimulatory molecules such as B7-H3. In this study, we attempted to establish mesenchymal stem cells from the human periodontal tissues, and investigated immunosuppressive function of the cells, especially focused on B7-H3.

2 Materials and Methods

A fibroblast-like cell line named H1C1 from extracted baby tooth periodontal tissue from a 6-year-old child was established by limiting dilution method and used for further research. The expression of immunosuppressive costimulatory molecules was analyzed by RT-PCR, immunocytochemistry, and FACSCalibur. Human PBMCs from healthy volunteers were cocultured with H1C1 for 6 days and the induction of regulatory T cells was investigated. For blocking of B7-H3, monoclonal antibody of B7-H3 was added to the coculture wells.

3 Results and Discussion

The cell line named H1C1 had abilities to differentiate into adipocytes, osteocytes, and chondrocytes; therefore, we termed H1C1 as the human mesenchymal stem cells. RT-PCR showed that H1C1 constantly expressed mRNA of costimulatory molecules B7-H1, B7-H2, and B7-H3, which are known as immunosuppressive molecules. When human PBMC were cocultured with H1C1, the population of Tregs was increased compared with monoculture of PBMC, and neutralization of B7-H3 suppressed this induction. These results suggest that the human oral mesenchymal stem cells such as H1C1 have the abilities to suppress immunoreactions by costimulatory molecules and increasing Tregs.

4 Conclusion

- We succeeded to establish the human mesenchymal stem cells named H1C1.
- H1C1 expressed B7-H1, B7-H2, and B7-H3.
- H1C1 had abilities to differentiate into adiopcytes, osteocytes, and chondrocytes.
- H1C1 induced regulatory T cells by cell-to-cell interactions.
- Neutralization of B7-H3 suppressed the induction of Tregs.

In conclusion, we succeeded to establish the human oral mesenchymal stem cells named H1C1, which may contribute to the oral tolerogenic environment by cell-to-cell interactions and promoting the induction of regulatory T cells by B7-H3 molecule.

References

1. Nagai Y, Aso H, Ogasawara H et al (2008) Anterior pituitary progenitor cells express costimulatory molecule 4Ig-B7–H3. J Immunol 181:6073–6081
2. Nagai Y, Ogasawara H, Taketa Y et al (2008) Bovine anterior pituitary progenitor cell line expresses interleukin (IL)-18 and IL-18 receptor. J Neuroendocrinol 20:1233–1241

A method for determining the profiles of biomass volume and glucan within dental plaque

Kazuo Kato, Kiyomi Tamura, Tran Thu Thuy, Haruo Nakagaki, and Takuichi Sato

Abstract. A method for determining the profiles of biomass volume and glucan within plaque was developed using depth-specific analysis of plaque. The profiles of biomass volume and glucan demonstrated a tendency to be higher in the plaque exposed to sucrose, suggesting that an evaluation of these two plaque indices would be important to see the cariogenicity in the diet.

Key words. depth-specific analysis, glucan, biomass volume, dental plaque

1 Introduction

It has been proposed that glucans synthesized from sucrose by *Streptococcus mutans* promote cariogenic effects of sugars by increasing the porosity of plaque and enhancing the diffusion of sugar substrates in the biofilms [1]. This crossover trial was carried out to clarify density profiles of biomass volume and glucan throughout the plaque, which was formed under a periodical sucrose exposure, using a depth-specific analysis.

2 Materials and Methods

Seven consenting subjects (19–22 years) rinsed their mouths with 15% sucrose (experimental) or placebo (control) solutions six times a day for 4 days. They wore in situ plaque-generating devices in their upper molars to allow plaque to form

K. Kato (✉), K. Tamura, T.T. Thuy, and H. Nakagaki
Department of Preventive Dentistry and Dental Public Health, School of Dentistry, Aichi-Gakuin University, 1-100 Kusumoto-cho, Chikusa-ku, Nagoya 464-8650, Japan
e-mail: kazkato@dpc.aichi-gakuin.ac.jp

T. Sato
Department of Oral Biology, Tohoku University Graduate School of Dentistry, 4-1 Seiryo-machi, Aoba-ku, Sendai 980-8575, Japan

T. Sasano et al. (eds.), *Interface Oral Health Science 2009*,
DOI 10.1007/978-4-431-99644-6_61, © Springer 2010

within the devices by abstaining from tooth brushing. The sample was separated into 4–6 layered fractions (180 μm thick) by serial sectioning (2 at 2 μm, 2 at 4 μm, then 4 at 6 μm) using a method reported elsewhere [2]. Thinner sections were used for area measurement to evaluate sample volume. Genomic DNA was isolated from middle thick sections to amplify the 16S rRNA gene sequences by PCR with *S. mutans*-specific primers [3]. Thicker sections were heated with 25 μl of 1M sulfuric acid in a microwave oven (250 W, 30 min) to produce hydrolysate, in which glucose concentrations were determined using an immobilized glucose oxidase membrane to estimate the total amount of glucans.

3 Results and Discussion

The profiles of plaque mass volume demonstrated a tendency to be higher in the outer region of the experimental group (Fig. 1). The profiles of glucan within plaque in both groups demonstrated a similar pattern, indicating that this short-term sucrose exposure failed to promote glucan production within the plaque (Fig. 2). Although detection patterns of positive fractions for the cariogenic bacteria varied among the subjects irrespective of the treatment, they tended to be found in the layer with richer mass volume. Significant negative relationships were found between the amount of glucan and mass volume per plaque layer in both groups. This might be related to, if any, a mirror-image pattern of the glucan and the mass volume profiles within the plaque. Further studies will be needed to check whether the negative relationship between the amount of glucan and mass volume would be one of the depth-specific properties of the plaque.

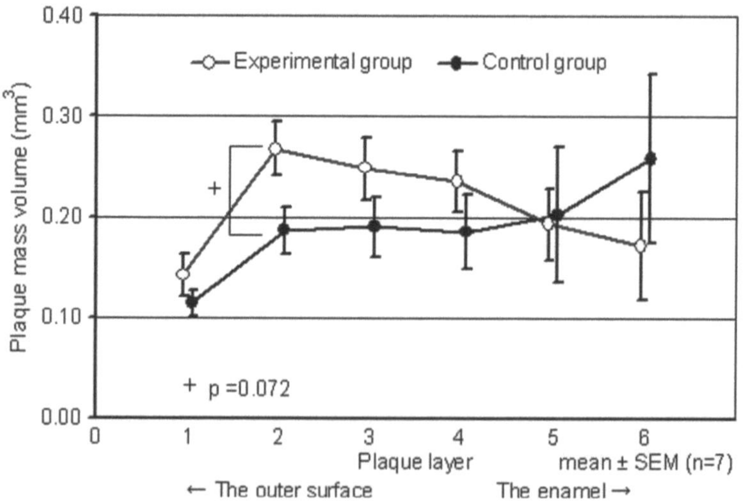

Fig. 1. Profiles of glucan concentrations per unit of plaque volume within plaque

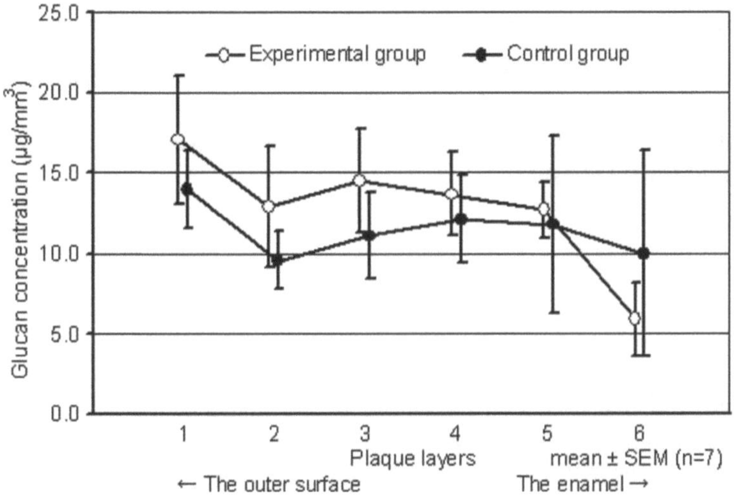

Fig. 2. Profiles of biomass volume within plaque

Acknowledgments This study was supported by KAKENHI (C) No. 20592477, Japan.

References

1. Dibdin GH, Shellis RP (1988) Physical and biochemical studies of *Streptococcus mutans* sediments suggest new factors linking the cariogenicity of plaque with its extracellular polysaccharide content. J Dent Res 67:890–895
2. Kato K, Sato T, Takahashi N et al (2004) A method for mapping the distribution pattern of cariogenic streptococci within dental plaque in vivo. Caries Res 38:448–453
3. Rupf S, Merte K, Eschrich K et al (2001) Peroxidase reaction as a parameter for discrimination of *Streptococcus mutans* and *Streptococcus sobrinus*. Caries Res 35:258–264

Porphyromonas gingivalis is widely distributed in subgingival plaque biofilm of elderly subjects

Yuki Abiko, Takuichi Sato, Kenji Matsushita, Reiko Sakashita, and Nobuhiro Takahashi

Abstract. The frequency of periodontal diseases appears to increase with age. *Porphyromonas gingivalis* is widely regarded as major periodontal pathogens. This study aimed to quantify *P. gingivalis* in subgingival plaque biofilm of elderly subjects by real-time polymerase chain reaction (PCR). Subgingival plaque was obtained from 198 periodontally healthy (mean age, 70.3 years) and 176 subjects with periodontitis (70.6 years). Quantification of total bacteria and *P. gingivalis* was performed by real-time PCR using universal and *P. gingivalis*-specific primers based on 16S rRNA genes, respectively. Both the detection frequency and mean proportion of *P. gingivalis* were significantly higher in subjects with periodontitis than in periodontally healthy subjects (p<0.0001). Nevertheless, *P. gingivalis* was detected frequently both from subjects with periodontitis and periodontally healthy subjects. These results suggest that *P. gingivalis* is widely distributed in subgingival plaque biofilm of elderly subjects.

Key words. elderly subjects, plaque biofilm, *Porphyromonas gingivalis*, quantitative polymerase chain reaction, 16S ribosomal RNA

1 Introduction

Qualitative and quantitative changes of the subgingival plaque biofilm microflora in periodontal pockets are thought to be associated with the development and progression of periodontitis. The frequency of periodontal diseases appears to increase

Y. Abiko(✉), T. Sato, and N. Takahashi
Division of Oral Ecology and Biochemistry, Tohoku University Graduate School of Dentistry, Sendai 980-8575, Japan
e-mail: abiko@mail.tains.tohoku.ac.jp

K. Matsushita
Department of Oral Disease Research, National Center for Geriatrics and Gerontology, Obu, Aichi 474-8522, Japan

R. Sakashita
Nursing Foundation, College of Nursing Art and Science, University of Hyogo, Akashi 673-8588, Japan

T. Sasano et al. (eds.), *Interface Oral Health Science 2009*,
DOI 10.1007/978-4-431-99644-6_62, © Springer 2010

with age. However, few studies have been investigated on subgingival microflora in elderly subjects. *Porphyromonas gingivalis* is widely regarded as major periodontal pathogens. This study aimed to quantify *P. gingivalis* in subgingival plaque biofilm of elderly subjects by real-time polymerase chain reaction (PCR).

2 Relationship Between Periodontal Status and *P. gingivalis* in Subgingival Plaque Biofilm of Elderly Subjects

Subgingival plaque was obtained from independent elderly subjects (60 years and over, $n=374$) from gingival crevices with the deepest probing depths. Gender, probing depth, bleeding on probing (BOP) and wearing of denture were examined and recorded. Quantification of total bacteria and *P. gingivalis* was performed by real-time PCR using universal and *P. gingivalis*-specific primers based on 16S rRNA genes [1], respectively. According to the deepest probing depths, 198 (mean age, 70.3 years) were considered as periodontally healthy, while 176 (70.6 years) were considered as subjects with periodontitis. Both the detection frequency and mean proportion of *P. gingivalis* were significantly higher in subjects with periodontitis than in periodontally healthy subjects ($p<0.0001$). The proportion of *P. gingivalis* was significantly higher in subjects with BOP-positive ($p<0.05$), but not related to subjects' gender, wearing of denture and age.

3 Inhabitation of *P. gingivalis* in Subgingival Plaque of Elderly Subjects

It has been reported that *P. gingivalis* is generally detected from subjects with periodontitis [1, 2]. However, in this study, *P. gingivalis* was detected frequently both from subjects with periodontitis and periodontally healthy subjects, suggesting that the inhabitation of *P. gingivalis* is a specific feature of subgingival plaque biofilm of elderly subjects. Amano et al. [3] reported that the pathogenicity of *P. gingivalis* was different among strains, suggesting the possibility that low pathogenic *P. gingivalis* strains colonized in periodontally healthy elderly subjects in this study.

4 Clinical Implication

This study suggests that the inhabitation of *P. gingivalis* in subgingival plaque biofilm of elderly subjects is one of the risk factors of periodontitis, as well as the decrease of host defense mechanism with age, although further study is required to elucidate the pathogenicity of *P. gingivalis* found in periodontally healthy elderly subjects. The findings of this study support the necessity and importance of periodontal control, especially in elderly subjects.

References

1. Abiko Y, Sato T, Mayanagi G et al (2010) Profiling of subgingival plaque biofilm microflora from periodontally healthy subjects and from subjects with periodontitis using quantitative real-time PCR. J Periodontal Res 45 (in press)
2. Kuboniwa M, Amano A, Kimura RK et al (2004) Quantitative detection of periodontal pathogens using real-time polymerase chain reaction with TaqMan probes. Oral Microbiol Immunol 19:168–176
3. Amano A, Nakagawa I, Kataoka K et al (1999) Distribution of *Porphyromonas gingivalis* strains with *fimA* genotypes in periodontitis patients. J Clin Microbiol 37:1426–1430

Profiling of dental plaque microflora on root caries lesions and the protein-degrading activity of these bacteria

Kazuhiro Hashimoto, Takuichi Sato, Hidetoshi Shimauchi, and Nobuhiro Takahashi

Abstract. This study aimed to profile plaque microflora on root caries lesions and to examine the protein-degrading activity of isolated bacteria. Plaque samples on root caries lesions (R) and from healthy supragingival sites (S) of six subjects were collected and cultured anaerobically on blood agar plates. The isolated bacteria were identified by 16S rRNA sequencing and examined for their protein-degrading activity, using the skim-milk plates, and for their acidogenicity, using the FAB broth. *Propionibacterium*, *Actinomyces*, *Streptococcus*, *Lactobacillus*, and *Bifidobacterium* were predominant in R. The skim-milk plates distinguished between protein-degrading and protein-coagulating bacteria, which comprised 7 and 33% of microflora in R, respectively. These results suggest that protein-coagulating bacteria demineralize hydroxyapatite and denature proteins of root dentin, and that protein-degrading bacteria may degrade the denatured proteins.

Key words. microflora, plaque biofilm, protein-degrading activity, root caries

1 Introduction

Root caries is thought to be initiated by decalcification by bacterial acids of plaque formed on root surface. In addition, degradation of dentinal proteins may be related to root caries, since cementum and dentin of root surface consist of not only

K. Hashimoto and H. Shimauchi
Division of Periodontology and Endodontology, Tohoku University Graduate School of Dentistry, 4-1 Seiryo-machi, Aoba-ku, Sendai 980-8575, Japan

K. Hashimoto, T. Sato (✉), and N. Takahashi
Division of Oral Ecology and Biochemistry, Tohoku University Graduate School of Dentistry, 4-1 Seiryo-machi, Aoba-ku, Sendai 980-8575, Japan
e-mail: tak@mail.tains.tohoku.ac.jp

T. Sasano et al. (eds.), *Interface Oral Health Science 2009*,
DOI 10.1007/978-4-431-99644-6_63, © Springer 2010

hydroxyapatite but also the proteins such as collagen. However, protein-degrading activity of plaque bacteria on root caries has not been reported yet. Therefore, the aims of this study were to profile plaque microflora on root caries lesions, and to examine the protein-degrading activity of these bacteria.

2 Predominant Bacteria in Plaque on Root Caries

Plaque samples on root caries (R), as well as from healthy supragingival sites (S), of six subjects (48–73 years) were cultured anaerobically on CDC blood agar plates, and isolated bacteria were identified by 16S rRNA sequencing. *Propionibacterium*, *Actinomyces*, *Streptococcus*, *Lactobacillus*, and *Bifidobacterium* were predominant in plaque on R. Among them, *Actinomyces*, *Streptococcus*, and *Lactobacillus* have been reported to be predominant in plaque on root caries [1, 2], in accordance with this study. On the other hand, *Propionibacterium* and *Bifidobacterium* have been reportedly often detected from caries dentin [3], suggesting that plaque microflora on root caries lesions is similar to that of caries dentin.

3 Protein-Degrading Bacteria in Plaque on Root Caries

The skim-milk plates distinguished between protein-degrading and protein-coagulating bacteria, which comprised 7 and 33%, and 26 and 0% of plaque microflora in R and S, respectively. In addition, protein-coagulating bacteria produced enough organic acids to denature proteins, i.e., alter protein conformation, since they lowered the pH of the medium to approximately 4.0 where the skim milk was found to be coagulated in the skim milk plates.

4 Conclusions

These results suggest that protein-coagulating bacteria demineralize hydroxyapatite and denature proteins of root dentin, and that protein-degrading bacteria may degrade the denatured proteins, thus resulting in the onset and progression of root caries.

References

1. Brailsford SR, Shah B, Simons D et al (2001) The predominant aciduric microflora of root-caries lesions. J Dent Res 80:1828–1833
2. van Houte J, Lopman J, Kent R (1994) The predominant cultivable flora of sound and carious human root surfaces. J Dent Res 73:1727–1734
3. Hoshino E (1985) Predominant obligate anaerobes in human carious dentin. J Dent Res 64:1195–1198

Characterization of glucosyltransferases synthesizing (1→6)-α-D-glucan from *Streptococcus sobrinus* and *Streptococcus downei*

Hideaki Tsumori, Atsunari Shimamura, Kazuo Yamakami, and Yutaka Sakurai

Abstract. Glucosyltransferase-T genes from *Streptococcus sobrinus* (serotype *g*) and *Streptococcus downei* (*h*) were cloned and the sequences were compared with those of *Streptococcus criceti* (*a*) and *S. sobrinus* (*d*). Deduced amino acid sequences of glucosyltransferase-T (GTF-T) from the four species revealed 83–85% homology. Glucans synthesized by purified GTF-Ts from sucrose without primer were (1→6)-α-glucan. Endodextranase digestion of the glucans reduced the size up to 50–70%, and remaining parts of the glucan were mainly composed of (1→3)-α-glucosidic linkage. The GTF-Ts from the four species of streptococci are highly similar at the amino acid level, and their catalytic properties are also similar. The GTF-Ts from these streptococci would trigger the formation of biofilm in vivo.

Key words. mutans streptococci, glucosyltransferase, glucan

1 Introduction

Dental caries continues to pose an important health problem worldwide. Mutans group of oral streptococci are strongly implicated as the primary causative agent in human dental caries [1]. *Streptococcus criceti* (serotype *a*), *Streptococcus sobrinus* (*d* and *g*), and *Streptococcus downei* (*h*) secrete four glucosyltransferases (GTFs, GTF-T, GTF-I, GTF-S, and GTF-U) in vivo. These GTFs synthesize a diversity of glucans that vary in glucosidic linkages, solubilities, and degree of blanching [2]. By their cooperative action, the GTFs synthesize adhesive water-insoluble glucan

H. Tsumori (✉) and A. Shimamura
Department of Chemistry, National Defense Medical College, 3-2 Namiki, Tokorozawa, Saitama 359-8513, Japan
e-mail: htsumori@ndmc.ac.jp

K. Yamakami and Y. Sakurai
Department of Preventive Medicine and Public Health, National Defense Medical College, 3-2 Namiki, Tokorozawa, Saitama 359-8513, Japan

T. Sasano et al. (eds.), *Interface Oral Health Science 2009*,
DOI 10.1007/978-4-431-99644-6_64, © Springer 2010

from dietary sucrose, which is closely related to biofilm formation on smooth tooth surface and their virulence. In the understanding of pathogenesis of GTF, we focused on GTF-Ts and *gtfT* genes from some species of mutans streptococci.

2 Experiments

2.1 Purification and Properties of GTF-T

GTF-Ts were purified from culture supernatants of four species of mutans streptococci. The p*I* values of the purified GTF-Ts from *S. sobrinus* (*d* and *g*) and *S. downei* (*h*) were estimated to be 5.3–5.8. The *Mr* values of the GTF-Ts from four strains were estimated to be 160 kDa by SDS-PAGE. The GTF-Ts were immunologically almost identical with each other by immunoprecipitation analysis.

3 Molecular Properties of GTF-T

Nucleotide sequence of *gtfT* gene of *S. sobrinus* (*g*, AB476745) and *S. downei* (*h*, AB476746) and the previous data of *S. criceti* (*a*) and *S. sobrinus* (*d*) revealed a ranging from 4,500 to 4,600 open reading frame encoding a protein with an N-terminal signal peptide. The deduced amino acid sequences of GTF-T from four species showed over the 84% identity and the *Mr* and p*I* values of the mature GTF-T was calculated to be approximately 160 kDa and 5.2–5.4, respectively.

4 Linkage Analysis of Glucans Synthesized by GTF-T

The glucans synthesized by the purified GTF-Ts were (1→6)-α-glucan (61–72 mol%) with (1→3)-α- and (1→3,6)-α-glucosyl residues. Over the 50% of these glucans were hydrolyzed by endodextranase (Merck, Whitehouse Station, NJ). Residual glucans were (1→3)-α-glucan with (1→6)-α- and (1→3,6)-α-glucosyl residues.

5 Conclusion

Mutans group of oral streptococci are implicated as the primary causative agent in human dental caries [1]. In this group, *S. criceti* (*a*), *S. sobrinus* (*d* and *g*), and *S. downei* (*h*) secrete four GTFs in vivo [2]. GTF-Ts synthesize water-soluble glucan from sucrose without primer that are necessary for activation of other GTFs [3]. Therefore, we purified GTF-Ts and cloned *gtf*T genes from the three species of mutans streptococci (four serotypes) and compared the properties of the enzymes. We found that glucans synthesized by the purified enzymes mainly consisted of α-(1→6) linked glucose. The GTF-Ts among mutans streptococci are highly similar at the amino acid sequence and catalytic properties are very alike. The findings should indicate that GTF-T is one of a virulence factor for dental caries, because the GTF-Ts are necessary for the initiation of biofilm formation.

References

1. Loesche WJ (1986) Role of *Streptococcus mutans* in human dental decay. Microbiol Rev 50:353–380
2. Monchois V, Willemot RM, Monsan P (1999) Glucansucrases: mechanism of action and structure-function relationships. FEMS Microbiol Rev 23:131–151
3. Hijum SAFT, Kralj S, Ozimek LK et al (2006) Structure-function relationships of glucansucrase and fructansucrase enzymes from lactic acid bacteria. Microbiol Mol Biol Rev 70:157–176

Profiling of dental plaque biofilm on first molars with orthodontic bands and brackets

Ryo Komori, Takuichi Sato, Teruko Takano-Yamamoto, and Nobuhiro Takahashi

Abstract. This study aimed to profile plaque microflora on first molars with orthodontic bands (Ba), brackets (Br), and without appliances (C). The mean bacterial numbers of plaque (logCFUs/mg) on the molars with Ba, Br, and C were 6.6, 6.7, and 6.9, respectively. *Actinomyces*, *Streptococcus,* and *Veillonella* were predominant in Br and C, while the proportions of *Actinomyces* and *Veillonella* were low in Ba. Periodontitis-associated bacteria including *Eubacterium*, *Fusobacterium*, *Porphyromonas*, and *Prevotella* were isolated in Br, but virtually not detected in Ba and C, suggesting that supragingival plaque biofilm of teeth with Br carries bacteria related to periodontitis. These findings may provide a helpful suggestion for self-care and regular professional plaque control in clinical orthodontics.

Key words. microflora, orthodontic appliances, plaque biofilm, polymerase chain reaction

1 Introduction

The aims of this study were to profile the microflora on upper and lower first molars with orthodontic bands (Ba), brackets (Br), and without appliances (C). Bacteria were counted, isolated, and identified using anaerobic culture and molecular biological techniques.

R. Komori and T. Takano-Yamamoto
Division of Orthodontics and Dentofacial Orthopedics, Tohoku University Graduate School of Dentistry, 4-1 Seiryo-machi, Aoba-ku, Sendai 980-8575, Japan

R. Komori, T. Sato (✉), and N. Takahashi
Division of Oral Ecology and Biochemistry, Tohoku University Graduate School of Dentistry, 4-1 Seiryo-machi, Aoba-ku, Sendai 980-8575, Japan
e-mail: tak@mail.tains.tohoku.ac.jp

T. Sasano et al. (eds.), *Interface Oral Health Science 2009*,
DOI 10.1007/978-4-431-99644-6_65, © Springer 2010

2 Profiling of Plaque Biofilm on Orthodontic Appliances

Supragingival plaque samples on the surface of upper and lower first molars with Ba, Br, and C of six subjects (aged 11–30) were cultured anaerobically on CDC blood agar plates, and isolated bacteria were identified by 16S rRNA sequencing, as described previously [1, 2]. The mean bacterial numbers (logCFUs/mg) were 6.6, 6.7, and 6.9 from Ba, Br, and C, respectively, indicating that there were no significant differences among Ba, Br, and C. These findings were in accordance with a previous study by Persson et al. [3] that there were no differences in the bacterial amounts of plaque on different surfaces, that is, enamel and composite resin. *Actinomyces* (44 and 40%), *Streptococcus* (24 and 35%), and *Veillonella* (9 and 10%) were predominant in Br and C, respectively, in accordance with a previous study by Hoshino et al. [4], which reported the composition of plaque on the enamel, while the proportions of *Actinomyces* (18%) and *Veillonella* (2%) were low in Ba. Periodontitis-associated bacteria including *Eubacterium* (3%), *Fusobacterium* (1%), *Porphyromonas* (3%), and *Prevotella* (2%) were isolated in Br, but virtually not detected both in Ba and in C.

3 Clinical Implication

These results suggest that supragingival plaque biofilm of teeth with Br carries bacteria related to periodontitis. These findings may provide a helpful suggestion for self-care and regular professional plaque control in clinical orthodontics.

References

1. Washio J, Sato T, Koseki T et al (2005) Hydrogen sulfide-producing bacteria in tongue coating and their relationship with oral malodour. J Med Microbiol 54:889–895
2. Sato R, Sato T, Takahashi I et al (2007) Profiling of bacterial flora in crevices around titanium orthodontic anchor plates. Clin Oral Implants Res 18:21–26
3. Persson A, Claesson R, van Dijken JWV (2005) Levels of mutans streptococci and lactobacilli in plaque in aged restorations of an ion-releasing and a universal hybrid composite resin. Acta Odontol Scand 63:21–25
4. Hoshino E, Sato M, Sasano T et al (1989) Characterization of bacterial deposits formed *in vivo* on hydrogen-ion-sensitive field-effect transistor electrodes and enamel surfaces. Jpn J Oral Biol 31:102–106

Hydrogen-sulfide production from various substrates by oral *Veillonella* and effects of lactate on the production

Jumpei Washio, Yoko Sakuma, Yuko Shimada, and Nobuhiro Takahashi

Abstract. Oral *Veillonella* is one of the dominant bacteria in the tongue coating that produces hydrogen sulfide (H_2S) known as one of the cause of oral malodor. This study aimed to examine various cysteine-containing substrates on the H_2S production by *Veillonella*, and the effects of oral environmental factors, such as pH and lactate, on the production. Three *Veillonella* species were grown and the cell suspensions were incubated with cysteine, cysteinyl-glycine, glutathione, cystine, and tryptone at pH 7 in the presence or absence of lactate. The amounts of H_2S produced were measured by methylene blue method. The H_2S production was the highest from cysteine and cysteinyl-glycine. In addition, lactate increased the H_2S production. These results suggested that *Veillonella* can produce H_2S from cysteine-containing substrates available in the oral cavity, and that the production is regulated by oral environmental factors such as lactate.

Key words. *Veillonella*, hydrogen sulfide, lactate, oral malodor

1 Introduction

Oral malodor is due to the metabolic products of bacteria in the oral cavity, particularly those living on the dorsum of the tongue [1].Oral *Veillonella* is one of the dominant bacteria in the tongue coating that produces hydrogen sulfide (H_2S) from cysteine [2]. However, the concentration of a single amino acid, cysteine, in the oral cavity is considered to be not high; therefore, the first purpose of this study is to examine various cysteine-containing substrates on the H_2S production by *Veillonella*.

J. Washio (✉), Y. Sakuma, Y. Shimada, and N. Takahashi
Division of Oral Ecology and Biochemistry, Tohoku University Graduate School of Dentistry, 4-1 Seiryo-machi, Aoba-ku, Sendai 980-8575, Japan
e-mail: j-washio@mail.tains.tohoku.ac.jp

T. Sasano et al. (eds.), *Interface Oral Health Science 2009*,
DOI 10.1007/978-4-431-99644-6_66, © Springer 2010

Moreover, the environmental conditions in the oral cavity fluctuates continuously, especially the concentration of acids such as lactate that is produced by bacteria from sugar. Therefore, the second purpose is to examine the effects of oral environmental factors, such as pH and lactate, on the production.

2 H_2S Production from Various Sulfur Compounds and Effects of Lactate on H_2S Production

The type strains of three *Veillonella* species, *Veillonella atypica, Veillonella dispar,* and *Veillonella parvula,* were anaerobically grown in tryptone–yeast extract medium containing 1.8% sodium lactate, and the bacterial cell suspensions were incubated with substrates, 1 mM cysteine, cysteinyl-glycine, glutathione, cystine, or 0.5% tryptone, at pH 7 in the presence or absence of 10 mM sodium lactate. After 3-h incubation, the amounts of H_2S produced were measured by methylene blue method.

The H_2S were produced from all substrates, especially cysteine and cysteinyl-glycine effectively, followed by tryptone, glutathione, and cystine. Lactate increased greatly the H_2S production from cysteine, cysteinyl-glycine, glutathione, and cystine, but had no effect on the production from tryptone.

These results suggest that *Veillonella* can produce an odorous compound H_2S from various cysteine-containing substrates available in the oral cavity and that lactate in tongue coating affects the H_2S production. *Veillonella* species are asaccharolytic, so the activation of H_2S production by lactate may occur under the coexistence with lactate-producing oral bacteria, such as *Streptococcus, Actinomyces and Lactobacillus.* It is suggested that the H_2S production may be regulated by interaction between bacterial members in the oral cavity, and thus the etiology of oral malodor should be considered in the entire oral microbial ecosystem.

Acknowledgments This study was supported partly by a Grant-in-Aid (no. 20791632 to JW and no. 19390539 to NT) from the JSPS.

References

1. Kazor CE, Mitchell PM, Paster BJ et al (2003) Diversity of bacterial populations on the tongue dorsa of patients with halitosis and healthy patients. J Clin Microbiol 41:558–563
2. Washio J, Sato T, Takahashi N et al (2005) Hydrogen sulfide-producing bacteria in tongue biofilm and their relationship with oral malodour. J Med Microbiol 54:889–895

Denture plaque removal efficacy of denture cleansing device utilizing radical disinfection ability of activated low concentration H_2O_2

Taro Kanno, Eisei Hayashi, Hiroyo Ikai, Keisuke Nakamura, Takayuki Mokudai, Masahiro Kohno, and Keiichi Sasaki

Abstract. The purpose of this study was to evaluate the plaque removal and disinfection effect of an experimental denture cleaning device applying radical disinfection system. An experimental denture cleaning device, which would be able to produce the hydroxyl ramdical by means of irradiation of light-emitting diodes (wavelength: 405 nm) to low concentration hydrogen peroxide was made.

Difference in cleaning capability between the experimental denture cleaning device and commercial cleaning agent was apparent. The result of the present study indicated that the experimental denture cleaning device using radical disinfection system had an ability of effectively cleaning dentures.

Key words. disinfection, low concentration of H_2O_2, light-emitting diodes, visible light, hydroxyl radical

1 Introduction

To clean the denture, sterilizing and cleaning capabilities that are effective against high drug-resistance pathogens, such as *Candida albicans*, are required. However, use of highly-concentrated agent for a home care product involves various risk factors. Therefore, development of an innovative denture cleaning system that enhanced bactericidal capability of low-concentrated agent by using a dedicated device is demanded.

T. Kanno(✉), E. Hayashi, H. Ikai, and K. Nakamura
Division of Fixed Prosthodontics, Tohoku University Graduate School of Dentistry, 4-1 Seiryo-machi, Aoba-ku, Sendai 980-8575, Japan
e-mail: tarou@mail.tains.tohoku.ac.jp

K. Nakamura, T. Mokudai, and M. Kohno
New Industry Creation Hatchery Center,Tohoku University, 6-6-10 Aoba, Aramaki, Aoba-ku, Sendai 980-8579, Japan

K. Sasaki
Division of Advanced Prosthodontic Dentistry, Tohoku University Graduate School of Dentistry, 4-1 Seiryo-machi, Aoba-ku, Sendai 980-8575, Japan

T. Sasano et al. (eds.), *Interface Oral Health Science 2009*,
DOI 10.1007/978-4-431-99644-6_67, © Springer 2010

We previously reported about a new bacteria killing system, which enhanced the radical bactericidal activity included in low-concentrated hydrogen peroxide with LED at a specific wavelength effectively killed oral bacteria [1, 2].

The purpose of this study was to evaluate the plaque removal and disinfection effect of an experimental denture cleaning device applying the new radical disinfection system.

2 Materials and Methods

Twenty partial removable dentures, which were being used by 20 patients, were used in this study. In test group, the experimental device was used, while in control group, a commercially available denture cleaning agent was used. The removable dentures were cleaned for 30 min in both groups. Before and after cleaning in both groups, microbiological analysis was performed using a kit for bacterial culturing to evaluate colony forming unit (CFU). The amount of hydroxyl radical generated from both the experimental device and the denture cleaning agent was measured using electron spin trap device and DMPO.

3 Results

The experimental denture cleaning device was found to reduce the bacteria in the denture plaque by 10^6 CFU/ml on average. On the other hand, the commercial cleaning agent decreased the bacteria only by 10^2–10^3 CFU/m. There was also statistically significant difference of CFU between test and control group after the cleaning.

The experimental denture cleaning device turned out to create about 0.2 mM hydroxyl radical in 1 min. The hydroxyl radical was continuously generated after 30 min. Although the commercial cleaning agent generated a bit of hydroxyl radical, it did not increase over time.

4 Discussion

Difference in cleaning capability between the experimental denture cleaning device and the commercial cleaning agent was apparent. The result of the present study indicated that the experimental denture cleaning device using radical disinfection system had an ability of effectively cleaning dentures.

The hydroxyl radical generated from the experimental denture cleaning device was found to increase in proportion to the cleaning time. This is a major contributor to the cleaning capability of the system. Creating the hydroxyl radical effectively, supposedly enhances cleaning and bactericidal capabilities.

References

1. Kanno T, Nakamura K, Ikai H et al (2008) Radical sterilization properties of hydrogen peroxide activated by LED (386 nm) irradiation for oral bacterium. A pilot study. Prosthodont Res Pract 7:138–140
2. Ikai H, Kanno T, Hayashi E et al (2008) New irradiation system for clinical dental treatment advanced using low density hydrogen peroxide and laser diode. J Jpn Prosthodont Soc 117th Special Issue 52:194

Detection of herpes simplex virus type 1 in human cadaver trigeminal ganglia

Yuko Monma, Hisako Motani, Hirotaro Iwase, and Satoshi Fukumoto

Abstract. Our previous study suggested that the genotyping of latent herpes simplex virus type 1 (HSV-1) in trigeminal ganglion in forensic autopsies were useful for tracing the origins of unidentified cadavers. Here, we tried to isolate HSV-1 from cadaver trigeminal ganglia. Trigeminal ganglia were cultured with Vero cells. DNA was extracted from the culture medium and from the remaining half of each ganglion samples. HSV-1 DNA detection rate of the culture medium was higher than those of the extract taken directly from trigeminal ganglia. These results suggest that latent HSV-1 in trigeminal ganglia may be reactivated.

Key words. HSV-1, trigeminal ganglion, cadaver

1 Introduction

Herpes simplex virus type 1 (HSV-1) produces the typical symptoms in the oropharyngeal region and can establish latent infection in human trigeminal ganglia. Previously, we identified HSV-1 DNA in human cadaver trigeminal ganglia in forensic autopsies and suggested that the genotyping for variable region (V1) were useful for tracing the origins of unidentified cadavers [1]. However, previous studies have not examined whether the latent virus detected in both ganglia was same strain or not. Here, we tried to isolate HSV-1 in cadaver trigeminal ganglia to gain reactivated-virus.

Y. Monma (✉) and S. Fukumoto
Division of Pediatric Dentistry, Department of Oral Health and Development Sciences, Tohoku University Graduate School of Dentistry, 4-1 Seiryo-machi, Aoba-ku, Sendai 980-8575, Japan
e-mail: monma@pedo.dent.tohoku.ac.jp

H. Motani and H. Iwase
Department of Legal Medicine, Graduate School of Medicine, Chiba University, Chiba, Japan

T. Sasano et al. (eds.), *Interface Oral Health Science 2009*,
DOI 10.1007/978-4-431-99644-6_68, © Springer 2010

2 Methods

Six trigeminal ganglia were obtained bilaterally from three cadavers. Half of each ganglion was placed on Vero cell monolayers to isolate HSV-1 virus. After 2 days, the culture medium of these cells was collected. The DNA was extracted from the culture medium samples and the remaining half of each ganglion samples. Using these DNA, two-positions in the stable RL2 region (147 bp) and V1 region (666 bp) were amplified twice with PCR methods [1]. PCR products were separated by electrophoresis using 2% Agarose gel and then visualized by ethidium bromide.

3 Results and Discussion

Cytopathic effect of HSV-1 virus was not observed in any trigeminal ganglia samples cultured with Vero cells. In DNA samples from culture medium, all were RL2 region, and half of them were V1 region positive (Fig. 1). In contrast, in DNA extracts taken directly from trigeminal ganglia, five samples were RL2 and two were V1 region positive. In brief, HSV-1 DNA detection rate of the culture medium was higher than those of extract directly from trigeminal ganglia. These results suggested that the latent virus in trigeminal ganglia could be reactivated and HSV-1 fragment might be released into the culture medium. To measure the quantity of virus in the trigeminal ganglia the culture medium may be necessary for future examination.

Reference

1. Motani H, Sakurada K, Ikegaya H et al (2006) Detection of herpes simplex virus type 1 DNA in bilateral human trigeminal ganglia and optic nerves by polymerase chain reaction. J Med Virol 78:1584–1587

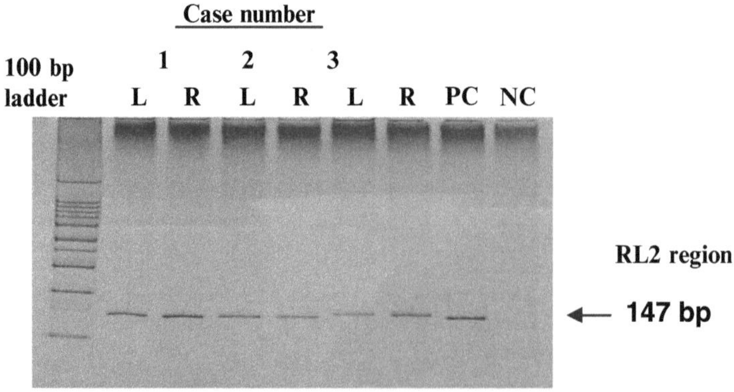

Fig. 1. The detection by PCR using DNA from culture medium

Transmitted laser beam power of the resin washed by experimental washing machine for dentures

Eisei Hayashi, Mika Tada, Taro Kanno, Hiroyo Ikai, Keisuke Nakamura, and Masahiro Kohno

Abstract. We had developed the experimental washing machine for denture, and had examined the cleansing property of the machine. The washing machine was using hydrogen peroxide as cleansing solution activated by LED (375 nm) and ultrasonic energy. However, the effect of long-term use of this machine was not examined. The purpose of this study was to examine the color alteration using measurement of transmitted laser beam. This study showed that there was not a change in the measurement of transmitted laser beam after 100 h for washing. If the color alteration in the surface of the samples had occurred because of washing, it was thought that absorption and dispersion of the laser light were influenced. Consequently, this study indicated that this washing machine might not change on the surface of the resin.

Key words. transmitted laser beam power, bactericidal effect, resin

We succeeded in developing a prototype device, which cleaned denture plaque on denture impression surface, by exciting the low-concentrated hydrogen peroxide water with energy of LED light and ultrasonic wave and generating hydroxyl radical efficiently, and investigated the cleaning effects of the device. The results of our investigations proved that the prototype denture cleaning device had significantly higher plaque removing effects compared with the commercial denture cleaning agent. However, we had not inspected how a long-term use of the prototype device would affect dentures. Thus, in this study, an acrylic resin for a denture plate was repeatedly cleaned with the same method as the actual denture cleaning method, and transmitted intensity of the laser beam irradiated was compared in order to investigate if color change occurred on the sample before and after the cleaning.

E. Hayashi (✉), T. Kanno, and H. Ikai
Division of Fixed Prosthodontics, Department of Restorative Dentistry, Tohoku University Graduate School of Dentistry, 4-1 Seiryo-machi, Aoba-ku, Sendai 980-8575, Japan
e-mail: eisei@mail.tains.tohoku.ac.jp

M. Tada, K. Nakamura, and M. Kohno
New Industry Creation Hatchery Center, Tohoku University, 6-6-10 Aoba, Aramaki, Aoba-ku, Sendai 980-8579, Japan

T. Sasano et al. (eds.), *Interface Oral Health Science 2009*,
DOI 10.1007/978-4-431-99644-6_69, © Springer 2010

One hundred hours of cleaning time in the experiment is converted into 200 days of the cleaning data, assuming that the dentures are cleaned once a day for 30 min before bedtime. There was no grossly-visible change found on the sample resin plates used for the cleaning test. This experiment made use of excellent directivity and convergence of the laser beam, but significant difference was not found in the ment, we inspected the change of the transmitted laser light intensity and did not directly measure the change of the resin surface color. Although change in characteristics and color of the resin surface due to the cleaning is thought to influence absorption and scattering of the laser beam, we can conclude that 100-h cleaning on the acrylic resin for the denture plate with the prototype denture cleaning device does not cause the color change, which affects transmission of the laser beam from the experiment results.

The evaluation of the dental disinfection device with low concentration of H_2O_2 and laser diode

Hiroyo Ikai, Taro Kanno, Keisuke Nakamura, Eisei Hayashi, Akihito Kudo, and Masahiro Kohno

Abstract. The purpose of the present study was to evaluate the experimental dental disinfection device, which is equipped with laser diode (LD) emitting visible radiation of wavelength (405 nm) and would be able to kill dental bacteria by effectively producing the hydroxyl radical by means of irradiation of LD to a low concentration (0.125–1 M) of the hydrogen peroxide (H_2O_2). The fungous suspension of *Candida albicans* (*C. albicans*) was exposed to the LD with an energy dose from 120 J/cm^2 (5 min at 50 mW) to 720 J/cm^2 (5 min at 300 mW). The results of this study indicate that the H_2O_2, even if the concentration is low, can effectively kill *C. albicans* when it is exposed to LD with a wavelength of 405 nm and an energy dose of over 360 J/cm^2.

Key words. disinfection, low concentration of H_2O_2, laser diode, visible light, hydroxyl radical

1 Introduction

It is well known that hydroxyl radical has the highest reactivity among reactive oxygen species or free radicals [1].

The purpose of the present study was to evaluate the ability of the experimental dental disinfection device, which would be able to kill *Candida albicans* (*C. albicans*)

H. Ikai (✉), T. Kanno, K. Nakamura, and E. Hayashi
Division of Fixed Prosthodontics, Department of Restorative Dentistry, Tohoku University Graduate School of Dentistry, 4-1 Seiryo-machi, Aoba-ku, Sendai 980-8575, Japan
e-mail: ikai@mail.tains.tohoku.ac.jp

K. Nakamura and M. Kohno
New Industry Creation Hatchery Center, Tohoku University, 6-6-10 Aoba, Aramaki, Aoba-ku, Sendai 980-8579, Japan

A. Kudo
PAX Ltd, 6-6-3 Minamiyoshinari, Aoba-ku, Sendai 989-3204, Japan

T. Sasano et al. (eds.), *Interface Oral Health Science 2009*,
DOI 10.1007/978-4-431-99644-6_70, © Springer 2010

by effectively producing the hydroxyl radical by means of irradiation of laser diode (LD) to a low concentration of hydrogen peroxide (H_2O_2).

2 Materials and Methods

An experimental dental disinfection device equipped with LD emitting visible light with a wavelength of 405 nm was made. ATP of the fungous suspension was measured by bioluminescent method. The cell numbers of the fungal suspension of *C. albicans* were adjusted to the Relative Light Unit (RLU) of 1×10^5. The suspension was mixed with the same volume of H_2O_2 resulting in a final concentration of 0.125, 0.25, 0.5, and 1 M. Immediately after mixing, the LD was irradiated to the suspension with an energy dose from 120 J/cm^2 (5 min at 50 mW) to 720 J/cm^2 (5 min at 300 mW). The survival fraction of *C. albicans* was evaluated by the RLU. In each condition, three samples were measured.

3 Results

When the power of the LD was used at more than 250 mW with 0.25 M H_2O_2 and more than 150 mW with 0.5 or 1.0 M H_2O_2, the decreasing rate of *C. albicans* was over 99%.

4 Discussion

We have reported that there is a close relationship between the amount of ATP and the number of bacteria [2]. Therefore, we used the bioluminescent method, which could be instantly performed in order to screen many conditions.

The results of this study indicate that H_2O_2, even if the concentration is low; can effectively kill *C. albicans* when it is exposed to the LD with a wavelength of 405 nm and an energy dose of over 360 J/cm^2.

C. albicans is a eukaryotic cell having a nuclear membrane and the larger cell size. Hence, it was very significant that this disinfection system could kill *C. albicans*.

This experimental device was developed to assume the treatment of oral infection such as caries and periodontitis. The target microorganisms of this device will be *Streptococcus mutans*, *Porphyromonas gingivalis,* etc., which might have less resistance to this disinfection system than *C. albicans*.

Therefore, we considered that the results of the present study are so significant to kill the dental bacteria by effectively producing the hydroxyl radical by means of irradiation of LD to a low concentration of the H_2O_2.

References

1. Chapple IL (1997) Reactive oxygen species and antioxidants in inflammatory diseases. J Clin Periodontol 24:287–296
2. Ikai H, Kanno T, Hayashi E et al (2008) New irradiation system for clinical dental treatment advanced using low density hydrogen peroxide and laser diode. J Jpn Prosthodont Soc 117th Special Issue 52:194

Rapid identification of HACEK group bacteria using 16S rRNA gene PCR-RFLP

Minoru Sasaki, Shihoko Tajika, Yoshitoyo Kodama, Yu Shimoyama, and Shigenobu Kimura

Abstract. HACEK bacteria (*Haemophilus* spp., *Aggregatibacter actinomycetemcomitans*, *Cardiobacterium hominis*, *Eikenella corrodens* and *Kingella* spp.), classified as fastidious and slow-growing Gram-negative coccobacilli, inhabit human oral cavity, and can cause infective endocarditis. The identification of the HACEK group of bacteria in blood samples from the infective endocarditis patients is known to be rather difficult and occasionally inconclusive by conventional culture methods, because the biochemical characteristics resemble each other. In this study, we developed a rapid and highly sensitive identification method for the HACEK group bacteria by means of 16S rRNA gene PCR amplification followed by restriction fragment length polymorphism analysis (PCR-RFLP) with *Hin*fI and *Msp*I. The 16S rRNA genes were successfully amplified from all the five representative strains of HACEK bacterial species, and typical restriction patterns of the five bacteria were obtained. The restriction patterns were readily distinguishable from each other and are also different from those of other 15 causative pathogens of infective endocarditis.

Key words. HACEK bacteria, identification, 16S rRNA, PCR-RFLP, endocarditis

Although infective endocarditis due to the HACEK group bacteria is a rare occurrence, the identification of the organisms is still important diagnostically for the specific antimicrobial therapy. The HACEK group bacteria inhabit human oral cavity and are classified as fastidious Gram-negative coccobacilli, requiring increased CO_2 tension for optimal growth. Further, the biochemical characteristics resemble each other. Thus, the identification of the HACEK group of bacteria has been rather difficult and sometimes inconclusive. In this study, we developed a rapid and highly sensitive identification method for the HACEK group bacteria by PCR-RFLP.

M. Sasaki, S. Tajika, Y. Kodama, Y. Shimoyama, and S. Kimura (✉)
Department of Oral Microbiology, Iwate Medical University School of Dentistry,
1-3-27 Chuo-dori, Morioka, Iwate 020-8505, Japan
e-mail: kimuras@iwate-med.ac.jp

T. Sasano et al. (eds.), *Interface Oral Health Science 2009*,
DOI 10.1007/978-4-431-99644-6_71, © Springer 2010

Reference strains (*Haemophilus aphrophilus* ATCC 33389, *Aggregatibacter actinomycetemcomitans* ATCC 33384, *Cardiobacterium hominis* ATCC 16826, *Eikenella corrodens* ATCC 23834, *Kingella kingae* ATCC 23330) and seven clinical isolates from the blood samples of endocarditis patients were used. The DNA samples were extracted as described previously [1]. After PCR amplification using the primers corresponding to *Escherichia coli* 16S rRNA gene, the PCR products were digested with 4 U of either *Hin*fI or *Msp*I at 37°C for 1.5 h. The samples were then separated on 1.8% agarose gel, and the restriction patterns were recorded.

Typical restriction patterns among five species of the HACEK bacteria were observed by combing use of *Hin*fI and *Msp*I digestions. The RFLP patterns of five species of the HACEK bacteria obtained by combing use of *Hin*fI and *Msp*I digestions were readily distinguished from each other and from the other pathogens of infective endocarditis, including viridans streptococci (Table 1). Furthermore, the PCR-RFLP method yielded a definitive identification of *C. hominis* from one of the blood samples of the patients with infective endocarditis in which causative pathogens could not be unidentified by biochemical identification kits. The result was confirmed by a 16S rRNA gene sequence analysis of the isolate.

Table 1. The of DNA fragments of the 16S rRNA gene PCR products cleaved with *Hin*fI and *Msp*I[a]

Strain	DNA fragments cleaved with *Hin*fI (bp)	DNA fragments cleaved with *Msp*I (bp)
H. aphrophilus ATCC 33389^T	**663**[b], **275, 155**, 121[c], 120	**545, 309**, 168, 130, [127][d], 109
A. actinomycetemcomitans ATCC 33384^T	**939, 209, 127**, 120	**546, 315, 211, 169**, [126], 110
C. hominis ATCC 16826^T	**508, 491, 208**, [153], 109	**490, 305, 242, 130**, [124], 110
E. corrodens ATCC 23834^T	**1,000, 337**	**547, 408, 130**, [126], 110
K. kingae ATCC 23330^T	**1,000, 327**	**657, 324, 169**, 130, [124]
S. mutans ATCC 25175^T	976, 245, 173	[562], 303, 288, 122, [113]
S. anginosus NCTC 10713^T	486, 404, 230, [173]	[340], 316, 222, 163, 125, 120
S. intermedius ATCC 27335	891, [344], 172	[563], 317, 163, 127, 125, 120
S. sanguinis ATCC 10556^T	891, [345], [173]	[555], 316, 163, 125, 120
S. gordonii ATCC 10558^T	894, [344], [173]	[564], 317, 211, 163, 125
S. mitis NCTC 3165	891, [344], [172]	[553], 317, 163, 125, 120
S. salivarius ATCC 7073^T	486, 412, 336, 172	561, 316, 211, 163, 125, 113
S. aureus ATCC 35844	978, [337], [172]	608, 388, 211, [156]
S. epidermidis ATCC 14990	978, [337], [172]	608, 388, 211, [156]
S. haemolyticus ATCC 29970^T	978, [337], [172]	608, 388, 211, [156]
A. defectiva ATCC 41976^T	783, 185, 173, [172], 133	606, 564, 163, [111]
G. adiacens ATCC 41975^T	545, 362, 225, [163], 146	537, 407, 164, [116]
G. elegans DSM 11693^T	545, 371, 363, 178	556, 538, 163, 133
P. endodontalis ATCC 35406	509, 336, 292, [173], 164	399, 337, 232, [127]
P. gingivalis ATCC 33277	802, 375, 133	451, 248, 232, 113, 110

[a]The DNA fragments of less than 99 bp are not shown in this table

[b]The boldface number indicates the DNA fragment that was found in the present experiments among the deduced DNA fragments

[c]The lightface number indicates the deduced DNA fragment that was not found in the present experiments

[d]The number in brackets is the deduced fragment containing either uncloned 5′ or 3′ ends

Consequently, the 16S rRNA gene PCR-RFLP analysis developed in this study was a rapid and highly sensitive identification method, and could be applicable for a definitive diagnostic discrimination of the HACEK group bacteria.

Reference

1. Ohara-Nemoto Y, Tajika S, Sasaki M et al (1997) Identification of *Abiotrophia adiacens* and *Abiotrophia defectiva* by 16S rRNA gene PCR and restriction fragment length polymorphism analysis. J Clin Microbiol 35:2458–2463

A novel aspartate-specific dipeptidylpeptidase produced from *Porphyromonas endodontalis*

Shigenobu Kimura, Hiroshi Haraga, Yuko Ohara-Nemoto, Takayuki K. Nemoto, Yu Shimoyama, Sachimi Agato, and Minoru Sasaki

Abstract. *Porphyromonas endodontalis*, a black-pigmented anaerobe, is a predominant pathogen of human periapical periodontitis with acute symptoms. In contrast to *P. gingivalis*, a major pathogen of chronic periodontitis, the pathogenic factors of *P. endodontalis* have been poorly characterized. In this study, the proteases as the pathogenic factors of *P. endodontalis* were investigated. The hydrolyzing profile toward various synthetic MCA peptides indicated that the fraction had a spectrum of proteolytic activities distinct from that of *P. gingivalis*. The chromatographic analysis revealed that aspartate- and alanine-specific proteolytic activities were substantially present in the fraction. The MALDI-TOF MS analyses of the chromatographically purified aspartate-specific protease fraction on the substrate preference using neuromedin B and the analogs indicated that *P. endodontalis* possesses a novel aspartate-specific dipeptidylpeptidase in the extracellular fraction, which could be involved in an important etiologic process in *P. endodontalis*-induced human periapical periodontitis.

Key words. *Porphyromonas endodontalis*, Asp-specific dipeptidylpeptidase

Porphyromonas endodontalis is an important pathogenic organism in human periapical periodontitis with acute symptoms. Like *P. gingivalis*, a major pathogen of chronic periodontitis, *P. endodontalis* is asaccharolytic and forms black-pigmented colonies on enriched blood agar plates. However, pathogenic factors of this microbe have been poorly characterized. In this study, the proteolytic activities of *P. endodontalis* were examined in comparison to those of *P. gingivalis* since the proteases are recognized as major virulent factors especially in these asaccharolytic organisms.

P. endodontalis ATCC 35406 and *P. gingivalis* ATCC 33277 were cultured on ABCM agar plates seated with sterilized dialysis membrane as described previously

S. Kimura (✉), H. Haraga, Y. Shimoyama, S. Agato, and M. Sasaki,
Department of Oral Microbiology, Iwate Medical University School of Dentistry,
1-3-27 Chuo-dori, Morioka, Iwate 020-8505, Japan
e-mail: kimuras@iwate-med.ac.jp

Y. Ohara-Nemoto and T.K. Nemoto
Department of Oral Molecular Biology, Nagasaki University Graduate School of Biomedical Sciences, 1-7-1 Sakamoto, Nagasaki 852-8588, Japan

T. Sasano et al. (eds.), *Interface Oral Health Science 2009*,
DOI 10.1007/978-4-431-99644-6_72, © Springer 2010

Table 1. The degradation of the synthetic peptides by the DEAE-purified fraction

Compound	Sequence	Degradation[a]
neuromedin B	Gly-Asn-Leu-Trp-Ala-Thr-Gly-His-Phe-Met-NH_2	5%
Asp_1-neuromedin B	Asp-Gly-Leu-Trp-Ala-Thr-Gly-His-Phe-Met-NH_2	<5%
Asp_2-neuromedin B	Gly-Asp-Leu-Trp-Ala-Thr-Gly-His-Phe-Met-NH_2	100%
Asp_3-neuromedin B	Gly-Ser-Asp-Trp-Ala-Thr-Gly-His-Phe-Met-NH_2	<5%
Asp_5-neuromedin B	Gly-Ser-Leu-Trp-Asp-Thr-Gly-His-Phe-Met-NH_2	<5%

[a]The percent degradation was calculated from the percent increase of the desired dipeptidyl fragments in the mass spectrums

[1]. After cultivation, the bacterial cells were suspended in sterilized ice-cold PBS (pH 7.4) and centrifuged (6,300×*g*, 15 min). The obtained supernatant was filtrated and then used as the extracellular fraction to measure the proteolytic activity. The proteolytic activities were examined by the hydrolyzing profile toward various synthetic MCA peptides. To purify the specific protease(s) of *P. endodontalis*, the extracellular fraction was subjected to ion-exchange chromatography and the aspartate-specific activity was obtained (designated as the DEAE-purified aspartate-specific protease fraction). The MALDI-TOF MS analyses of the fraction on the substrate preference were performed using neuromedin B and a series of synthetic neuromedin B analogues (Table 1).

The results on the protease activities in the *P. endodontalis* extracellular fraction indicated that the extracellular fraction had a spectrum of proteolytic activities distinct from that of *P. gingivalis*, and that aspartate- and alanine-specific proteolytic activities were substantially present in the extracellular fraction. Then, the extracellular fraction was subjected to ion-exchange chromatography, and the aspartate-specific proteolytic activity was eluted from a DEAE-Sephacel column separately from the alanine-specific one. The MALDI-TOF MS analyses of the chromatographically purified aspartate-specific protease fraction clearly indicated that *P. endodontalis* possesses a novel aspartate-specific dipeptidylpeptidase in the extracellular fraction (Table 1).

Reference

1. Ohara-Nemoto Y, Ikeda Y, Kobayashi M, Sasaki M, Tajika S, Kimura S (2002) Characterization and molecular cloning of glutamyl endopeptidase from *Staphylococcus epidermidis*. Microb Pathog 33:33–41

Short-term effect of single NaF-mouthrinse on glucose-induced pH fall in dental plaque

Kazuko Nakajo, Tomofumi Asanoumi, Akinobu Shibata, Yoko Yagishita, Kazuo Kato, and Nobuhiro Takahashi

Abstract. This study aimed to evaluate the short-term effect of a single fluoride-mouthrinse containing different concentrations of sodium fluoride (NaF) on glucose-induced pH fall and fluoride retention within dental plaque. The NaF-mouthrinse inhibited decreases in plaque-pH in short-term, probably due to fluoride retention within plaque. Thus, it is expected that plaque acid production by bacterial sugar metabolism may be inhibited by NaF-mouthrinse in vivo.

Key words. dental plaque, fluoride, pH fall

1 Introduction

Sodium fluoride (NaF)-mouthrinse is widely used as a beneficial agent for preventing dental caries. The major effects of fluoride on preventing dental caries are the enhancement of acid resistance of enamel and the promotion of remineralization of enamel. Dental plaque contains oral bacteria that metabolize dietary fermentable carbohydrates to produce acidic end products, such as lactic, formic, and acetic acids. These acids can diffuse and demineralize the tooth surface. It is known that fluoride, in concentrations as low as 8-ppm, can inhibit acid production by oral streptococci at acidic pH in vitro [1]. Therefore, the inhibition of bacterial acid production by fluoride should also be an important approach for prevention of dental caries. However, in the previous study, NaF-mouthrinse (225-ppm F) could not inhibit acid production by dental plaque sampled more than 30-min after rinsing in vivo [2]. The discrepancy between results obtained in vivo and in vitro suggests that NaF-mouthrinse has a potential for inhibition of acid production by the plaque in short-term. Thus, this study aimed to evaluate the short-term effect of a single fluoride-mouthrinse

K. Nakajo(✉), T. Asanoumi, A. Shibata, Y. Yagishita, and N. Takahashi
Division of Oral Ecology and Biochemistry, Tohoku University Graduate School of Dentistry, 4-1 Seiryo-machi, Aoba-ku, Sendai, Japan
e-mail: nakajo@mail.tains.tohoku.ac.jp

K. Kato
Department of Preventive Dentistry and Dental Public Health, School of Dentistry, Aichi-Gakuin University, 1-100 Kusumoto-cho, Chikusa-ku, Nagoya, Japan

T. Sasano et al. (eds.), *Interface Oral Health Science 2009*,
DOI 10.1007/978-4-431-99644-6_73, © Springer 2010

containing different concentrations of NaF on glucose-induced pH fall of dental plaque and fluoride retention within dental plaque.

2 NaF-Mouthrinse Inhibits Plaque Acid Production

Volunteers were asked to accumulate plaque for 24-h and rinse with 10-mL NaF solution for 1-min. The plaque pH collected before NaF-mouthrinse continued to fall to 4.3 (control), while those at 15-min after the rinse with 250-, 500-, and 900-ppm ceased to fall at 5.0. Similar results were obtained for the plaque collected 30-min after the rinse except for the pH value with 250-ppm rinse, which returned to the control level.

3 Fluoride Retained within Dental Plaque Can Inhibit Plaque Acid Production in Acidic Environment

NaF-mouthrinse can inhibit short-term plaque acid-production, probably due to fluoride retention within plaque, since amounts of fluoride detected from dental plaque after NaF-mouthrinse were higher than those detected from the control. The termination of pH decrease around pH 5 suggests that plaque-bound fluorides efficiently suppress bacterial sugar metabolism at acidic pH. Fluoride is known to penetrate the cell membrane of plaque bacteria in acidic environment [3] and subsequently inhibit the glycolytic enzymes including enolase [4]. It is expected that plaque acid production by bacterial sugar metabolism may be inhibited by NaF-mouthrinse in vivo.

References

1. Maehara H, Iwami Y, Mayanagi H et al (2005) Synergistic inhibition by combination of fluoride and xylitol on glycolysis by mutans streptococci and its biochemical mechanism. Caries Res 39:521–528
2. Giertsen E, Emberland H, Scheie AA (1999) Effects of mouth rinses with xylitol and fluoride on dental plaque and saliva. Caries Res 33:23–31
3. Gutknecht J, Walter A (1981) Hydrofluoric and nitric acid transport through lipid bilayer membranes. Biochim Biophys Acta 644:153–156
4. Curran TM, Buckley DH, Marquis RE (1994) Quasi-irreversible inhibition of enolase of *Streptococcus mutans* by fluoride. FEMS Microbiol Lett 119:283–288

Short-time effect of fluoride on acid production by *Streptococcus mutans*

Hitomi Domon, Kazuko Nakajo, Jumpei Washio, Harumi Miyasawa-Hori, Satoshi Fukumoto, and Nobuhiro Takahashi

Abstract. This study aimed to evaluate the effects of a short-time exposure of fluoride (250, 450, and 950 ppm) on glucose-induced pH fall of *Streptococcus mutans* and fluoride retention of *S. mutans* cells. Short-time fluoride exposure inhibited the pH fall by *S. mutans* probably due to fluoride binding to cell components of *S. mutans*. Thus, it is expected that a short-time fluoride rinse can reduce the risk of caries in terms of inhibitory effect on acid production by plaque bacteria in vivo.

Key words. fluoride, short-term effect, pH fall, *Streptococcus mutans*

1 Introduction

Daily mouth-rinse with a low concentration of fluoride is effective for caries prevention by stimulating remineralization of tooth surface. Recently, oral care products containing a high-concentration of fluoride became available. It is known that the acid production by oral bacteria is inhibited in the presence of fluoride [1]. Besides, this inhibitory effect has been previously estimated by concomitant administration of bacterial cell and fluoride, but it is not realistic that fluoride is retained in the oral environment, because oral cavity is constantly flushed by saliva. Nevertheless, it was recently reported that fluoride mouth-rinse could inhibit acid production by dental plaque for a short time [2] suggesting that fluoride may bind to dental plaque for a short time and inhibit bacterial acid production. Therefore, we aimed to examine effects of a short-time exposure of fluoride on one of the most acidogenic oral bacteria, *Streptococcus mutans*, in an environment similar to the oral cavity in vitro.

H. Domon and S. Fukumoto
Division of Pediatric Dentistry, Tohoku University Graduate School of Dentistry, 4-1 Seiryo-machi, Aoba-ku, Sendai, Japan

H. Domon, K. Nakajo, J. Washio, H. Miyasawa-Hori, and N. Takahashi (✉)
Division of Oral Ecology and Biochemistry, Tohoku University Graduate School of Dentistry, 4-1 Seiryo-machi, Aoba-ku, Sendai, Japan
e-mail: nobu-t@mail.tains.tohoku.ac.jp

T. Sasano et al. (eds.), *Interface Oral Health Science 2009*,
DOI 10.1007/978-4-431-99644-6_74, © Springer 2010

2 Short-Time Fluoride Exposure Inhibits pH Fall by *Streptococcus mutans*

The pH falls by glucose fermentation of *S. mutans* without fluoride exposure reached 3.5 (control), while the pH falls ceased completely 4.0–5.0 by *S. mutans* which had been exposed to potassium fluoride at a concentration of 250, 450 and 950 ppm for 10 min and washed twice with 2 mM potassium phosphate buffer (pH 7.0). These findings indicate that a short-time fluoride exposure can inhibit acid production by *S. mutans* probably due to fluoride binding to cell components of *S. mutans*. It was confirmed that amounts of fluoride detected from *S. mutans* cells that were exposed to fluoride and washed were higher than those detected from the control cells. Fluoride uptake by oral streptococci after a fluoride exposure for 30 min was reported previously [3].

3 Conclusion

The present study demonstrated that amounts of fluoride retained to streptococcal cells during a short-time fluoride exposure were sufficient to inhibit acid production by oral streptococci. The short-time use of fluorides, such as mouth-rinse, may prevent dental caries through inhibition of acid production in dental plaque in vivo.

References

1. Curran TM, Buckley DH, Marquis RE (1994) Quasi-irreversible inhibition of enolase of *Streptococcus mutans* by fluoride. FEMS Microbiol Lett 119:283–288
2. Nakajo K, Asanoumi T, Shibata A et al. (2009) Short-term effect of single NaF-mouthrinse on glucose-induced pH fall in dental plaque. In: T. Sasano et al. (eds.) Interface oral health science 2009. Springer, New York (See the chapter by K. Nakajo, this volume)
3. Kashket S, Rodriguez VM (1976) Fluoride accumulation by a strain of human oral *Streptococcus sanguis*. Arch Oral Biol 21:459–464

Real-time PCR analysis of cariogenic bacteria in supragingival plaque biofilm microflora on caries lesions of children

Junko Matsuyama, Takuichi Sato, Yuki Abiko, Ayako Hasegawa, Kazuo Kato, and Etsuro Hoshino

Abstract. The aim of this study was to quantify cariogenic bacteria, *Streptococcus mutans*, in supragingival plaque biofilm microflora on caries lesions of children. After informed consent was obtained from each subject, supragingival plaque on caries lesion was obtained from 10 children (3–10 years). Healthy supragingival plaque was also obtained from the same subject, as a control. Total bacteria and *S. mutans* were quantified by real-time PCR using universal and *S. mutans*-specific primers based on 16S ribosomal RNA genes, respectively, and the proportion of *S. mutans* in each sample was calculated. Total bacterial DNA levels in carious plaque and healthy plaque were found to be similar, though the proportion of *S. mutans* in carious plaque was higher than that in healthy plaque. The findings of this study suggest that amount of cariogenic bacteria in supragingival plaque biofilm may be associated with the status of dental caries.

Key words. children, dental caries, mutans streptococci, plaque biofilm, quantitative polymerase chain reaction

J. Matsuyama (✉)
Division of Pediatric Dentistry, Niigata University Graduate School of Medical and Dental Sciences, Niigata 951-8514, Japan
e-mail: junko@dent.niigata-u.ac.jp

T. Sato, Y. Abiko, and A. Hasegawa
Division of Oral Ecology and Biochemistry, Tohoku University Graduate School of Dentistry, Sendai 980-8575, Japan

K. Kato
Department of Preventive Dentistry and Dental Public Health, School of Dentistry, Aichi-Gakuin University, Nagoya, Japan

E. Hoshino
Division of Oral Ecology in Health and Infection, Niigata University Graduate School of Medical and Dental Sciences, Niigata 951-8514, Japan

T. Sasano et al. (eds.), *Interface Oral Health Science 2009*,
DOI 10.1007/978-4-431-99644-6_75,

1 Introduction

The plaque biofilm microflora is considered healthy when it is composed of indigenous bacteria. Numerous changes in the plaque biofilm environment, including alterations in pH and nutrition, may lead to an accumulation of cariogenic bacteria on the surface of teeth, resulting in the initiation of dental caries. The aim of this study was to quantify cariogenic bacteria, *Streptococcus mutans*, in supragingival plaque biofilm microflora on caries lesions of children.

2 Relationship Between the Status of Dental Caries and *S. mutans* in Supragingival Plaque

After informed consent was obtained from each subject, supragingival plaque on caries lesion was obtained from 10 children (3–10 years). Healthy supragingival plaque was also obtained from the same subject as a control. Genomic DNA was extracted from the plaque biofilm with the QIAamp DNA Stool Mini Kit (Qiagen) according to the manufacturer's instructions. Total bacteria and the target species, *S. mutans*, were quantified by real-time PCR using universal and species-specific primers, respectively [1–3], and its proportion was calculated. Total bacterial DNA levels in carious plaque and healthy plaque were found to be similar. While, the mean proportion of *S. mutans* in carious plaque was higher than that in healthy plaque. The samples with deep dental caries had large amounts of *S. mutans* in carious plaque, but those with shallow dental caries did not. The proportion of *S. mutans* in supragingival plaque appears to be associated with the status of human dental caries. The findings of the present study suggest that quantification of cariogenic bacteria, including *S. mutans* in supragingival plaque biofilm, may be an appropriate tool for the diagnosis of dental caries of children.

Acknowledgments This study was supported in part by Grants-in-Aid for Scientific Research (20592220) from the JSPS and Niigata University Young Researchers Encouragement Program (to JM) in 2008 and 2009.

References

1. Matsuyama J, Sato T, Takahashi N et al (2007) Real-time PCR analysis of genera *Veillonella* and *Streptococcus* in healthy supragingival plaque biofilm microflora of children. In: Watanabe M, Okuno O (eds) Interface oral health science 2007. Springer, Tokyo, pp 255–256
2. Matsuyama J, Sato T, Washio J et al (2005) PCR for detection of mutans streptococci in human dental plaque. Int Congr Ser 1284:158–162
3. Sato T, Matsuyama J, Mayanagi G et al (2007) Nested PCR for the sensitive detection of cariogenic bacteria. Cariol Today 3–4:17–20

Involvement of cough reflex impairment and silent aspiration of oral bacteria in postoperative pneumonia: a model of aspiration pneumonia

Takuichi Sato, Yasushi Hoshikawa, Takashi Kondo, Kazuhiro Hashimoto, Yuki Abiko, Ayako Hasegawa, Junko Matsuyama, and Nobuhiro Takahashi

Abstract. Postoperative pneumonia (PP) occurs in elderly subjects whose cough reflex (CR) is impaired postoperatively. This finding indicates that silent aspiration of oral bacteria may be involved in PP. This study aimed to clarify the involvement of CR impairment and silent aspiration of oral bacteria in PP. Intraoperative bronchial aspirates (BA) from 22 subjects aged over 60 years undergoing lung resection were cultured anaerobically on blood agar plates. The CR of all subjects was measured on 1st postoperative day. Eleven subjects showed impaired CR, whereas 11 showed normal. The bacterial amounts in intraoperative BA of subjects with postoperatively impaired CR were found to be higher than those with normal CR. PP occurred in 2 out of 11 subjects with impaired CR but not in 11 subjects with normal CR. These results suggest that impairment of the CR and silent aspiration of oral bacteria may be associated with the development of PP in elderly subjects.

Key words. bronchial aspirates, cough reflex, microflora, oral bacteria, postoperative pneumonia

---

T. Sato (✉), Y. Abiko, A. Hasegawa, and N. Takahashi
Division of Oral Ecology and Biochemistry, Tohoku University Graduate School of Dentistry, Sendai 980-8575, Japan
e-mail: tak@mail.tains.tohoku.ac.jp

Y. Hoshikawa and T. Kondo
Department of Thoracic Surgery, Institute of Development, Aging, and Cancer, Tohoku University, Sendai 980-8575, Japan

K. Hashimoto
Division of Periodontology and Endodontology, Tohoku University Graduate School of Dentistry, Sendai 980-8575, Japan

J. Matsuyama
Division of Pediatric Dentistry, Niigata University Graduate School of Medical and Dental Sciences, Niigata 951-8514, Japan

T. Sasano et al. (eds.), *Interface Oral Health Science 2009*,
DOI 10.1007/978-4-431-99644-6_76, © Springer 2010

1 Introduction

Postoperative pneumonia (PP) occurs in elderly subjects whose cough reflex (CR) is impaired postoperatively. This finding indicates that silent aspiration of oral bacteria may be involved in PP. The aim of the present study was to clarify the involvement of CR impairment and silent aspiration of oral bacteria in PP.

2 Relationship Between Involvement of Cough Reflex Impairment and Silent Aspiration of Oral Bacteria

After informed consent was obtained from each subject, intraoperative bronchial aspirates (BA) from 22 subjects aged over 60 years undergoing lung resection were sampled and cultured anaerobically [1] on CDC blood agar plates. CR of all subjects was measured on first postoperative day. Eleven subjects showed impaired CR, whereas 11 showed normal CR. The bacterial amounts in intraoperative BA of subjects with postoperatively impaired CR were found to be higher than those with normal CR. PP occurred in 2 out of 11 subjects with impaired CR but not in 11 subjects with normal CR. The findings of the present study suggest that impairment of the CR and silent aspiration of oral bacteria may be associated with the development of PP in elderly subjects.

Acknowledgments This study was supported in part by Grants-in-Aid for Scientific Research (20591661 to YH; 20592220 to TS) from the Japan Society for the Promotion of Science.

Reference

1. Sato R, Sato T, Takahashi I et al (2007) Profiling of bacterial flora in crevices around titanium orthodontic anchor plates. Clin Oral Implants Res 18:21–26

The production of secretory leukocyte protease inhibitor from gingival epithelial cells in response to *Porphyromonas gingivalis* lipopolysaccharides

Taichi Ishikawa, Yuko Ohara-Nemoto, Shihoko Tajika, Minoru Sasaki, and Shigenobu Kimura

Abstract. Secretory leukocyte protease inhibitor (SLPI) has been recognized as not only a protease inhibitor but also an important defense component in mucosal secretory fluids. To elucidate the functional role in innate immunity in gingival crevices, the SLPI production from a gingival epithelial cell line, GE1, with and without a stimulation of *Porphyromonas gingivalis* lipopolysaccharides (Pg-LPS), and the inhibitory effect on *P. gingivalis* proteases were investigated. The unstimulated GE1 cells showed low, but significant, levels of SLPI mRNA expression, which increased after stimulation with Pg-LPS. The upregulation of SLPI mRNA expression in GE1 cells was accompanied by the inductions of IL-6, TNF-α, and IL-1β mRNA expressions. Further experiments using rSLPI indicated that SLPI showed a direct inhibitory effect on the *P. gingivalis* protease of Lys-gingipain. Thus, it was suggested that gingival epithelial cells could be a substantial producer of SLPI that functions inhibitory to the pathogenic *P. gingivalis* protease in gingival crevices.

Key words. *Porphyromonas gingivalis*, secretory leukocyte protease inhibitor, gingival epithelial cells, lipopolysaccharides

Secretory leukocyte protease inhibitor (SLPI) is reported to participate in the mucosal defense by reducing inflammation, however, the protective roles as well as the mechanism in oral cavity have not been elucidated. In this study using GE1 cells, the potential role of gingival epithelial cells as an SLPI producer in response

T. Ishikawa, S. Tajika, M. Sasaki, and S. Kimura (✉)
Department of Oral Microbiology, Iwate Medical University School of Dentistry, 1-3-27 Chuo-dori, Morioka, Iwate 020-8505, Japan
e-mail: kimuras@iwate-med.ac.jp

Y. Ohara-Nemoto
Department of Oral Molecular Biology, Nagasaki University Graduate School of Biomedical Sciences, 1-7-1 Sakamoto, Nagasaki 852-8588, Japan

DOI 10.1007/978-4-431-99644-6_77, © Springer 2010

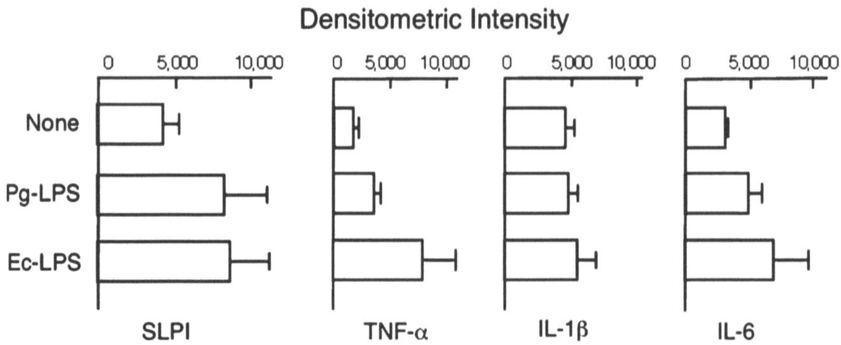

Fig. 1. The mRNA levels of SLPI and inflammatory cytokines in GE1 cells. After 6 h-incubation with Pg-LPS or Ec-LPS, total RNA was purified and mRNA levels were measured by RT-PCR

to Pg-LPS and the inhibitory effect of SLPI on *P. gingivalis* proteases were investigated.

The Pg-LPS and *Escherichia coli* LPS (Ec-LPS) were prepared as described previously [1]. The expressions of SLPI and inflammatory cytokine mRNAs in GE1 cells were examined by RT-PCR. The inhibitory effect of SLPI on the *P. gingivalis* proteases was assessed by the reduction in hydrolyzing activities in the *P. gingivalis* extracellular fraction toward synthetic MCA peptides (Bz-Arg-MCA for Arg-gingipain and Z-His-Glu-Lys-MCA for Lys-gingipain).

The unstimulated GE1 cells showed low, but significant, levels of SLPI mRNA expression, which increased after stimulation with Pg-LPS (10 μg/ml) as well as Ec-LPS (1 μg/ml) for 6 h (Fig. 1). The Pg-LPS also induced proliferative responses and the mRNA expression of inflammatory cytokines (TNF-α, IL-1β, and IL-6) in GE1 cells (Fig. 1). The kinetics analysis suggested that the upregulation of SLPI production in GE1 cells could not be a second response to the inflammatory cytokines induced by the stimulant, but a direct response to Pg-LPS. The experiments using rSLPI indicated that SLPI inhibited the Lys-gingipain activity of the *P. gingivalis* extracellular fraction, but not on the Arg-gingipain protease activity. These results suggest that gingival epithelial cells could be a substantial producer of SLPI that functions inhibitory to the pathogenic *P. gingivalis* protease in gingival crevices. It was also suggested that the SLPI production could increase in response to *P. gingivalis* through the stimulation with its pathogenic constituents.

Reference

1. Kimura S, Tamamura T, Nakagawa I et al (2000) CD14-dependent and independent pathways in lipopolysaccharide-induced activation of a murine B-cell line, CH12.LX. Scand J Immunol 51:392–399

Analysis of antigen incorporating and processing cells in sublingal immunotherapy

Daisuke Shiraishi, Yasuhiro Nagai, Yasuo Endo, Hidetoshi Shimauchi, and Shunji Sugawara

Abstract. For analysis of allergen incorporation mechanism, OVA solution was applied under tongue of the mice, and immunostaining was performed on the sublingual tissues using anti-OVA antibody as primary antibody in immunohistology. We found that the OVA antigen localized around the sublingual lamina propria.

To establish of sublingual immunotherapy for anaphylactic shock, sensitized mice that onset anaphylactic shock were vaccinated by sublingual route. Anaphylactic symptom tended to improve by sublingual vaccination.

Key words. sublingual immunotherapy, oral tolerance, anaphylactic shock

Antigen-specific sublingual immunotherapy (SLIT) is safe and efficient in treating type I allergies like pollen hypersensitivity in both adults and children in Europe.

Mechanism of SLIT is reported that induction of oral tolerance and antigen specific IgG or secretory IgA protect against such allergies. Recent study suggests that the antigen administered in sublingual compartment is absorbed through the sublingual mucosa and taken up by local dendritic cells or oral langerhans cells, and these cells with unique tolerogenic properties are present in the sublingual mucosa.

In this study, we analyzed the mechanism of the allergen incorporation treated in sublingual compartment and what kind of cells is involved in these phenomenon by immunohistology. Next, we tried to establish the sublingual tolerogenic of Balb/c mice, which onset anaphylactic shock.

D. Shiraishi (✉) and H. Shimauchi
Department of Periodontology and Endodontology, School of Dentistry, Tohoku University, Sendai 980-8575, Japan
e-mail: D-sk1173@m.tains.tohoku.ac.jp

D. Shiraishi, Y. Nagai, Y. Endo, and S. Sugawara
Department of Oral Immunology, Graduate School of Dentistry, Tohoku University, Sendai 980-8575, Japan

T. Sasano et al. (eds.), *Interface Oral Health Science 2009*,
DOI 10.1007/978-4-431-99644-6_78, © Springer 2010

OVA antigen located at the submucosal section after 0.25 and 0.5 hour, then gradually disappeared after 1 hour. It suggests that the antigen was processed by some cells in the sublingual tissue.

In 0.5 hour, some amount of OVA localized at the sublingual gland. In 1 and 2 hours, OVA was not detected. It suggests that the sublingual grand has a role in OVA processing in sublingual compartment.

Anaphylactic symptom tended to improve by sublingual vaccination. Further study is needed to establish SLIT for anaphylactic shock.

Acoustic mineral density measurement to evaluate clinical demineralized lesions

Jun Suzuki, Yudai Yamada, Sadao Omata, Emi Ito, Katsuhiko Taura, and Takeyoshi Koseki

Abstract. The aim of this study is to measure the acoustic characteristics of human extracted teeth with sound or dentine caries lesion by using an ultrasonic device and to compare the readouts to the diagnosis based on the DIAGNOdent (KaVo institution, Germany) and dyed with Caries-Check (Nishika institution, Tokyo).

Key words. dental caries, ultrasonic device, nondestructive measurement

The Acoustic Mineral Density (AMD) index indicates 1,372±98 in the sound dentine area, while in the dyed area, the average AMD index is 947±169. The average AMD index of the dentine caries area where DIAGNOdent indicates 30 or more is 674±291. The average AMD index of the sound dentine area where the score of DIAGNOdent is under 30, is 1,368±326. Our data clearly shows the significant difference of the AMD index between the sound dentine and the demineralised dentine.

The method is convenient and is applicable for assessing the process of the demineralization and remineralization of the tooth surface.

1 Objective

The aim of this study is to measure the acoustic characteristics in sound or dentine caries lesion of human extracted teeth by using an ultrasonic device and to compare the readouts to the diagnosis based on the prevalent methods in clinics.

J. Suzuki, Y. Yamada, S. Omata, E. Ito, K. Taura, and T. Koseki (✉)
Division of Preventive Dentistry, Tohoku University Graduate School of Dentistry, Sendai, Japan
e-mail: tkoseki@mail.tains.tohoku.ac.jp

T. Sasano et al. (eds.), *Interface Oral Health Science 2009*,
DOI 10.1007/978-4-431-99644-6_79, © Springer 2010

2 Methods

Human extracted teeth with dentine caries were embedded into acrylic polymer, and cut to expose the center of the demineralized dentine lesions by using a water-cooled saw. The specimens were scanned – the autofluorescence with DIAGNOdent (KaVo institution, Germany) and acoustic characteristics of the tooth surface with ultrasonic mineral density analyzer Ha-Ichigou (Tauza Research Institute, Koriyama). After the measurement, the specimens were dyed with Caries-Check (Nishika institution, Tokyo) to visualize the decayed lesions. The data were analyzed statistically with Dunnett's multiple comparison tests.

3 Results and Discussions

AMD index in sound dentine was 1,372 ± 98. The average AMD index in dyed area was 947 ± 169. The average AMD index of the dentine area where the score of DIAGNOdent was more than 30 was 674 ± 291, while the average AMD index in the dentine area where DAGNOdent indicated fewer than 30 was 1,368 ± 326.

There are multiple diagnosis methods used in clinics such as ocular inspection, palpation on root surface, dying of dental caries, and X-ray diagnosis, however, these methods are subjective depending upon the examiner's skills. Optical method is a useful tool to detect the predental caries legion before the beginning of real defect on dental surface. In QLF method, the reflection of fluorescence light releasing from sound tooth is diminished by the opaque lesion of primary dental caries. The QLF method can detect the change of early caries lesions, however, the image is just a shadow, not the exact mineral loss of lesions. Furthermore, the QLF method is unstable to evaluate root caries. Although there are multiple methods for inspecting the hardness of dentin, such as Micro-hardness test in vitro, all of them damage the test tooth. So those methods are unsuitable for the clinical diagnosis method. This present study proposes a new diagnosis method, AMD analyser, which keeps teeth intact. The score of the AMD analysis in dentin caries is related to the value of micro hardness. Thus our study suggests that the AMD analysis is useful for diagnosis of root caries in daily clinics.

AMD index indicated the significant difference between the sound dentine and the demineralised dentin lesions more clearly than DIAGNOdent. These results suggest that our method is more sensitive when compared with DIAGNOdent and is useful for measuring not only enamel caries but also dentin caries. The advantage of this AMD analyzer used in this study is noninvasive, nondestructive, pain-free, instant and exact measurement of the mineral density of the tooth surfaces. According to the results, the AMD analyzer showed the possibility to be a powerful clinical tool to diagnose early caries lesions of dentine and evaluate the process of the demineralization and remineralization of the tooth surfaces.

Session III
Biometrial Interface

Experimental Ti–Ag alloys inhibit biofilm formation

Masatoshi Takahashi, Kazuko Nakajo, Nobuhiro Takahashi, Keiichi Sasaki, and Osamu Okuno

Abstract. We prepared experimental Ti–Ag alloys and investigated their properties. The strength and hardness of the Ti–Ag alloys increased with the concentration of Ag. Moreover, the alloys exhibited sufficient elongation, making them suitable for dental applications. The machinability of the Ti–Ag alloys was superior to that of pure titanium. By carrying out the anode polarization test and immersion test, we found that the corrosion resistance of the Ti–Ag alloys was comparable to that of pure titanium. Further, we performed a biofilm formation test and found that the amount of biofilm formed on the experimental Ti–Ag alloys was less than that on pure titanium, pure silver, and a dental alloy. It was concluded that the experimental Ti–Ag alloys are new types of biomaterials that have an inhibitory effect on biofilm formation as well as excellent mechanical properties and outstanding machinability.

Key words. Ti–Ag alloy, dental alloy, biofilm, biomaterial

1 Introduction

As a part of our efforts to develop a new dental titanium alloy with enhanced machinability and mechanical properties, we prepared experimental Ti–Ag alloys and investigated their properties [1–3]. The strength and hardness values of Ti–Ag

M. Takahashi (✉) and O. Okuno
Division of Dental Biomaterials, Tohoku University Graduate School of Dentistry, 4-1 Seiryo-machi, Aoba-ku, Sendai 980-8575, Japan
e-mail: m-tak@mail.tains.tohoku.ac.jp

K. Nakajo and N. Takahashi
Division of Oral Ecology and Biochemistry, Tohoku University Graduate School of Dentistry, 4-1 Seiryo-machi, Aoba-ku, Sendai 980-8575, Japan

K. Sasaki
Division of Advanced Prosthetic Dentistry, Tohoku University Graduate School of Dentistry, 4-1 Seiryo-machi, Aoba-ku, Sendai 980-8575, Japan

T. Sasano et al. (eds.), *Interface Oral Health Science 2009*,
DOI 10.1007/978-4-431-99644-6_80, © Springer 2010

alloys with 5–20 mass% Ag were found to increase with the concentration of Ag. Moreover, the alloys exhibited sufficient elongation, making them suitable for dental applications [1]. The machinability of Ti–Ag alloys with 20 mass% Ag was superior to that of pure titanium [1]. By carrying out the anode polarization test and immersion test, we found that the corrosion resistance of Ti–Ag alloys with up to 25 mass% Ag was comparable to that of pure titanium [2, 3]. From these results, the experimental Ti–Ag alloys, especially those with ≥20 mass% Ag, are found to be suitable for use in dental prostheses and implants.

2 Biofilm Inhibition: An Additional Characteristic of Ti–Ag Alloys

Microorganisms comprising normal oral microfloras are not necessarily harmful to humans. However, biofilms formed by microorganisms on oral surfaces cause oral diseases. Therefore, the inhibition of biofilm formation is important for the prevention of oral diseases and improvement of oral health. Titanium is used in dentures and dental implants. If titanium is alloyed with silver, which is known to exhibit antibacterial activity, the obtained alloy may inhibit biofilm formation.

The polished plates of experimental Ti–Ag alloys (with 20 and 25 mass% Ag), pure titanium, pure silver, and a dental alloy (Castwell MC, GC, Tokyo, Japan) were ultrasonically cleaned and then sterilized in an autoclave. *Streptococcus mutans* ATCC 31989 was cultured anaerobically in a complex culture medium containing 0.5% sucrose in the presence of each plate. After 12 h, the bacteria accumulated on the plates were removed by gently vibrating the plates in water. Then, the biofilm formed on the plates was observed visually.

Bacterial adhesion, which consisted of biofilms and bacterial accumulations, was observed on all the plates after cultivation, and the amount of bacterial adhesion did not differ among specimens. However, after removing the accumulations, the amount of biofilm formed on the Ti–Ag alloys appeared to be less than that on the other metals and alloys. The amount of biofilm formed on pure titanium, pure silver, and the dental alloy was almost the same.

3 Conclusion

We have developed a new type of biomaterial – experimental Ti–Ag alloys – which exhibit inhibitory effect on biofilm formation in addition to excellent mechanical properties and machinability.

References

1. Takahashi M, Kikuchi M, Takada Y et al (2005) Grindability and mechanical properties of experimental Ti-Au, Ti-Ag and Ti-Cu alloys. In: Watanabe M, Takahashi N, Takada H (eds) Interface oral health science. Elsevier, Amsterdam, pp 326–327
2. Takahashi M, Kikuchi M, Takada Y et al (2006) Electrochemical behavior of cast Ti-Ag alloys. Dent Mater J 25:516–523
3. Takahashi M, Takada Y, Kikuchi M et al (2007) Released ions and microstructures of dental cast experimental Ti-Ag alloys. In: Watanabe M, Okuno O (eds) Interface oral health science 2007. Springer, New York, pp 311–316

Apatite formation from octacalcium phosphate with fluoride

Yukari Shiwaku, Yoshitomo Honda, Takahisa Anada, Keiichi Sasaki, and Osamu Suzuki

Abstract. This chapter focuses on the interaction between octacalcium phosphate (OCP) and fluoride. Synthetic OCP was incubated in 150 mM Tris buffer in the presence of fluoride at neutral pH and 37°C. Then, the effects of the fluoride ion concentrations on the process of OCP–HA conversion were investigated. The analyses showed that OCP–HA conversion advanced completely in a short time in the presence of fluoride up to 200 ppm. Further study is necessary to search for the optimum conditions so that the intermediates in the conversion of OCP to HA can be obtained.

Key words. octacalcium phosphate, fluoride, hydrolysis

1 Introduction

Octacalcium phosphate (OCP) is known as a precursor of biological apatite and confirmed to facilitate bone regeneration in vivo. Fluoride ion is capable of promoting conversion of OCP to HA in addition to osteoblast activation [1]. Controlling OCP–HA conversion by fluoride could be effective to enhance bone regenerative property of the original OCP. This is because the process of OCP–HA conversion itself is also suggested to stimulate osteoblastic cell differentiation [2, 3]. Creating a composite of OCP and fluoride could be a way to develop a potential substitute

Y. Shiwaku and K. Sasaki
Division of Advanced Prosthetic Dentistry, Tohoku University Graduate School of Dentistry, Sendai 980-8575, Japan

Y. Shiwaku, Y. Honda, T. Anada, and O. Suzuki (✉)
Division of Craniofacial Function Engineering, Tohoku University Graduate School of Dentistry, Sendai 980-8575, Japan
e-mail: suzuki-o@m.tains.tohoku.ac.jp

T. Sasano et al. (eds.), *Interface Oral Health Science 2009*,
DOI 10.1007/978-4-431-99644-6_81,

material. However, it is still unclarified how fast OCP–HA conversion progresses depending on the fluoride ion concentrations. The aim of this chapter is to review the character of the hydrolysis in OCP and the products in fluoride.

2 Hydrolysis of OCP with Fluoride

The detail of the hydrolysis of OCP with fluoride was in part described previously [1]. Briefly, synthetic OCP was hydrolyzed by incubating the crystals in 150 mM Tris buffer (at neutral pH) including fluoride (as NaF) at 37°C. Fluoride concentrations up to 200 ppm were used. After the stirring up to 60 min, the OCP slurry was centrifuged, and the precipitates were washed several times with distilled water. The recovered samples were dried in an oven around 100°C overnight. The structures of hydrolyzates with fluoride were examined by X-ray diffraction (XRD) and Fourier transform infrared spectroscopy (FTIR).

3 The character of the Hydrolyzed Product

In XRD analysis, the characteristic (100) reflection of OCP decreased with the advancement of hydrolysis. In FTIR analysis, the center peak at 1,078 cm^{-1} in three distinct bands around 1,030–1,130 cm^{-1} became obscured, as reported in the previous study [1]. These results suggest that OCP was completely converted to fluoridated hydroxyapatite (FHA) in the presence of fluoride even in a short incubation time. Fluoride ion in the Tris buffer (at neutral pH) could be consumed rapidly during apatite formation. These results suggest the necessity to conduct further study searching for the optimum conditions so that the intermediates in the conversion of OCP to HA can be obtained.

References

1. Suzuki O, Yagishita H, Yamazaki M et al (1995) Adsorption of bovine serum albumin onto octacalcium phosphate and its hydrolyzates. Cell Mater 5(1):45–54
2. Suzuki O, Kamakura S, Katagiri T et al (2006) Bone formation enhanced by implanted octacalcium phosphate involving conversion into Ca-deficient hydroxyapatite. Biomaterials 27(13):2671–2681
3. Anada T, Kumagai T, Honda Y et al (2008) Dose-dependent osteogenic effect of octacalcium phosphate on mouse bone marrow stromal cells. Tissue Eng Part A 14(6):965–978

Effect of topography of the octacalcium phosphate granule surfaces on its bone regenerative property

Yoshitomo Honda, Takahisa Anada, Shinji Kamakura, and Osamu Suzuki

Abstract. We have previously reported that synthetic octacalcium phosphate (OCP) enhanced differentiation of osteoblasts and osteoclasts in vitro and also the bone regeneration in vivo. While the conversion of OCP into hydroxyapatite (HA) has been considered to be involved in the stimulatory capacity of OCP, little is known about the effect of topography of the OCP granule surface on its osteogenic capability. The present study reviews whether the surface topography of OCP granules, aggregates consisting of randomly oriented plate-like crystals, affects its intrinsic bone regenerative property in vivo. Two OCP granules, showing similar conversion velocity but a different surface topography, induced the distinct bone-forming ability in the calvaria defect of mouse. These results suggest that special attention should be paid for the surface microstructure as well as the conversion velocity in OCP granules to prepare the potent bone regenerative scaffold.

Key word. octacalcium phosphate, topography, bone, microstructure, bone regeneration

Reproduced from Honda et al. [1] with permission from Mary Ann Liebert, Inc., publishers. Supported by Grants-in-Aid (17076001, 19390490, 21659452) from the Ministry of Education, Science, Sports and Culture of Japan.

Y. Honda, T. Anada, and O. Suzuki (✉)
Division of Craniofacial Function Engineering (CFE), Tohoku University Graduate School of Dentistry, 4-1 Seiryo-machi, Aoba-ku, Sendai 980-8575, Japan
e-mail: suzuki-o@m.tains.tohoku.ac.jp

S. Kamakura
Division of Bone Regenerative Engineering, Department of Regenerative Biomedical Engineering, Tohoku University Graduate School of Biomedical Engineering, 2-1 Seiryo-machi, Aoba-ku, Sendai 980-8574, Japan

T. Sasano et al. (eds.), *Interface Oral Health Science 2009*,
DOI 10.1007/978-4-431-99644-6_82, © Springer 2010

1 Introduction

Implanted synthetic octacalcium phosphate (OCP) has a potential to enhance bone formation by osteoblasts coupled with its own biodegradation by osteoclast-like cells. On the other hands, OCP spontaneously and irreversibly converts to bone-like apatite at the physiological conditions in vivo and in vitro. On the basis of these results, we have assumed that the conversion process might be involved in the osteogenic capability of OCP. In fact, the process was associated with stimulating the osteoblast differentiation in vitro, corresponding to the acceleration of bone formation by OCP in vivo. However, little attention has been given to the topography of OCP granule surface on its osteogenic capability. This brief chapter outlines the results of comparing the osteogenic capability of two OCPs that had a different surface topography but similar conversion velocity in vivo.

2 Characterization of OCP Granules and Implant Procedure

Two types of OCP granules composed of different topographical surfaces were prepared from aggregation with distinct dimensional crystals. They were hereafter referred to as fine OCP (F-OCP) granules and coarse OCP (C-OCP) granules. Two distinct dimensional OCP crystals (4.0 and 26.6 μm) were prepared by the modified synthesis methods previously reported [2]. The crystals were characterized by X-ray diffraction (XRD) and Fourier transform infrared spectroscopy. The crystal microstructures were analyzed by scanning electron microscopy (SEM), and the crystal sizes were evaluated using a laser diffraction particle size analyzer. All animal handling and surgical procedures were approved by the Animal Research Committee of Tohoku University. Granules were implanted into critical sized defects (4.2 mm) in 9–10-week-old ICR mice calvaria for up to day 14. Osteogenic capability of the implanted OCP granules was determined by histological observations and histomolphometrical analysis. To evaluate the conversion velocity of OCP granules in vivo, both OCP granules implanted were retrieved from the implantation sites and examined by XRD analysis.

3 Effect of Surface Topography for the Osteogenic Capability of OCP Granules

From the SEM observation, it was apparent that the two types of OCP granules had different surface topography. Profile of both the granules showed that OCP tended to convert an apatite structure in a similar conversion velocity (about 50% from the original OCP granules in 6 days). Nevertheless, in the histomorphometrical analysis,

F-OCP granules displayed the osteoconduction in the bone defects, while C-OCP granules suppressed a new bone formation markedly relative to the F-OCP up to day 14. These results suggested that their bone regenerative properties are affected by not only conversion process but also topography of OCP granules. Overall, this may provide a new insight as to how OCP can be prepared to raise the bone regenerative property.

References

1. Honda Y, Anada T, Kamakura S et al (2009) The effect of microstructure of octacalcium phosphate on the bone regenerative property. Tissue Eng Part A 15:1965–1973
2. Suzuki O, Nakamaura M, Miyasaka Y et al (1991) Bone formation on synthetic precursors of hydroxyapatite. Tohoku J Exp Med 164:37–50

The influence of sericin solution on wettability and antifungal effect of resin surface

Guang Hong, Taizo Hamada, Takeshi Maeda, Sadayuki Yuda, Hideyuki Yamada, Kazuhisa Tsujimoto, and Shinsuke Sadamori

Abstract. This study is aimed at analyzing the antifungal effect of sericin solution and its influence on wettability of resin surface. The results suggest that the sericin solution improved the wettability of the resin surface significantly and showed antifungal effect on *Candida albicans*, *Candida tropicalis*, and *Candida glabrata*.

Key words. sericin solution, wettability, antifungal, *Candida albicans*, *Candida tropicalis*, *Candida glabrata*

1 Background

The wettability of denture in the oral cavity is kept by saliva. However, in elderly persons, reduction of salivary secretion with age, various symptoms including xerostomia due to various drugs, and dysfunctions occur on a daily basis. The purpose of this study was to develop a safe and user-friendly denture lubricant that stabilizes dentures and maintains wettability in the oral cavity.

G. Hong (✉) and T. Hamada
Department of Oral Health Care Promotion, Graduate School of Dentistry, Tohoku University, Sendai, Japan
e-mail: hong@m.tains.tohoku.ac.jp

T. Maeda and S. Sadamori
Department of Prosthetic Dentistry, Graduate School of Biomedical Sciences, Hiroshima University, Hiroshima, Japan

S. Yuda
Namitec Co., Osaka, Japan

H. Yamada and K. Tsujimoto
Seiren Co. Ltd., Fukui, Japan

T. Sasano et al. (eds.), *Interface Oral Health Science 2009*,
DOI 10.1007/978-4-431-99644-6_83, © Springer 2010

2 Material and Methods

In the wettability test, heat-polymerizing denture base resin (Acron, GC Co., Ltd., Japan) was used. The samples were polymerized according to the manufacturer's instruction to the rectangular blocks. The surfaces of test specimens were abraded with 280 grit waterproof abrasive paper. The three samples were immersed into the air (Control), distilled water; trial material-A & B (Stock sericin solution, 1%), and trial material-A/10 & B/10 (ten times diluents sericin solution) for 14 days, and the contact angle of each sample was measured after 0, 1, 3, 7, and 14 days of immersion. In the antifungal test of sericin solution, *Candida albicans*, *Candida tropicalis*, and *Candida glabrata* were used. Preincubate the Candida in Sabourand liquid medium for 24 h at 37°C. 100 μl of fungal culture (1.8×10^6 cells/ml) and 100 μl of distilled water; or trial material-A & B (Stock sericin solution, 1%) or trial material-A/10 & B/10 (ten times diluents sericin solution) were mixed, and 40 μl of mixed culture were inoculate to Sabouraud agar medium, and cultivate at 37°C. After 24 h, fungal colonies were counted. Statistical analysis was performed using 1-way analysis of variance (ANOVA), Student–Newman–Keuls multiple comparison tests (S–N–K test; $\alpha = 0.05$).

3 Results and Discussion

A significant difference ($p < 0.05$, ANOVA) was found among different immersed solutions for contact angle of denture base resin. After 1 day of immersion, the samples of immersed into the trial material-B/10 exhibited lowest values of contact angle with 40.1°. The samples of immersed in distilled water and air exhibited higher values of contact angle in all immersion periods. The values of contact angle of immersed trial material-A & B and trail material-A/10 & B/10 samples significantly decreased with an increase in time ($p < 0.05$, ANOVA), except immersed in air and distilled water samples (Fig. 1). A significant difference ($p < 0.05$, ANOVA) was found among different solutions for antifungal effect on *Candida albicans*, *Candida tropicalis*, and *Candida glabrata*.

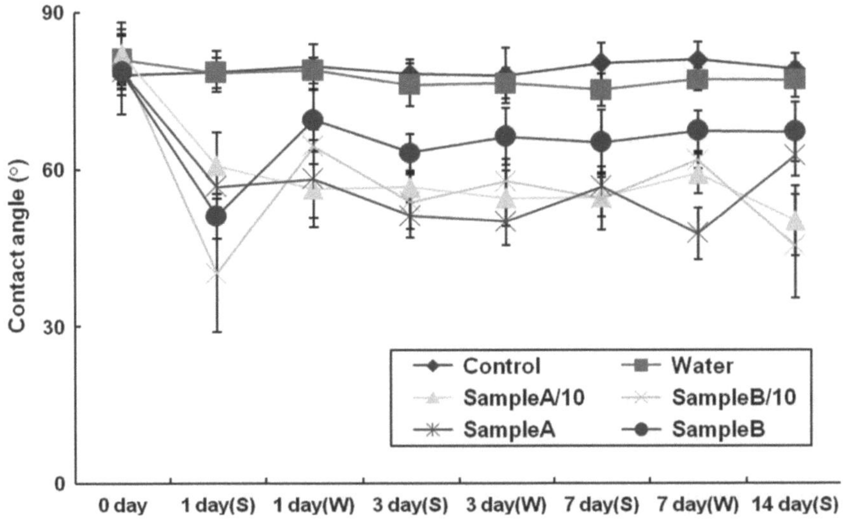

Fig. 1. The changes of contact angle with time. Each specimen were immersed in each solution for 12 h at 23±2°C [day(S)], washed thoroughly with tap water and distilled water, and then immersed in distilled water for 12 h at 37°C [day(W)]. This procedure was repeated each day for 14 days

4 Conclusions

The sericin solution improved the wettability of the acrylic denture base resin significantly and showed antifungal effect on *Candida albicans*, *Candida tropicalis* and *Candida glabrata*.

Adhesives and resin composites as functional units

Masafumi Kanehira, Werner J. Finger, and Masashi Komatsu

Abstract. This study evaluated the interaction between adhesives and resin composites and determined significant parameters influencing marginal adaptation in dentin cavities. The three one-step adhesives FluoroBond Shake One, iBond, and an experimental self-etching adhesive were combined with three resin composites. Shear bond strength on human dentin was determined 15 min after light activation. Marginal adaptation of the nine adhesive–composite combinations was evaluated in butt-joint dentin cavities. Polymerization contraction was measured using the bonded disk method by Watts & Cash. Polymerization shrinkage stress was evaluated by photoelasticity using the first order isochromatic line in cylindrical Araldit cavities for calculation of stress. In conclusion, there is no relationship between marginal cavity adaptation and bond strength, yet a distinct correlation with polymerization contraction and contraction stress exists.

Key words. dental adhesive, marginal adaptation, composite resin, polymerization contraction, contraction stress

1 Introduction

Despite the rapid progress made in adhesive bonding of direct resin composite restorations, the effectiveness of simplified one-step self-etch adhesives is reportedly still inadequate [1]. Clinical failures are mostly consequences of incomplete sealing of the tooth/restoration interface and thus closely related to polymerization contraction and contraction stress of the resin composite material.

Interactions between adhesives and different resin composites are rarely acknowledged. The stress generated in a polymerizing resin composite loading the

M. Kanehira (✉), W.J. Finger, and M. Komatsu
Division of Operative Dentistry, Department of Restorative Dentistry, Tohoku University Graduate School of Dentistry, 4-1 Seiryo-machi, Aoba-ku, Sendai 980-8575, Japan
e-mail: kane@mail.tains.tohoku.ac.jp

T. Sasano et al. (eds.), *Interface Oral Health Science 2009*,
DOI 10.1007/978-4-431-99644-6_84,

adhesive interface is not only dependent on the curing contraction of the material per se. Cavity size and geometry, expressed by the configuration factor, application, processing, curing techniques, and mechanical characteristics of the resin composite are presumably equally important parameters.

Aim of this study was to evaluate interactions between adhesives and resin composites in order to determine significant parameters influencing marginal adaptation in dentin cavities.

2 Shear Bond Strength to Dentin

The three one-step adhesives FluoroBond Shake One (Shofu), iBond, and an experimental self-etching material (Heraeus Kulzer) were combined with three resin composites, namely Beautifil (Shofu), Venus, and an experimental composite NEUN (Heraeus Kulzer). Shear bond strength was determined on peripheral dentin of extracted human molars. For each adhesive–composite combination, eight specimens were produced. The early dentin bonds strengths (MPa) mediated with each individual adhesive were not significantly different for the three resin composites [2]. Two-way ANOVA showed that only the adhesives were significant determinants.

3 Marginal Gap Width and Polymerization Contraction Stress

For determination of the marginal adaptation in cylindrical butt-joint cavities, the nine adhesive–composite combinations were tested. After exposing the entire cavity margin, maximum gap widths were measured using a light microscope.

Marginal cavity adaptation was best with NEUN. The resin composite was a significant determinant, whereas the adhesive system had no significant impact.

The polymerization contraction 5 min after light activation was measured using the bonded disk method by Watts & Cash, and the polymerization shrinkage stress was determined using a photoelastic method. The shrinkage stress was calculated from the first order isochromatic line.

The marginal cavity adaptation of the nonbonded restorations and the polymerization contraction data showed no uniform tendency. Beautifil produced the widest marginal gaps. Although the polymerization contraction of Venus was larger, the marginal gaps of the nonbonded restorations were significantly smaller than that for Beautifil. NEUN produced the smallest gaps. The polymerization contraction stress, loading the opposing cavity walls during and after polymerization of the resin composite, followed the tendency of the marginal gap data for nonbonded and bonded restorations, i.e., the stress exerted by Beautifil was significantly larger than the stress produced by Venus. Comparisons of the data of cavity adaptation, polymerization contraction, and contraction stress demonstrate that cavity adaptation depends on different parameters and not only on the free polymerization contraction [2].

4 Conclusions

Cavity adaptation is related to polymerization contraction and shrinkage stress of the resin composite used and not related to bond strength. Therefore, bond strength data are not suitable to predict the clinically more relevant adaptation of bonded restorations in cavities.

References

1. Van Landuyt K, De Munck J, Snauwaert J et al (2005) Monomer-solvent phase-separation in one-step self-etch adhesives. J Dent Res 84:183–188
2. Kurokawa R, Finger WJ, Hoffmann M et al (2007) Interactions of self-etch adhesives with resin composites. J Dent 35:923–929

Effects of bisphosphonates on bone marrow stromal cells

En Luo, Guozhu Yin, Xiaohui Zhang, Xian Liu, and Jing Hu

Abstract. Bisphosphonates are known to prevent bone resorption through their effects on osteoclastic activity. At the same time, there is increasing evidence that bisphosphonates also interact with bone marrow stromal cells (BMSCs), which are able to differentiate along multiple lineages and can be used as precursor cells for tissue reconstitution. This chapter discusses the effects of bisphosphonates on BMSCs in different ways and the extent of it.

Key words. bisphosphonate, BMSCs, differentiation

1 Introduction

Bisphosphonates, the synthetic analogs of pyrophosphate, are potent inhibitors of bone resorption and have been successfully used with increasing frequency in the treatment of osteoporosis, fibrous dysplasia, osteoarthritis, and tumor [1]. Bisphosphonates are well-known potent inhibitors of osteoclast activity. Recent evidence from in vitro and in vivo studies indicates that bisphosphonates may additionally promote osteoblastic bone formation [2]. Bone marrow stromal cells (BMSCs) are biologically important cells for tissue reconstitution and could differentiate along multiple lineages, such as chondrocytes, osteocytes, adipocyte and astrocytes. More and more reports discuss the effects of bisphosphonate on BMSCs.

E. Luo, G. Yin, X. Zhang, X. Liu, and J. Hu (✉)
West China Stomatology College, State Key Laboratory of Oral Disease, Sichuan University, No. 14, Section 3, Renmin South Avenue, Chengdu, Sichuan, P.R. China
e-mail: donow3000@scu.edu.cn

T. Sasano et al. (eds.), *Interface Oral Health Science 2009*,
DOI 10.1007/978-4-431-99644-6_85, © Springer 2010

2 Bisphosphonate Zoledronate

Zoledronate is a bone seeking specific inhibitor of protein farnesylation and geranylgeranylation. The studies show that zoledronate in vitro causes inhibition of proliferation and induction of apoptosis in hMSC when applied in concentrations of 20 and 50 μM for more than 24 h, which can be rescued by treatment with 10 μM geranylgeranyl pyrophosphate. These data set the stage for the future dissection of the effects of zoledronate and other aminobisphosphonates on cells beyond osteoclasts, with respect to cell differentiation in benign metabolic and to antitumor efficacy in metastatic bone diseases, as well as adverse events due to putative substance accumulation in bone during long-term treatment [3].

3 Bisphosphonate Clodronate

The bisphosphonate clodronate, having a high affinity for hydroxyapatite (HA), is proved to be a potent inhibitor of bone resorption and an effective accelerator of bone regeneration. BMSCs were isolated from Sprague-Dawley rat bone marrow and then cocultured with both HA and clodronate-HA. SEM, MTT, and ALP of BMSCs cultured on modified HA, and HA demonstrated no significant differences between two groups. BMSCs could be differentiated into adipocyte, osteocyte, and myocyte after being cocultured with both pure HA and clodronate-HA. The studies show that clodronate on the HA has no obvious inhibition to the proliferation and activities of BMSCs on the composites and evidently no negative effect on multidirectional capability of the BMSCs [4].

4 Bisphosphonates (Alendronate, Risedronate, or Zoledronate)

The studies evaluated the effects of three FDA-approved and clinically utilized bisphosphonates on the proliferation and osteogenic differentiation of hBMSC. Cells were obtained from patients undergoing primary total hip arthroplasty for end-stage degenerative joint disease. Cells were treated with or without a bisphosphonate (alendronate, risedronate, or zoledronate) and analyzed over 21 days of culture. Steady-state mRNA levels of key genes involved in osteogenic differentiation, such as BMP-2, bone sialoprotein-II, and cbfa1, were generally increased by bisphosphonate treatment in a type- and time-dependent manner. The studies demonstrate that bisphosphonates enhance proliferation of BMSC and initiate osteoblastic differentiation [5].

Acknowledgments Supported by grants from the financial support of the National Science Foundation of China (30700950 and 30973346), the Doctoral Specialty Foundation of High School, China (20070610061).

References

1. Siris ES, Selby PL, Saag KG et al (2009) Impact of osteoporosis treatment adherence on fracture rates in North America and Europe. Am J Med 122:S3–S13
2. Chaplet M, Detry C, Deroanne C et al (2004) Zoledronic acid up-regulates bone sialoprotein expression in osteoblastic cells through Rho GTPase inhibition. Biochem J 384:591–598
3. Ebert R, Zeck S, Krug R et al (2009) Pulse treatment with zoledronic acid causes sustained commitment of bone marrow derived mesenchymal stem cells for osteogenic differentiation. Bone 44:858–864
4. Song GD, Bao CY, Hu J et al (2008) Effect of clodronate modifying hydroxyapatite on mesenchymal stem cells multi-directional differentiation. J Med Res 37:42–45
5. Von Knoch F, Jaquiery C, Kowalsky M et al (2005) Effects of bisphosphonates on proliferation and osteoblast differentiation of human bone marrow stromal cells. Biomaterials 26:6941–6949

Tooth shape reconstruction from dental micro CT images

**Shin Kasahara, Shinichiro Omachi, Hirotomo Aso,
Kousuke Saito, and Satoshi Yamada**

Abstract. In this study, we propose a reconstruction method for obtaining a 3-dimensional shape of a tooth from dental micro CT images. An initial outline of a tooth is obtained from a CT image of a tooth that erupted from the alveolar bone. The next outline is determined in subsequent images by searching the positions with strong edges, and a closed spline curve is used to represent the contour. The concept and method of this research will be useful in dental research and diagnoses for dental treatment.

Key words. 3-dimensional tooth shape, reconstruction, CT image, spline curve

1 Introduction

3-dimensional information regarding teeth is useful for dental treatment and research. However, it is difficult to acquire 3-dimensional information of teeth in vivo without physically extracting them.

In this study, we propose a method for the reconstruction of the 3-dimensional shape of a tooth in a living body using images acquired from a dental micro CT. In the proposed method, the shape of a tooth is obtained by propagating a closed curve representing an outline of a tooth between images using the active outline with a spline curve.

S. Kasahara (✉)
Division of Fixed Prosthodontics, Department of Restorative Dentistry,
Graduate School of Dentistry, Tohoku University, Sendai, Japan
e-mail: kasahara@mail.tains.tohoku.ac.jp

S. Omachi, H. Aso, and K. Saito
Department of Electrical and Communication Engineering, Graduate School
of Engineering, Tohoku University, Sendai, Japan

S. Yamada
School of Dentistry, Ohu University, Koriyama, Japan

T. Sasano et al. (eds.), *Interface Oral Health Science 2009*,
DOI 10.1007/978-4-431-99644-6_86, © Springer 2010

2 Materials and Methods

Edge images are obtained from these CT images with a Sobel operator (a digital operator used in image processing, particularly within edge detection algorithms). In the proposed method, an image in which the teeth are not in contact with each other and are not buried in the alveolar bone is selected first. Namely, choosing one tooth in a figure of CT image, the initial outline is obtained from the edge image. For the adjacent CT images of the same tooth, a close curve that represents the outline of the tooth is obtained from each image, and the shape of the tooth is duplicated by integrating these closed curves. The active outline is used to obtain the closed curve that lies where the edge is strong. The closed curve is represented by a spline curve, and a similar closed curve is duplicated.

For comparison, the outlines of a tooth are drawn on each CT image manually, and the 3-dimensional shape of the tooth is reconstructed.

3 Results

Figure 1a displays the 3-dimensional shape of a tooth reconstructed from these outlines obtained using the proposed method. Figure 1b displays a manually detected 3-dimensional shape.

4 Discussion and Conclusion

The experimental results show that the shape of the tooth was accurately duplicated using the proposed method. From comparison of these images, we obtain a precise outline of a tooth from the incisal edge to the middle of root. We were unable to obtain a precise duplicate of the area near the root apex.

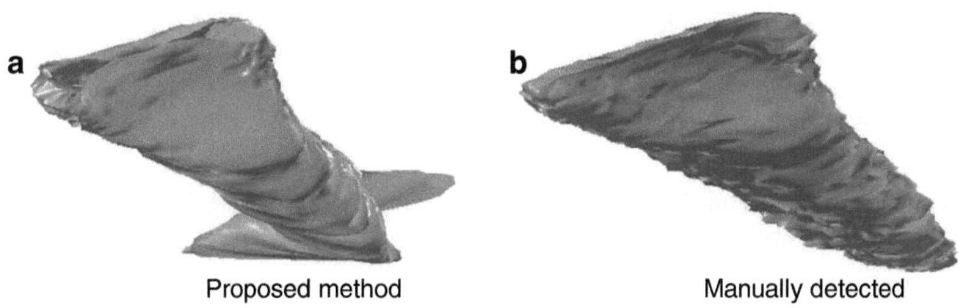

Fig. 1. Reconstructed 3-dimensional shape of the tooth (upper central incisor)

Formation of hydroxyapatite film on tooth using powder-jet-deposition

Ryo Akatsuka, Mohamamd Saeed Sepasy Zahmaty, Miyoko Noji, Takahisa Anada, Tsunemoto Kuriyagawa, Osamu Suzuki, and Keiichi Sasaki

Abstract. Using a powder-jet-deposition process, a thick hydroxyapatite (HAp) film can be created on a human tooth surface. Two different types of HAp particles, calcinated at 1,200 and 1,300°C, were used. The HAp particle was mixed with nitrogen as the carrier gas to form an aerosol flow and then accelerated and blasted from the nozzle onto the enamel substrate at room temperature and atmospheric pressure. HAp particles in the deposited film were tightly packed. There was no gap between the HAp film and the enamel substrate. The bonding strength of the HAp film is almost the same as the composite resin on the enamel.

Key words. tooth-biomaterial interface, hydroxyapatite, powder-jet-deposition, bonding strength, ultrafine particle

1 Introduction

The possibility of applying a hydroxyapatite (HAp) film as a new dental restorative material with compositional and mechanical properties corresponding to the tooth substance is presented. Using a powder-jet-deposition (PJD) process, a thick HAp

R. Akatsuka (✉), M. Noji, and K. Sasaki
Division of Advanced Prosthetic Dentistry, Tohoku University Graduate School of Dentistry, 4-1 Seiryo-machi, Aoba-ku, Sendai 980-8575, Japan
e-mail: akatsuka@mail.tains.tohoku.ac.jp

T. Anada and O. Suzuki
Division of Craniofacial Function Engineering, Tohoku University Graduate School of Dentistry, 4-1 Seiryo-machi, Aoba-ku, Sendai 980-8575, Japan

T. Kuriyagawa
Department of Nanomechanics, Tohoku University Graduate School of Engineering, 6-6-1 Aramakiaza-aoba, Aoba-ku, Sendai 980-8579, Japan

M.S.S. Zahmaty
Department of Mechanical Engineering, University of Tehran, Tehran 11365-4563, Iran

T. Sasano et al. (eds.), *Interface Oral Health Science 2009*,
DOI 10.1007/978-4-431-99644-6_87, © Springer 2010

film can be created on a human tooth surface [1]. In this chapter, microstructure and mechanical characteristics of the HAp film are investigated.

2 Materials and Methods

Two different types of HAp particles, calcinated at 1,200 and 1,300°C, were used. The HAp particle was mixed with nitrogen as the carrier gas to form an aerosol flow and then was accelerated and blasted from the nozzle on the enamel substrate at room temperature and atmospheric pressure. Acceleration pressure was 0.5 MPa. The surface and cross-section of the HAp films were observed by a scanning electron microscope (SEM). The 3D-profile of HAp films were measured by 3D non-contact measurement systems. Hardness of HAp films were measured by ultramicro Vickers hardness test. The load was 200 mN. The bonding strength between the HAp films and the enamel substrate was evaluated through a microtensile test.

3 Results

HAp particles were densely packed, and the shape of each HAp particle deposited on the enamel substrate has been highly deformed from the original one. There was no gap between the HAp film and the enamel substrate. The deposition-area was over 3×4 mm. The maximum film thickness was more than 40 μm, and the average film thickness was about 30 μm.

Micro-Vickers hardness of the enamel substrate and two types of deposited HAp films, using particles calcinated at 1,200 and 1,300°C, were 542.6 ± 52.7, 572.2 ± 68.3, and 557.4 ± 35.8, respectively, and there were no significant difference ($p > 0.05$, Steel–Dwass's multiple comparison test).

When a microtensile test was evaluated, particles calcinated at 1,200°C were used. There was no sample destroyed at the interface between the enamel and the HAp film. The bonding strength between the HAp films and enamel substrate was 14.0 ± 5.7 MPa and that of a composite resin was 15.8 ± 1.5 MPa. There was no significant difference ($p > 0.05$, Steel–Dwass's multiple comparison test).

4 Discussions

Through conducting a microtensile test on a deposited HAp film with tooth at its substrate, we confirmed two fracture patterns: (1) internal fracture of the HAp film and (2) fracture at both HAp film and enamel part. The latter showed a high value of tensile strength while the former was lower. Furthermore, SEM observation indicated that fracture pattern of HAp film at the interface is similar to that of stud side (reverted image). There were also no sample with fracture at the interface of the enamel and the HAp film, indicating a well deposition process.

5 Conclusions

The microstructure of the deposited HAp film was dense and the hardness was the same as that of human enamel. The bonding strength of the HAp film was almost the same as that of a composite resin which is normally used in dental treatment.

Reference

1. Noji M et al (2009) Characteristics of the hydroxyapatite film deposited on human enamel: deposition of a ceramic film by powder jet deposition technique. Int J Abrasive Technol 2:83–96

Electrodeposition of apatite onto titanium substrates under pulse current

Masakazu Kawashita, Zhixia Li, Tomoyasu Hayakawa, Gikan Takaoka, and Toshiki Miyazaki

Abstract. Apatite films were deposited onto titanium (Ti) metal substrates by an electrodeposition method in metastable calcium phosphate solution under a pulse current. We used a current-on time (T_{ON}) equal to current-off time (T_{OFF}) of 10 ms, 100 ms, 1 s, and 15 s. The adhesive strength between apatite and Ti substrates were relatively high at $T_{ON} = T_{OFF} = 10$ ms. It is believed that electrodeposition of apatite under pulse current would be useful for inducing bone-bonding ability to metallic materials.

Key words. Ti, electrodeposition, apatite, simulated body fluid

1 Introduction

Titanium (Ti) metal and its alloys are widely used for orthopedic and dental applications. However, they cannot directly bond to living bones. In order to improve its bone-bonding property, hydroxyapatite (HA) coating is applied on Ti implants. The electrodeposition method is a process that uses aqueous solutions at low temperatures; this method hardly affects the implant structure and can be applied to complex shapes. The HA layer can be rapidly formed on Ti substrates by electrodeposition

M. Kawashita (✉) and Z. Li
Graduate School of Biomedical Engineering,
Tohoku University, Aoba-ku, Sendai 980-8579, Japan
e-mail: m-kawa@ecei.tohoku.ac.jp

T. Hayakawa and G. Takaoka
Photonics and Electronics Science and Engineering Center, Graduate School of Engineering,
Kyoto University, Nishikyo-ku, Kyoto 615-8510, Japan

T. Miyazaki
Graduate School of Life Science and Systems Engineering, Kyushu Institute of Technology,
Wakamatsu-ku, Kitakyushu 808-0196, Japan

T. Sasano et al. (eds.), *Interface Oral Health Science 2009*,
DOI 10.1007/978-4-431-99644-6_88, © Springer 2010

in a metastable calcium phosphate solution. This chapter reviews our recent researches on electrodeposition of apatite onto Ti metal in metastable calcium phosphate solution by using various pulse widths.

2 Electrodeposition of Apatite onto Ti Substrates

The working electrode was a pure Ti substrate, and the counter electrode was a platinum plate. Before the electrodeposition, the Ti substrates were etched in H_2SO_4 solution. The current-on time (T_{ON}) was equal to the current-off time (T_{OFF}) irrespective of the pulse width. The electrolyte was 1.5SBF (simulated body fluid) without magnesium (Mg^{2+}) ions at 36.5°C. Deposition time was 90 min.

3 Structure of Apatite Electrodeposited on Ti Substrates

Dense HA layers were formed on the surfaces of all specimens by electrodeposition irrespective of the pulse width. It was found that the HA crystals on all specimens had cracks. The width of the cracks decreased with T_{ON}. Also, it was confirmed that the HA layers contained carbonate (CO_3^{2-}) ions. The amount of CO_3^{2-} in HA decreased with T_{ON}. However, the acid etching hardly affected the crystal structure of HA.

4 Adhesion Between Apatite and Ti Substrates

Figure 1 shows the SEM photographs of the unetched and etched Ti substrates subjected to electrodeposition at various T_{ON} followed by the peeling test using Scotch® tape. The deposited HA was easily peeled off by the Scotch® tape for the etched and unetched Ti substrates subjected to electrodeposition at $T_{ON}=1$ and 15 s. However, it did not peel off (with the glue of the Scotch® tape remaining on the surface) from the etched Ti substrates subjected to electrodeposition at $T_{ON}=10$ and 100 ms and the unetched Ti substrates subjected to electrodeposition at $T_{ON}=10$ ms. The glue of the Scotch® tape that remained on the surface indicates that the adhesive strength of the HA coating of these specimens was relatively high.

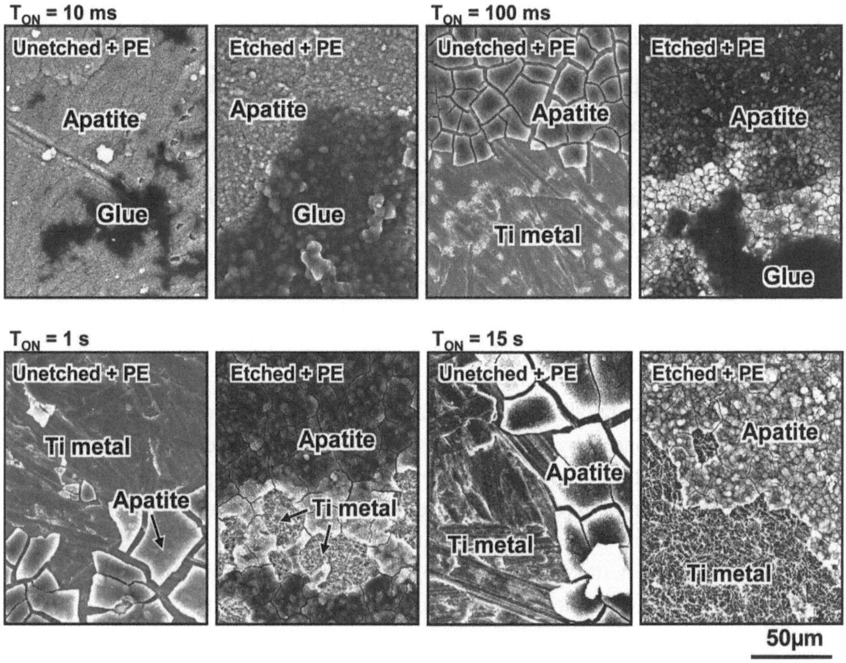

Fig. 1. SEM photographs of unetched and etched Ti substrates subjected to electrodeposition at various T_{ON} followed by peeling test using Scotch® tape

Alginate/octacalcium phosphate composites enhance bone formation in critical-sized mouse calvaria defects

Takeshi Fuji, Takahisa Anada, Yoshitomo Honda, Yukari Shiwaku, Keiichi Sasaki, and Osamu Suzuki

Abstract. Alginate/octacalcium phosphate (Alg/OCP) composites with different pore sizes were prepared by centrifuging Alg gels precipitated by OCP crystals. We investigated whether Alg/OCP promotes osteoblastic cell proliferation in vitro and bone regeneration in vivo. The analyses showed that cell proliferation and bone regeneration increased with an increase in the pore size. The results suggest that Alg/OCP provides a better scaffold for osteoblasts to attach and proliferate in its three-dimensional environment and promotes bone regeneration.

Key words. octacalcium phosphate, alginate, scaffold, bone regeneration

Reproduced from Fuji et al. (2009, Ref. 3) with permission from Mary Ann Liebert, Inc., publishers. Supported by Grants-in-Aid (17076001, 19390490, 21659452) from the Ministry of Education, Science, Sports and Culture of Japan.

1 Introduction

Our previous studies showed that synthetic octacalcium phosphate (OCP) is capable of promoting osteoblastic cell differentiation in vitro and has a potential to enhance bone regeneration [1, 2]. In this study, alginate (Alg) was used as the matrix to test the distribution effect of OCP in three-dimensional environment. We established a method to prepare Alg/OCP composites with different pore sizes

T. Fuji, T. Anada, Y. Honda, Y. Shiwaku, and O. Suzuki (✉)
Division of Craniofacial Function Engineering, Tohoku University Graduate School of Dentistry, Sendai 980-8575, Japan
e-mail: suzuki-o@mail.tains.tohoku.ac.jp

T. Fuji, Y. Shiwaku, and K. Sasaki
Division of Advanced Prosthetic Dentistry, Tohoku University Graduate School of Dentistry, Sendai 980-8575, Japan

T. Sasano et al. (eds.), *Interface Oral Health Science 2009*,
DOI 10.1007/978-4-431-99644-6_89, © Springer 2010

utilizing the centrifugation technique [3]. We hypothesized that osteoblastic cells in the pores may encounter the three-dimensionally arranged OCP crystals precipitated in the Alg matrix, which would disclose the intrinsic osteoconductive property of OCP in a three-dimensional noncell interactive matrix.

2 Materials and Methods

2.1 Preparation and Characterization of Alg/OCP Composites

OCP slurries were prepared according to the method previously reported [1], using phosphate solution containing various amounts (0, 0.2, 0.4 wt%) of Alg. Then each slurry was centrifuged at various speed (3,000–15,000 rpm). After supernatant removal and lyophilization, composites with different pore sizes were formed. Finally, disks (4 mm in diameter, 0.5 mm in thickness) of Alg/OCP were sterilized. The composites were characterized by Fourier transform infrared spectroscopy and X-ray diffraction. The porosity and pore size were measured by the mercury intrusion technique.

2.2 Cell Culture and Assay of Cell Proliferation in the Alg/OCP Disks

Mouse bone marrow stromal ST-2 cells were cultured on Alg/OCP disks for 3 days. The number of cells on the disks was determined using the colorimetric assay (cell counting kit-8).

2.3 Implantation Procedure

Alg/OCP disks and sodium Alg disks were implanted into critical-sized defects (4 mm) in the calvarium of ICR mice for 21 days ($N=6$). The samples were fixed and stained with hematoxylin–eosin and tartrate-resistant acid phosphatase (TRAP).

3 Results and Discussions

The pore size in the composites decreased with increasing centrifugal speed. The cell number of the 0.4Alg/OCP composites was higher than those of 0.2Alg/OCP. The 0.4Alg/OCP (3,000 rpm) composite with the largest pore size showed a higher n-Bone% than other groups. TRAP-positive cells were observed on the newly formed bone throughout the defect in the 0.4Alg/OCP (3,000 rpm) composite. The results suggest that OCP could be used in improving the osteogenic condition of noncell-interactive polymeric scaffold, such as Alg. We concluded that the cell proliferation in vitro and bone regeneration could be controlled by the pore size of the Alg/OCP composites.

References

1. Suzuki O, Nakamura M, Miyasaka Y et al (1991) Bone formation on synthetic precursors of hydroxyapatite. Tohoku J Exp Med 164:37–50
2. Suzuki O, Kamakura S, Katagiri T et al (2006) Bone formation enhanced implanted octacalcium phosphate involving conversion into Ca-deficient hydroxyapatite. Biomaterials 27:2671–2681
3. Fuji T, Anada T, Honda Y et al (2009) Octacalcium phosphate-precipitated alginate scaffold for bone regeneration. Tissue Eng PartA 15:3525–3535

Strength of porcelain fused to Ti-20% Ag alloy made by CAD/CAM

Ryoichi Inagaki, Masanobu Yoda, Masafumi Kikuchi, Kohei Kimura, and Osamu Okuno

Abstract. Titanium and titanium alloys are difficult to machine. This problem arises when milling, using dental CAD/CAM systems. In a previous study, an experimental binary titanium alloy with 20 mass% Ag showed good grindability. In this study, the fracture strength of porcelain fused to a Ti-20 mass% Ag alloy crown made using a CAD/CAM (GN-1, GC, Japan) system is investigated. As controls, similar pure titanium (JIS grade II) samples made using cast and using the CAD/CAM system were also examined. The crowns were made assuming a maxillary left central tooth. The fracture strengths were statistically analyzed using one-way ANOVA followed by Tukey pairwise tests. There was no significant difference in the fracture strength of porcelain fused to metal crowns between the Ti-20 mass% Ag alloy frame crowns and the pure titanium frame crowns.

Key words. Ti-20 mass% Ag alloy, CAD/CAM system, porcelain fused to metal crown, fracture strength

1 Introduction

In recent years, many dental computer-aided design and manufacturing (CAD/CAM) systems have been commonly used to fabricate dental prostheses. The ease of machining (cutting or grinding) a material is referred to as its "machinability." Machinability is the most important machining operation employed in many dental

R. Inagaki (✉), M. Yoda, and K. Kimura
Division of Fixed Prosthodontics, Department of Restorative Dentistry, Tohoku University Graduate School of Dentistry, 4-1 Seiryo-machi, Aoba-ku, Sendai 980-8575, Japan
e-mail: inagaki@mail.tains.tohoku.ac.jp

M. Kikuchi and O. Okuno
Division of Dental Biomaterials, Department of Restorative Dentistry, Tohoku University Graduate School of Dentistry, 4-1 Seiryo-machi, Aoba-ku, Sendai 980-8575, Japan

T. Sasano et al. (eds.), *Interface Oral Health Science 2009*,
DOI 10.1007/978-4-431-99644-6_90, © Springer 2010

CAD/CAM systems. Further development of new dental materials suited for machining is desired. On the other hand, commercially pure titanium (CP-Ti) has been evaluated for use in the fabrication of the framework for porcelain fused to a metal crown because of its excellent characteristics, such as its biocompatibility, corrosion resistance, and mechanical properties. However, CP-Ti is difficult to machine due to its intrinsic properties. In a previous study [1, 2], an experimental binary titanium alloy with 20 mass% Ag showed good machinability. This study investigated the fracture strength of porcelain fused to a Ti-20 mass% Ag alloy crown made using a CAD/CAM system.

Ti–Ag blocks were machined to fabricate the framework using a CAD/CAM system (GN-1, GC, Japan). As controls, similar pure titanium (JIS grade II) samples made using cast and using the CAD/CAM system were also examined. The crowns were made assuming a maxillary left central tooth. Commercial porcelain for titanium (VITA-Titankeramik, VITA, Germany) was applied to the frames. Each completed crown was luted to a metal abutment using zinc phosphate cement. The fracture strengths were evaluated by applying a static load at the incisal edge of the crown and statistically analyzed using one-way ANOVA followed by Tukey pairwise tests ($\alpha=0.05$). There was no significant difference in the fracture strength of porcelain fused to metal crowns between the Ti-20 mass% Ag alloy frame crowns and the pure titanium frame crowns.

A more precise clinical examination needs to be conducted before porcelain fused to titanium can be used for clinical applications. However, there are no definite regulations for the method to evaluate the fracture strength. Therefore, the examination for the strength of porcelain fused to metal crown is examined using various methods. In particular, regarding the application of CAD/CAM, there are some reports on the measurement of all ceramic crowns but none on porcelain fused to metal crowns. We independently devised a method to test the fracture strength by applying a static load on the incisal edge of the crown and reported the fracture strength of the porcelain fused to metal crowns.

There are many reports about the measurement of the occlusal force. Consequently, the occlusal force of a maxillary central tooth was researched with 100–250 N. Ti–Ag alloy crowns and CP-Ti crowns, which showed a much stronger value than the occlusal force, can be considered to have enough strength. Therefore, acceptable fracture strength could be achieved by Ti-20 mass% Ag alloy frame crowns.

References

1. Kikuchi M, Okuno O (2004) Machinability evaluation of titanium alloys. Dent Mater J 23:37–45
2. Kikuchi M, Takahashi M, Okuno O (2008) Machinability of experimental Ti-Ag alloys. Dent Mater J 27:216–220

Effect of various solutions to exudation of internal fluids from dentinal tubules

Hiromi Sasazaki and Masashi Komatsu

Abstract. Effect of various solutions to the exudation of internal fluids from dentinal tubules on cut dentin surfaces was observed by SEM using precision replica technique. Labial enamel of freshly extracted bovine teeth was removed and dentin surfaces were exposed. The right half of these surfaces was treated with various solutions. These treated surfaces were immediately impressed with silicone impression material. Epoxy resin was poured into these impressions. Precision replicas were coated with platinum and observed by SEM. Exudation of internal fluids from dentinal tubules was observed on the Super bond D liner-treated dentin surfaces. Exudation was slightly observed on the 21% $K_2C_2O_4$+3% KHC_2O_4 and 15.5% $Fe_2(SO4)_3$-treated dentin surfaces. Exudation was not observed on the Super bond D primer + D liner and Clearfil LB primer-treated dentin surfaces. Round crystal was observed on the $Ag(NH_3)_2F$-treated dentin surfaces. 21% $K_2C_2O_4$+3% KHC_2O_4, 15.5% $Fe_2(SO_4)_3$, Super bond D liner primer + D liner and Clearfil LB Primer were effective to stop the exudation of internal fluids from dentinal tubules.

Key words. internal fluid, exudation, dentinal tubules, replica, SEM

1 Introduction

With removal of enamel and exposure of dentin, it is known that tissue fluid exudes from dentinal tubules. The flow of dentinal fluid has been widely investigated in relation to tooth pain. Many in vitro studies concerning dentinal fluid have been reported. However, morphological observation of fluid accumulation on dentin surfaces has not been reported. We reported the precision replica technique to observe

H. Sasazaki (✉) and M. Komatsu
Division of Operative Dentistry, Department of Restorative Dentistry, Tohoku University Graduate School of Dentistry, 4-1 Seiryo-machi, Aoba-ku, Sendai 980-8575, Japan
e-mail: sasazaki@ddh.tohoku.ac.jp

T. Sasano et al. (eds.), *Interface Oral Health Science 2009*,
DOI 10.1007/978-4-431-99644-6_91, © Springer 2010

the exudation of dentinal fluid from dentinal tubules [1]. The purpose of this study was to investigate the effect of various solutions to the exudation of internal fluids from dentinal tubules on cut dentin surfaces using precision replica technique.

2 Materials and Methods

Labial enamel of freshly extracted bovine tooth was removed with emery paper (#400: Marumoto) and dentin surfaces were exposed. The right half of these surfaces was treated with Super bond D liner primer (Sun medical), Super bond D liner primer + D liner, $Ag(NH_3)_2F$ (Saforide: Bee brand medico dental), 23.1% $K_2C_2O_4$ + 2.9% KHC_2O_4 (Wako), Clearfil LB Primer (Kuraray), 15.5% $Fe_2(SO_4)_3$ (Astringedent: Ultradent). These treated surfaces were immediately covered with a silicone impression material (Silascon RTV 501: Dow corning). Epoxy resin (Epon 815 4 ml, Quetol-812 1 ml, DDSA 7 ml, DMP-30 0.1 ml: Nissin), degassed in a vacuum for 20 min, was poured into these impressions and cured in an oven at 50°C for 48 h. Precision replicas that were obtained were sputter coated with platinum and observed by SEM.

3 Results

Exudation of internal fluids from dentinal tubules was observed on the Super bond D liner primer-treated dentin surfaces. Exudation of internal fluids from dentinal tubules was slightly observed on the 23.1% $K_2C_2O_4$ + 2.9% KHC_2O_4 and 15.5% $Fe_2(SO_4)_3$-treated dentin surfaces. Exudation of internal fluids from dentinal tubules was not observed on the Super bond D liner primer + D liner and Clearfil LB primer-treated dentin surfaces. Round crystal was observed on the $Ag(NH_3)_2F$-treated dentin surfaces.

4 Discussion and Conclusion

23.1% $K_2C_2O_4$ + 2.9% KHC_2O_4, 15.5% $Fe_2(SO_4)_3$, Super bond D liner primer + D liner and Clearfil LB Primer were effective to stop the exudation of internal fluids from dentinal tubules. These solutions were effective to cure a patient with hyperesthesia teeth.

Reference

1. Sasazaki H, Okuda R (1994) Observation of the exudation of internak fluids in dentinal tubules, the Japanese. J Cons Dent 37(4):1164–1171

Evaluation of retentive force of β-type Ti–6Mo–4Sn alloy wire to apply for the abutment tooth of removable partial denture

Nobuhiro Yoda, Masayoshi Yokoyama, Takahiro Chiba, Genki Adachi, Masatoshi Takahashi, and Keiichi Sasaki

Abstract. The retentive force of wire clasp made of a low Young's modulus β-type Ti–6Mo–4Sn alloy (β-Ti-alloy) was evaluated using a piezoelectric transducer to determine an appropriate undercut (UC) for the retainer of removable partial dentures. There were no significant differences between the retentive force of a β-Ti-alloy wire with a 0.50 mm UC and that of a cobalt–chrome alloy (Co–Cr-alloy) wire with a 0.25 mm UC or between a β-Ti-alloy wire with a 0.75 mm UC and a Co–Cr-alloy wire with a 0.50 mm UC. As for the maximum magnitudes of the lateral forces, there were no significant differences between the β-Ti-alloy wire with the 0.50 mm UC and the Co–Cr-alloy wire with the 0.25 mm UC. The β-Ti-alloy wire could be applicable for abutment teeth with large amount of UCs.

Key words. Ti–6Mo–4Sn alloy wire, high elastic module, wire clasp, removable partial denture, retentive force

1 Introduction

Recently, a β-type Ti–6Mo–4Sn alloy (β-Ti-alloy) wire that has outstanding characteristics, such as a low Young's modulus, a high elastic limit, and biocompatibility, was developed. This study aimed to compare the retentive force of bent wires made of β-Ti-alloy and those made of a cobalt-chrome alloy (Co–Cr-alloy), so as to determine the appropriate amount of undercut (UC) for using β-Ti-alloy wire as a wire clasp.

N. Yoda(✉), M. Yokoyama, T. Chiba, G. Adachi, and K. Sasaki
Division of Advanced Prosthetic Dentistry, Tohoku University Graduate School of Dentistry, 4-1 Seiryo-machi, Aoba-ku, Sendai 980-8575, Japan
e-mail: n-yoda@umin.ac.jp

M. Takahashi
Division of Dental Biomaterials, Tohoku University Graduate School of Dentistry, 4-1 Seiryo-machi, Aoba-ku, Sendai 980-8575, Japan

T. Sasano et al. (eds.), *Interface Oral Health Science 2009*,
DOI 10.1007/978-4-431-99644-6_92, © Springer 2010

2 Materials and Methods

β-Ti-alloy wire (Neo-Titanium Wire, Yamahachi Dental Mfg., Co.) and Co–Cr-alloy wire (Sun-Cobalt Wire-Clasp, Dentsply Sankin K.K.) were evaluated. Seven wire clasps were bent along the abutment as a single arm clasp, for both materials, each with different UC depth (0.25, 0.5, and 0.75 mm). A piezoelectric force transducer [1] (Z18400, Kistler Instruments AG) was used to measure the retentive force. A force measuring device that consisted of the transducer and a spherical abutment model made of stainless steel was developed. The device enabled us to measure the three-dimensional forces when the bent wires were removed from the abutment by the universal testing machine.

3 Results

The forces on three orthogonal axes were recorded when the clasps were removed by the universal testing machine. There were no significant differences between the retentive force of β-Ti-alloy wire with the 0.50 mm UC and that of the Co–Cr-alloy wire with the 0.25 mm UC or between the β-Ti-alloy wire with the 0.75 mm UC and the Co–Cr-alloy wire with the 0.50 mm UC. As for the maximum magnitudes of the lateral forces, there were no significant differences between the β-Ti-alloy wire with the 0.50 mm UC and the Co–Cr-alloy wire with the 0.25 mm UC or between the β-Ti-alloy wire with the 0.75 mm UC and the Co–Cr-alloy wire with the 0.25 mm UC.

4 Discussion

The difference in the retentive force between the clasps of β-Ti-alloy wire and of Co–Cr-alloy was thought to be caused by differences in the Young's modulus [2, 3]. The β-Ti-alloy wire would be able to apply the abutment tooth with a larger amount of UC than the Co–Cr-alloy wire applied, even if a sufficient holding element of the RPD retainer, currently accepted, were used. Furthermore, β-Ti-alloy wire is thought to be advantageous for partial edentulous patients who have a metallic allergy because of its biocompatibility.

5 Conclusions

RPD wire-clasps made of β-Ti-alloy wire could be applicable for abutment teeth with large amounts of UC.

References

1. Kawaguchi T, Kawata T, Kuriyagawa T et al (2007) In vivo 3-dimensional measurement of the force exerted on a tooth during clenching. J Biomech 40:244–251
2. Waldmeier MD, Grasso JE, Norberg GJ et al (1996) Bend testing of wrought wire removable partial denture alloys. J Prosthet Dent 76:559–565
3. Mahmoud A, Wakabayashi N, Takahashi H et al (2005) Deflection fatigue of Ti-6Al-7Nb, Co-Cr, and gold alloy cast clasps. J Prosthet Dent 93:183–188

Medical application of magnesium and its alloys as degradable biomaterials

Yoshinaka Shimizu, Akiko Yamamoto, Toshiji Mukai, Yoko Shirai, Mitsuhiro Kano, Tadaaki Kudo, Hiroyasu Kanetaka, and Masayoshi Kikuchi

Abstract. The development of a degrading biomaterial is highly desired for use in new treatment modalities. In the search for degradable materials, a new category of biomaterials made from corroding metals was noticed. Magnesium and its alloys are biodegradable metals and exhibit improved mechanical properties and corrosion resistance compared to pure magnesium and industrial magnesium alloys. Previous research has demonstrated that magnesium alloys have acceptable biocompatibility as new biodegradable biomaterials.

Key words. magnesium alloy, biodegradation, biocompatibility, medical application

---

Y. Shimizu (✉), M. Kano, and M. Kikuchi
Division of Oral and Maxillofacial Anatomy, Tohoku University Graduate School of Dentistry, 4-1 Seiryo-machi, Aoba-ku, Sendai 980-8575, Japan
e-mail: shimizu@anat.dent.tohoku.ac.jp

A. Yamamoto and Y. Shirai
Biomaterials Center, National Institute for Materials Science, 1-1 Namiki, Tsukuba, Ibaraki 305-0044, Japan

T. Mukai
Structural Metals Center, National Institute for Materials Science, 1-2-1 Sengen, Tsukuba, Ibaraki 305-0047, Japan

T. Kudo
Division of Oral Physiology, Tohoku University Graduate School of Dentistry, 4-1 Seiryo-machi, Aoba-ku, Sendai 980-8575, Japan

H. Kanetaka
Department of Physical Medicine and Rehabilitation, Tohoku University Graduate School of Biomedical Engineering, 4-1 Seiryo-machi, Aoba-ku, Sendai 980-8575, Japan
Division of Advanced Prosthetic Dentistry, Tohoku University Graduate School of Dentistry, 4-1 Seiryo-machi, Aoba-ku, Sendai 980-8575, Japan

T. Sasano et al. (eds.), *Interface Oral Health Science 2009*,
DOI 10.1007/978-4-431-99644-6_93,

1 Introduction

Both metallic and polymeric materials are used in medical applications for osteosynthesis. Metals are more suitable for load-bearing applications but must be removed by a second surgical procedure after healing. On the other hand, polymeric materials are biodegradable, but are limited in their load-bearing capacity. Magnesium alloys exhibit better mechanical properties compared to polymeric materials and have been shown to degrade in vivo by corrosion. Recently, corrodible magnesium-based alloys have been introduced for use as coronary stents [1] and orthopedic devices [2].

2 Biodegradation of Magnesium Alloys

The corrosion of magnesium alloys is mainly dependent on their elemental compositions and the corrosive environment. Magnesium corrodes through reaction with aqua and chlorine ions. These ions produced from the corrosion of magnesium are nontoxic and can be readily excreted in the urine. However, the corrosion of magnesium and its alloys also produces hydrogen gas. A slowly corroding magnesium alloy showed no observable gas cavities since the slowly evolved hydrogen gas diffused into the surrounding tissues [3]. In order to limit the production of hydrogen gas, the previous study investigated the optimal elemental composition of the magnesium alloy.

3 Biocompatibility of Magnesium

The nontoxicity and corrosion resistance of magnesium alloys must be sufficiently investigated for medical application. Magnesium is essential to human metabolic functions and is the fourth most abundant cation in human body. In vitro cytotoxicity of pure and surface-treated magnesium metal showed positive cell proliferation and viability with no sign of growth inhibition [4]. The fracture toughness of magnesium is greater than that of ceramics, but pure magnesium corrodes too quickly in the physiological environment (pH 7.4–7.6), losing mechanical integrity before tissue healing. In an effort to maintain the mechanical integrity, more complex alloying compositions are necessary.

4 Medical Application of Magnesium Alloys

Recent developments in processing technology have led to the improvement of both the mechanical properties and corrosion resistance of magnesium alloys. Other advantages include their osteoconductive properties and the solubility of ions released during corrosion process in physiological media [3]. These improved magnesium alloys are currently being investigated for use as coronary stents, bone plates, and scaffolding for tissue regeneration and drug-eluting bioabsorbable materials [1].

References

1. Maeng M, Jensen LO, Falk E et al (2009) Negative vascular remodeling after implantation of bioabsorbable magnesium alloy stents in porcine coronary arteries: a randomized comparison with bare-metal and sirolimus-eluting stents. Heart 95:241–246
2. Staiger MP, Pietak AM, Huadmai J et al (2006) Magnesium and its alloys as orthopedic biomaterials: a review. Biomaterials 27:1728–1734
3. Witte F, Kaese V, Haferkamp H et al (2005) In vivo corrosion of four magnesium alloys and the associated bone response. Biomaterials 26:3557–3563
4. Li L, Gao J, Wang Y (2004) Evaluation of cyto-toxicity and corrosion behavior of alkali-heat-treated magnesium in simulated body fluid. Surf Coat Technol 185:92–98

Session IV
Social Interface

Difference between age generation of oral health examination in a rural town

Naoko Tanda, Kyoko Ikawa, Jumpei Washio, Yoshiko Shigihara, Yoshiro Shibuya, Masaki Iwakura, Megumi Haga, Yuhei Ogawa, Katsuhiko Taura, and Takeyoshi Koseki

Abstract. Oral health examinations were held in summer continuously for 2 years with general health checks in a rural town. We included oral health check, practice of oral cleaning, and measurement of oral malodor in the examination. Aim of this study was to examine the difference of impression in the oral health examination between younger and older participants by questionnaires. No difference was found between male and female ratio in each generation. Also, no difference was seen as for ratio of joiners of oral cleaning and ratio of their satisfaction after the practice. Significant difference between two generations was seen as for strong satisfaction ratio with explanation about the result of oral malodor, and effective motivation ratio for oral hygiene caused by measurement of oral malodor. The oral health examination held with general checks with measurement of oral malodor seemed more impressive to older generation.

Key words. general health checks, oral health checks, oral malodor

N. Tanda (✉), K. Ikawa, Y. Shigihara, Y. Shibuya, K. Taura, and T. Koseki
Division of Preventive Dentistry, Department of Oral Health and Development Sciences, Tohoku University Graduate School of Dentistry, 4-1 Seiryo-machi, Aoba-ku, Sendai 980-8575, Japan
e-mail: ntanda@mail.tains.tohoku.ac.jp

J. Washio
Division of Oral Ecology and Biochemistry, Department of Oral Biology, Tohoku University Graduate School of Dentistry, 4-1 Seiryo-machi, Aoba-ku, Sendai 980-8575, Japan

M. Iwakura and M. Haga
Department of Human Health and Nutrition, Faculty of Comprehensive Human Sciences, Shokei Gakuin College, 4-10-1Yurigaoka, Natori 981-1295, Japan

Y. Ogawa
Department of Dentistry and Oral Surgery, Akita University School of Medicine, 1-1-1 Hondo, Akita 010-8543, Japan

T. Sasano et al. (eds.), *Interface Oral Health Science 2009*,
DOI 10.1007/978-4-431-99644-6_94, © Springer 2010

1 Introduction

The WHO sees oral health as integral to general health and a determinant for quality of life. Promotion of oral health is essential in reducing inequalities in oral health care between urban and rural areas and/or between generations, in Japan and worldwide.

Operational research and efforts to translate science into practice in community-based oral health programs are necessary.

2 Aim

To examine the difference of impression, in the oral health examination, between younger and older participants by questionnaires.

3 Methods

We included oral health examinations with general health checks in a rural town of Japan. The oral health examination consisted of oral health checks, measurement of oral malodor by Breathtron™ (Cosmos Ltd. Osaka, Japan) [1], and practice of oral cleaning. Examinees responded to a questionnaire on their impressions after the examination. We analyzed statistical differences in impressions between younger (20–59) and older (60+) participants.

4 Result

No difference was found between male and female ratio in each generation. Also, no difference was seen as for ratio of joiners of oral cleaning and ratio of their satisfaction after the practice. Older participants showed significantly higher ratio as for strong satisfaction with explanation about the result of oral malodor and effective motivation for oral hygiene caused by measurement of oral malodor. As a result, older participants showed significant higher ratio of satisfaction than younger ones.

5 Conclusion

The oral health examination held with general checks and measurement of oral malodor seemed more impressive to older participants in rural area.

Reference

1. Tanda N, Washio J, Ikawa K et al (2007) A new portable sulfide monitor with a zinc-oxide semiconductor sensor for daily use and field study. J Dent 35:552–557

Impact of oral health status on healthy life expectancy in community-dwelling population: The AGES Project cohort study

Jun Aida, Miyo Nakade, Tomoya Hanibuchi, Hiroshi Hirai, Ken Osaka, and Katsunori Kondo

Abstract The aim of our study was to determine the association between oral health status and healthy life expectancy considering the socioeconomic status and health related variables. In this cohort study, self-administered questionnaires were mailed to 24,374 community-dwelling elderly living in 25 municipalities in 2003. A Cox's proportional hazard model was applied to assess the relationship between oral health and healthy life expectancy. Twelve thousand and thirty one people (49.4%) replied to the questionnaire. During the follow-up period of 3 years, 1,450 (12.6%) of 12,031 respondents died or received certification of long-term care requirement. After adjusting all covariables, respondents who answered "I cannot eat everything" had 1.21 (95% confidential interval: 1.01–1.45) times higher hazard ratio compared with those answered "I can eat anything." The results showed that perceived oral health status was significantly associated with healthy life expectancy even after being adjusted for socioeconomic and health related variables.

Key words. oral health, healthy life expectancy, Cox's proportional hazard model

J. Aida (✉) and K. Osaka
Department of International and Community Oral Health, Tohoku University Graduate School of Dentistry, 4-1 Seiryo-machi, Aoba-ku, Sendai 980-8575, Japan
e-mail: j-aida@umin.ac.jp

M. Nakade
Department of Nutrition and Food Science, Aichi Gakusen College, 28 Kamikawanari, Hegoshi-cho, Okazaki 444-8520, Japan

T. Hanibuchi
Research Center for Disaster Mitigation of Urban Cultural Heritage, Ritsumeikan University, 58 Komatsubara Kitamachi, Kita-ku, Kyoto 603-8341, Japan

H. Hirai
Center for Well-being and Society, Nihon Fukushi University, 5-22-35 Chiyoda, Naka-ku, Nagoya 460-0012, Japan

K. Kondo
Center for Well-being and Society, Nihon Fukushi University, Okuda, Mihama, Aichi 470-3295, Japan

T. Sasano et al. (eds.), *Interface Oral Health Science 2009*,
DOI 10.1007/978-4-431-99644-6_95, © Springer 2010

1 Introduction

A growing body of research has demonstrated the potential effect of oral health on general health [1–3]. The relationships between oral health and mortality have been reported in several studies. A few studies used the healthy life expectancy as an outcome variable. The aim of our study was to determine the association between oral health status and healthy life expectancy considering the socioeconomic status and health related variables.

2 Methods

In this cohort study, self-administered questionnaires were mailed to 24,374 community-dwelling elderly living in 25 municipalities in 2003 [4]. The questionnaire included information on perceived oral health, perceived general health, alcohol consumption, smoking status, exercise, and equivalent income. Mortality data for 3 years and certification as being in need of long-term care data were provided from each municipality. A Cox's proportional hazard model was applied to assess the relationship between oral health and healthy life expectancy.

3 Results

Twelve thousand and thirty one people (49.4%) replied to the questionnaire. In perceived oral health question, 44.6% of respondents answered "I can eat anything," 47.3% answered "I can eat almost anything," 5.8% answered "Because of oral health problem, I cannot eat everything," and 2.3% of them did not answer. During the follow-up period of 3 years, 1450 (12.6%) of 12,031 respondents died or received certification of long-term care requirement. After adjusting all seven covariables, respondents who answered "I cannot eat everything" had 1.21 (95% confidential interval: 1.01–1.45) times higher hazard ratio compared with those answered "I can eat anything."

4 Conclusion

The results showed that perceived oral health status was significantly associated with healthy life expectancy even after being adjusted for socioeconomic and health related variables.

Acknowledgments This study was supported by funding from the Ministry of Education, Culture, Sports, Science and Technology of Japan (the 21st Century Center of Excellence Program).

References

1. Nakanishi N, Fukuda H, Takatorige T et al (2005) Relationship between self-assessed masticatory disability and 9-year mortality in a cohort of community-residing elderly people. J Am Geriatr Soc 53:54–58
2. Fukai K, Takiguchi T, Ando Y et al (2007) Dental health and 15-year mortality in a cohort of community-residing older people. Geriatr Gerontol Int 7:341–347
3. Fukai K, Takiguchi T, Ando Y et al (2008) Mortalities of community-residing adult residents with and without dentures. Geriatr Gerontol Int 8:152–159
4. Kondo K (ed) (2007) Exploring "inequalities in health": a large-scale social epidemiological survey for care prevention in Japan (Japanese). Igaku-Shoin Ltd, Tokyo

Wireless magnetic motion capture system for medical use

Hiroyasu Kanetaka, Shin Yabukami, Syuichiro Hashi, and Ken-Ichi Arai

Abstract. The estimation of organic functions in the human body is important for both diagnosis and treatment in the medical field. In order to adequately assess these functions, a detection technique having an accuracy of better than 1 mm is required for body motion analysis. A wireless magnetic motion capture system is one such effective detection technique. Recently, the development of a new motion capture system using an LC resonant magnetic marker (LC marker) has been reported. This new system consists of both driving and pickup coils as well as small LC markers that do not have batteries or electric wires that may interfere with natural organic functions. It is suggested that this system is reliable and effective for diagnosis and treatment in the medical field, because it is a noninvasive system with six degrees of freedom that is able to capture magnetic markers in the human body.

Key words. wireless, motion capture, six degrees of freedom, LC marker, medical use

---

H. Kanetaka (✉)
Department of Physical Medicine and Rehabilitation, Tohoku University Graduate School of Biomedical Engineering, 4-1 Seiryo-machi, Aoba-ku, Sendai 980-8575, Japan
Division of Advanced Prosthetic Dentistry, Tohoku University Graduate School of Dentistry, 4-1 Seiryo-machi, Aoba-ku, Sendai 980-8575, Japan
e-mail: kanetaka@m.tains.tohoku.ac.jp

S. Yabukami
Department of Electrical Engineering and Information Technology, Tohoku Gakuin University, 1-13-1 Chuo, Tagajo 985-8537, Japan

S. Hashi
Research Institute of Electrical Communication, Tohoku University, 2-1-1 Katahira, Aoba-ku, Sendai 980-8577, Japan

K.-I. Arai
The Research Institute for Electrical and Magnetic Materials, 2-1-1 Yagiyama-minami, Taihaku-ku, Sendai 982-0807, Japan

T. Sasano et al. (eds.), *Interface Oral Health Science 2009*,
DOI 10.1007/978-4-431-99644-6_96, © Springer 2010

1 Introduction

Motion capture techniques are earnestly desired for minimally-invasive diagnosis and treatment without X-ray exposure. For medical use, a small marker and a 1-mm order of accuracy are required. Magnetic motion capture systems satisfy these requirements; however, conventional systems require that the magnetic object be large or that the marker contains electric wiring, in order to obtain a high signal-to-noise (S/N) ratio for the magnetic signal emitted by the marker [1, 2]. Therefore, a wireless magnetic motion capture system using a magnetically coupled small LC marker has been developed [3–5].

2 System Setup and Position Detection

The system consists of measurement equipment and a coil assembly comprising an exciting coil and an array of pick-up coils [3–5]. The pick-up coil array consists of 25 pick-up coils. The LC marker consists of a Ni–Zn ferrite core with 500 turns of wound coil, and a chip capacitor, representing an LC series circuit. The excitation field is generated by the driving coil, and the marker is strongly excited at its resonant frequency by electromagnetic induction [4, 5].

Both the position and orientation of the markers are obtained by solving an inverse problem; more than six values of the flux density at known locations are needed to determine both the position and orientation of the marker as the magnetic flux source.

3 Conclusion

LC markers are effective in improving the accuracy and reducing the size of the system. In addition, the system is capable of simultaneous multimarker detection when the LC markers are assigned individual resonant frequencies [3–5]. This system is suggested to be both reliable and effective for diagnosis and treatment in the medical field.

References

1. Paradiso AJ, Hsiao K, Stricken J et al (2000) Sensor systems for interactive surfaces. IBM Syst J 39:892–914
2. Yabukami S, Kikuchi H, Yamaguchi M et al (2000) Motion capture system of magnetic markers using three-axial magnetic field sensor. IEEE Trans Magn 36:3646–3648

3. Yabukami S, Hashi S, Tokunaga Y et al (2004) A development of position sensing system for a wireless magnetic marker. J Magn Soc Jpn 28:877–885
4. Hashi S, Toyoda M, Yabukami S et al (2008) Wireless magnetic motion capture system using multiple LC resonant magnetic markers with high accuracy. Sens Actuators A Phys 142:520–527
5. Hashi S, Yabukami S, Kanetaka H et al (2009) Numerical study on the improvement of detection accuracy for a wireless motion capture system. IEEE Trans Magn 45:2736–2739

Evaluation of the optimal time of the dental treatment for the elderly

Yoshinori Tamazawa, Masaaki Iwamatsu, Kaoru Tamazawa, Satoshi Yamaguchi, and Makoto Watanabe

Abstract. The purpose of this study was to evaluate the optimal dental treatment time for the elderly. The subjects were 100 outpatients aged 65 years old or older (elderly group) and 100 in their 20s (control group) after obtaining the informed consent. A questionnaire survey by hearing was performed about the desired dental treatment time and good/poor physical condition, systemic disease, and medicines. In the elderly group, the desired time for dental treatment was most often found to be 10:00–11:00 a.m. (42%), with 73% preferring treatment in the morning (9:00–12:00). In the control group, the desired time widely distributed from 9:00 to 21:00 h. In the elderly group, 56% complained of a poor physical condition immediately after waking up in the morning and from after lunch to the evening; however, very few complained between 9:00 and 12:00. These results suggested that the optimal time for dental treatment in elderly patients is 9:00–12:00 in the morning when their physical condition is stable.

Key words. elderly, dental treatment time, physical condition, systemic disease

---

Y. Tamazawa (✉)
Division of Infection Control, Tohoku University Hospital, Sendai 980-8575, Japan
e-mail: y-tama@mail.tains.tohoku.ac.jp

M. Iwamatsu
Division of Comprehensive Dentistry, Tohoku University Hospital, Sendai 980-8575, Japan

K. Tamazawa
Division of Periodontology and Endodontology, Tohoku University Graduate School of Dentistry, Sendai 980-8575, Japan

S. Yamaguchi and M. Watanabe
Division of Aging and Geriatric Dentistry, Tohoku University Graduate School of Dentistry, Sendai 980-8575, Japan

T. Sasano et al. (eds.), *Interface Oral Health Science 2009*,
DOI 10.1007/978-4-431-99644-6_97, © Springer 2010

1 Introduction

Humans have biorhythms, and sleeping [1, 2] and eating habits that are inconsistent with them are considered to impair health. In particular, in the elderly, decrease in reserve abilities for physical and mental activities [3] and a considerable decrease in the ability to adapt to environments have been suggested. Among medical treatments, dental treatment is particularly accompanied by physical pain and mental stress. The purpose of this study was to evaluate the optimal dental treatment time for the elderly.

2 Materials and Methods

The subjects were 100 outpatients aged 65 years old or older (elderly group), in the Tohoku University Dental Hospital, and 100 in their 20s (control group) after obtaining the informed consent. A questionnaire survey by hearing was performed about the desired dental treatment time and good/poor physical condition, a habit of napping, systemic disease, and medicines. In addition, vital signs were measured in the outpatient clinic, and the optimal dental treatment time was evaluated.

3 Results and Discussion

In the elderly group, the desired time for dental treatment was most often found to be 10:00–11:00 a.m. (42%), with 73% preferring treatment in the morning (9:00–12:00). In the control group, the desired time widely distributed from 9:00 to 21:00 h. The distribution of the desired treatment time significantly differed between the two groups (Mann–Whitney U-test, $p<0.001$). In the elderly group, 56% complained of a poor physical condition immediately after waking up in the morning and from after lunch to the evening; however, very few complained in the morning (9:00–12:00), and 23% had a habit of napping within a wide time zone between 11:00 and 17:00 h with a peak at 14:00. In the control group, 21% complained of a poor physical condition mostly immediately after waking up in the morning (14%). However, few controls had a habit of napping. Consequently, complaints of a poor physical condition showed significant differences in the number of subjects and the time distribution between the two groups (Mann–Whitney U-test, $p<0.001$). In the elderly group, 89% had medical treatment; however, neither systemic disease nor medicines affect dental treatment time.

In the elderly patients, the optimal treatment time results may be influenced by biorhythms and factors associated with their general condition.

4 Conclusions

These results suggested that the optimal time zone for dental treatment in elderly patients is 9:00–12:00 in the morning when their physical condition is stable.

Acknowledgment This study was supported by the JSPS grant (No.15659460).

References

1. Harrington JJ, Lee-Chiong T Jr (2007) Sleep and older patients. Clin Chest Med 28:673–684
2. Picarsic JL, Glynn NW, Taylor CA et al (2008) Self-reported napping and duration and quality of sleep in the lifestyle interventions and independence for elders pilot study. J Am Geriatr Soc 56:1674–1680
3. Martini B, Buffington AL, Welsh-Bohmer KA et al (2008) Time of day affects episodic memory in older adults. Neuropsychol Dev Cogn B Aging Neuropsychol Cogn 15:146–164

Educational effect on tooth preparation of visual feedback using computer graphics

Yayoi Okuyama, Toshinobu Abe, Shin Kasahara, and Masanobu Yoda

Abstract. This review article describes the educational effect on tooth preparation of visual feedback using computer graphics. The result is that in the feedback group, the average occlusal reductions showed significant differences at the three cusps, and the axial wall taper showed significant differences in the three regions. A series of studies indicates that visual feedback using computer graphics was effective for tooth preparation practices.

Key words. tooth preparation, educational effect, feedback, computer graphics, three-dimensional shape measuring system

1 Introduction

Tooth preparation for complete cast restorations is a routine clinical procedure performed intraorally prior to taking the impression. Preclinical students require training in the preparation technique, and quantitative evaluation of prepared artificial teeth permits students to evaluate their own skill level. In recent years, inlay cavities and teeth prepared for complete cast restorations have been quantitatively evaluated using a computer-aided system.

2 Research Projects

Okuyama et al. [1] reported the computerized calculation method designed for the visualization of the axial wall taper of abutment teeth immediately after tooth reduction. The shape of the prepared tooth was measured using a noncontact

Y. Okuyama (✉), T. Abe, S. Kasahara, and M. Yoda
Division of Fixed Prosthodontics, Department of Restorative Dentistry, Tohoku University Graduate School of Dentistry, 4-1 Seiryo-machi, Aoba-ku, Sendai 980-8575, Japan
e-mail: yayoi-o@mail.tains.tohoku.ac.jp

T. Sasano et al. (eds.), *Interface Oral Health Science 2009*,
DOI 10.1007/978-4-431-99644-6_98, © Springer 2010

three-dimensional shape measuring system (Surflacer VMS-100R, Unisn Inc, Osaka, Japan) with a minimum resolution of 0.035 mm. The data were combined with 3D Graphics Software (Imageware9, EDS PLM Solutions, Detroit, USA) to yield a profile of the whole surface. This shape measuring system showed that objective evaluation of abutment teeth is possible. Objective evaluation should be all-inclusive; that is, it should be combined with quantitative evaluation of other parameters – volume of occlusal reduction, margin shape, surface smoothness, and damage to adjacent artificial teeth. Moreover, objective assessment of damage to adjacent artificial teeth in tooth preparation practice was introduced [2]. The depth of the adjacent teeth damage could be calculated automatically from computer graphics and displayed as a rendering of a colored map consisting of contour lines. Applying this shape measuring system in actual preparation practices, Abe et al. [3] reported the educational effect of visual feedback using computer graphics on tooth preparation practices. A total of 27 dental preclinical students prepared an artificial tooth for complete cast restorations in a manikin head. After the first practice, these students were randomly divided into two groups. One group attended a 30-min visual feedback session involving their own preparation work. After feedback, this group (feedback group) prepared another artificial tooth. The other group (nonfeedback group) practiced a second preparation without any feedback. Between the first practice and the second practice, the quantity of occlusal reductions and axial wall taper were compared between the two groups. In the feedback group, occlusal reductions showed significant differences at the mesiobuccal cusp, distobuccal cusp, and distolingual cusp. The axial wall taper showed significant differences in the buccal, mesial, and distal region. In the nonfeedback group, there were no significant differences in occlusal reductions or the axial wall taper between the practices.

3 Conclusion

A series of studies indicates that visual feedback using computer graphics was effective for tooth preparation practices.

References

1. Okuyama Y, Kasahara S, Kimura K (2005) Quantitative evaluation of axial wall taper in prepared artificial teeth. J Oral Sci 47:129–133
2. Okuyama Y, Abe T, Kasahara S et al (2008) Objective assessment of damage to adjacent artificial teeth in tooth preparation education. J Jpn Dent Educ Assoc 24:170–174
3. Abe T, Okuyama Y, Kasahara S et al (2009) Educational effect on tooth preparation of visual feedback using computer graphics. Ann Jpn Prosthodont Soc 1:123–129

Japanese men OSAHS patient's anatomical features

Mau Okubo, Masaaki Suzuki, and Teruko Takano-Yamamoto

Abstract. The obstructive sleep apnea hypopnea syndrome (OSAHS) prevalence rate among Japanese presumes 1.7% (male 3.3%, female 0.5%). The obese subjects having body mass index of 30 kg/m^2 or higher were 20% of the population in United States and 2% of the population in Japan. However, in these two countries, there is no great difference in the prevalence rate of OSAHS. Even if the severity of obesity is low, the Japanese is guessed as the participation of abnormality of craniofacial morphology that OSAHS develops easily. Then, Japanese OSAHS patients' three-dimensional anatomical features are described in this chapter.

Key words. obstructive sleep apnea hypopnea syndrome (OSAHS), 3DMRI, divergence, internal length, area

1 Introduction

The primary structural risk factors for obstructive sleep apnea hypopnea syndrome (OSAHS) are either obesity or having an abnormal upper-airway anatomy. In this study, the morphologic characteristics of the mandible and upper-airway soft tissue were elucidated using 3D volumetric MRI reconstruction to clarify the anatomic risk factors for the upper airway in Japanese patients with OSAHS [1].

M. Okubo (✉) and T. Takano-Yamamoto (✉)
Division of Orthodontics and Dentofacial Orthopedics, Tohoku University Graduate School of Dentistry, 4-1 Seiryo-machi, Aobaku, Sendai 980-8575, Japan
e-mails: mau@m.tains.tohoku.ac.jp, t-yamamo@mail.tains.tohoku.ac.jp

M. Suzuki
Department of Otolaryngology, Teikyo University School of Medicine, Tokyo, Japan

DOI 10.1007/978-4-431-99644-6_99, © Springer 2010

2 Material and method

The study population consisted of 51 Japanese men matched for age and body mass index (31 patients with OSAHS and 20 healthy control subjects).

All the patients were newly diagnosed as having OSAHS. None of the control subjects had a history of snoring or excessive daytime sleepiness. Potential apnea was identified as a nearly flat airflow (<10% of baseline) and hypopnea as airflow <95% of the baseline for at least 10 s associated with either an oxygen desaturation of >3% or an arousal.

Sixty-four 2-mm-thick MRI slices were each obtained from the hard palate to the epiglottic vallecula, the base of the epiglottis. Using 3D imaging software, the mandible was trimmed for each slice from the bottom to the head. Five measurements of the mandible and three measurements of soft tissues were analyzed using 3D imaging software. This plane was defined as the mandibular base plane in this study. All mandibular parameters were based on the mandibular base plane. Figure 1 schematically shows the results of the axial morphologic analyses of the mandible on the mandibular base plane. The internal width was determined from the distance between the internal right gonion (IRG) and the internal left gonion (ILG). The mean distances between RG and IRG and between LG and ILG were defined as the bony thickness. The divergence was defined as the degree between the spina mentalis (SM)-IRG line and the SM-ILG line. The internal length was determined as the perpendicular distance from the SM to the line on both RG and LG (Fig. 1). The integration of the area within the internal mandible, hereafter referred to as the area, was digitally calculated on the lowest slice in which the entire corpus mandibulae appeared. The tongue, soft palate, and lateral pharyngeal walls were trimmed in each slice using 3D imaging software from the bottom to the head of each anatomic structure (Fig. 2).

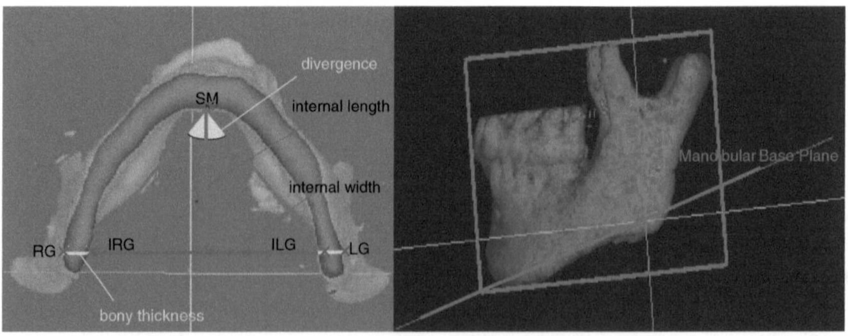

Fig. 1. Axial morphologic analyses of mandible on mandibular base plane

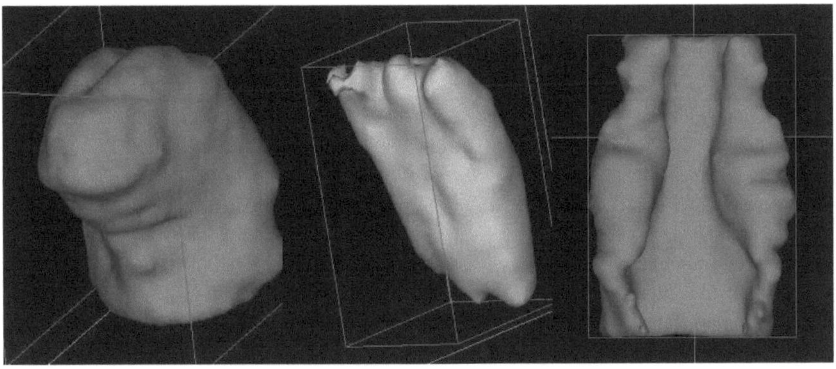

Fig. 2. Tongue volume, soft palate volume, lateral pharyngeal wall volume

Table 1 Comparison of 3D MRI reconstruction parameters from Japanese men with OSAHS and control

Variable	OSAHS	Control	*p* Value
Internal width (mm)	88.8±5.0	89.7±4.7	0.76
Divergence (°)	77.5±5.1**	73.5±5.1**	0.01**
Bony thickness (mm)	6.5±1.1	6.1±1.3	0.2
Internal length (mm)	53.9±4.2*	57.4±5*	0.04*
Area (cm^2)	32.7±4.6**	37.0±4.4**	0.002**
Tongue volume (cm^3)	78.1±11.9	77.1±11.6	0.91
Soft palate volume (cm^3)	6.1±2.3	5.9±1.2	0.86
Lateral pharyngeal wall volume (cm^3)	20.4±3.6	18.5±3.4	0.06

Mann-Whitney U-test
*$P<0.05$
**$P<0.01$

3 Results and discussion

1. The mandible in the patients with OSAHS had a wider divergence, a smaller internal length, and a smaller area at the mandibular base plane than that in the control subjects (Table 1).
2. There were no significant differences in the mandibular bone parameters between obese ($n=12$) and nonobese ($n=19$) patients with OSAHS (Table 2).
3. Tongue, soft palate, and lateral pharyngeal wall volumes were not significantly different between the OSAHS and control groups, although lateral pharyngeal wall volume correlated with AHI.

Japanese men patients with OSAHS have specific anatomic morphologies in the bottom part of the mandible. However, increased upper-airway soft-tissue volume was not proven to be an important risk factor for Japanese men with OSAHS.

Table 2 3D MRI reconstruction parameters were compared between obese and nonobese patients with OSAHS

Variable	BMI 25≧	BMI 25<	p Value
Internal width (mm)	88.9±5.6	88.7±4.1	0.69
Divergence (°)	77.8±5.2	77.1±5.3	0.73
Bony thickness (mm)	6.6±1.2	6.4±1.1	0.65
Internal length (mm)	54±4.9	53.7±3	0.58
Area (cm^2)	33±5.3	34.6±8.5	0.97
Tongue volume (cm^3)	79.1±13.2	76.4±9.7	0.69
Soft palate volume (cm^3)	6.2±2.3	5.9±2.4	0.57
Lateral pharyngeal wall volume (cm^3)	20.5±3.8	20.2±3.5	0.81

Investigators and clinicians must realize that ethnicity may modify the effects of obesity and abnormal craniofacial anatomy as risk factors for the pathogenesis of OSAHS.

Reference

1. Okubo M, Suzuki M, Horiuchi A et al (2006) Morphologic analyses of mandible and upper airway soft tissue by MRI of patients with obstructive sleep apnea hypopnea syndrome. Sleep 29:909–915

Association between periodontal disease and risk for atherosclerosis in hypertensive patients

Kaoru Tamazawa, Yoshinori Tamazawa, and Hidetoshi Shimauchi

Abstract. The purpose of this study is to investigate the relationship between the state of periodontal disease and the severity of atherosclerosis. The subjects were 71 hypertensive patients (aged 61.1 ± 9.6 years). The mean of the probing depth (PD) at all measured sites, the mean PD at deepest sites each tooth and percentage of sites with bleeding on probing (BOP) were 4.0 ± 1.0 mm, 6.2 ± 2.2 mm, and 43.6 ± 29.4% respectively, indicating markedly advanced periodontal disease. The pulse wave velocity (PWV) was significantly higher ($p<0.05$, t-test) in the high BOP group (BOP ≥ 40%) than in the low BOP group (BOP < 40%), and was higher in the large PD group (PD ≥ 4 mm) than in the small PD group (PD < 4 mm). Increased PWV values indicate an increased arterial stiffness. Therefore, these findings suggest that advanced periodontal disease in hypertensive patients is associated with increased atherosclerosis risk.

Key words. PWV, probing depth, bleeding on probing, atherosclerosis, hypertension

1 Introduction

Hypertension is a risk factor for coronary artery disease (CAD). Also periodontal disease has been reported as a potential risk factor for CAD. The purpose of this study is to investigate the relation between the periodontal status in hypertensive patients and the severity of atherosclerosis, intimately involved in CAD.

K. Tamazawa (✉) and H. Shimauchi
Division of Periodontology and Endodontology, Tohoku University Graduate School of Dentistry, 4-1 Seiryo-machi, Aoba-ku, Sendai 980-8575, Japan
e-mail: tamazawa@mail.tains.tohoku.ac.jp

Y. Tamazawa
Division of Infection Control, Tohoku University Hospital, Sendai, Japan

T. Sasano et al. (eds.), *Interface Oral Health Science 2009*,
DOI 10.1007/978-4-431-99644-6_100,

2 Materials and Methods

This study was carried out in 71 subjects diagnosed as hypertensive patients (61.1 ± 9.6 years) after obtaining informed consent. Primary evaluation items were the probing depth (PD), bleeding on probing (BOP), pulse wave velocity (PWV), and bacterial flora from subgingival plaques. PWV, which is an index of atherosclerosis, was measured using automatic waveform analyzer (BP-203RPE; Colin, Komaki, Japan). Subgingival plaques were cultured under anaerobic conditions for 96 h, and the bacteria were identified. The PWV value and plaque bacterial flora were compared between large PD group (PD ≥ 4 mm) and small PD group (PD < 4 mm), and compared between high BOP group (BOP ≥ 40%) group and low BOP group (BOP < 40%).

3 Results

1. The mean PWV was 1,650 ± 310 cm/s, elevated above normal values.
2. The mean PD of all measured sites was 4.0 ± 1.0 mm, the mean largest PD at six sites each tooth was 6.2 ± 2.2 mm, and percentage of sites with BOP was 43.6 ± 29.4%, indicating markedly advanced periodontal disease.
3. PD was significantly larger in the high BOP group than in the low BOP group ($p < 0.0001$), and the BOP rate was significantly higher in the large PD group than in the small PD group ($p < 0.0001$).
4. The PWV was significantly higher ($p < 0.05$, *t*-test) in the high BOP group than in the low BOP group. Similarly, it was higher in the large PD group than in the small PD group.
5. In bacterial flora, the detection rate of species of the genus *Prevotella* was significantly higher ($p < 0.01$, *t*-test) in the large PD group than in the small PD group, and similarly higher ($p < 0.05$, *t*-test) in the high BOP group than the low BOP group.

4 Discussion

We defined atherosclerosis as PWV ≥ 1,400 cm/s [1]. The mean PWV was 1,650 cm/s in our subjects, remarkably over normal value, and that was moreover higher in the high BOP group and in the large PD group. These results may explain that hypertensive patients with advanced periodontal disease should be assessed for atherosclerosis risk.

Also, the detection rate of species of the genus *Prevotella* was significantly higher in the group with advanced periodontal disease. *Prevotella* bacteria may play an active role in the progression of arterial lesions [2].

Acknowledgment This study was supported by the JSPS grant (No. 15209071)

References

1. Yamashina A, Tomiyama H, Arai T et al (2003) Brachial-ankle pulse wave velocity as a marker of atherosclerotic vascular damage and cardiovascular risk. Hypertens Res 26:615–622
2. Haraszthy VI, Zambon JJ, Trevisan M et al (2000) Identification of periodontal pathogens in athermanous plaques. J Periodontol 71:1554–1560

Leading a patient to a dental office: the evaluation of pain and stress during the dental treatment using an air-pad sensor system

Shigeru Shoji, Keiko Yamaki, Koji Hanawa, Terumi Takemoto, Fumio Obayashi, and Kazuo Yoshida

Abstract. To ensure the general health of people, every dentist should work toward maintaining the dental health of the public. But most people dislike going to a dental clinic to take treatment. To encourage patients to undergo dental treatment, we should provide them with painless treatment. The purpose of this study is to evaluate patient's pain and stress during dental treatment using an air-pad sensor to impart painless dental treatment. We have to measure the change of respiration and heartbeat using air-pad sensor. We obtain a low frequency (2–7 Hz) of body vibration using an air-pad sensor. We observe the change in height and width of heartbeat and respiration during the dental treatment. This would help us scientifically determine the pain and stress a patient undergoes during dental treatment.

Key words. dental treatment, pain, stress, air-pad sensor

1 Purpose

Many patients felt that because of the pain and dread caused by dental treatment they preferred not to go to a dental clinic to take treatment [1]. Hence, if we establish a painless dental treatment, several patients will stop hesitating to go to a dental clinic. However, it is very difficult to detect and prove pain and stress scientifically. The purpose of this study is to evaluate patient's pain and stress during dental treatment using an air-pad sensor to establish painless dental treatment.

S. Shoji (✉)
Department of Periodontics and Endodontics, Tohoku University Graduate School of Dentistry, Sendai, Japan
e-mail: shoji@mail.tains.tohoku.ac.jp

K. Yamaki, K. Hanawa, T. Takemoto, F. Obayashi and K. Yoshida
The Yoshida Dental Mfg. Ltd., Tokyo, Japan

T. Sasano et al. (eds.), *Interface Oral Health Science 2009*,
DOI 10.1007/978-4-431-99644-6_101, © Springer 2010

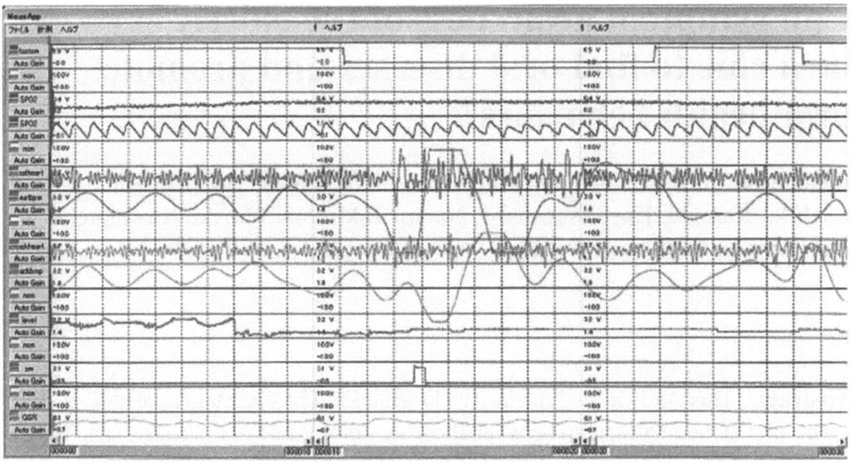

Fig. 1 Record of heartbeat and respiration using an air-pad sensor

2 Materials and Methods

We measure the change of respiration and heartbeat using an air-pad sensor system. This sensor can record from 0.1 Hz to 1 kHz of body vibration. To record and analyze the data, we use Labview system (National Instruments Co. Ltd.). With the consent of patients, we measure and record the vibration during scaling treatment using an ultrasonic scaler. The treatment is recorded by a CCD video camera.

3 Results

We obtain the low frequency (2–7 Hz) of body vibration using an air-pad sensor.

When a patient feels pain or stress, the interval of respiration is enlarged as R–R space in an electrocardiogram. We observe the change in height and width of heartbeat (Fig. 1). This helps us determine the pain and stress during the dental treatment.

Conclusion

The results of this study prove that it is possible to determine the pain and stress of a patient during dental treatment scientifically.

Reference

1. Streffer ML, Buchi S, Morgeli H et al (2009) PRISM (pictorial representation of illness and self measurement): a novel visual instrument to assess pain and suffering in orofacial pain patients. J Orofac Pain 23:140–146

Can symptom awareness of the elderly be a clue to find oral diseases and promote oral health behaviors?

Reiko Sakashita, Tomoko Miyashiba, Kumiko Otsuka, Takuichi Sato, Michiko Kamide, Kayo Watanabe, Naomi Takimoto, Mariko Kawaguchi, and Tomoko Nishihira

Abstract. This study aimed to clarify (1) what kind of symptoms the elderly were aware of, (2) the relationship between those symptoms and oral diseases, and (3) the relationship between those symptoms and oral health behaviors. Subjects consisted of 459 individuals 60 years and over, who were asked about subjective symptoms and oral health behaviors, and given an oral health examination. Findings were: (1) even though most subjects (75.2%) had the subjective symptoms, 55.7% of them did not think of them as health problems, (2) logistic regression analysis revealed that those who had subjective symptoms were at higher risk to have decayed teeth, periodontitis, and missing teeth ($p<0.01$–0.05), and (3) the elderly who had oral complaints or the subjective symptoms used an interdental brush or a dental floss much more often than those who did not ($p<0.05$). However, the elderly who had the oral complaint showed negative responses towards the visiting dentists ($p<0.05$).

Key words. the elderly, subjective symptom, oral care, oral diseases, health behaviors

---

R. Sakashita (✉), T. Miyashiba, K. Otsuka, and T. Nishihira
Nursing Foundation, College of Nursing Art and Science, University of Hyogo, Kitaoujicho 13-71, Akashi 673-8588, Japan
e-mail: sakashita@cnas.u-hyogo.ac.jp

T. Sato
Division of Oral Ecology and Biochemistry, Tohoku University Graduate School of Dentistry, Sendai, Japan

M. Kamide and K. Watanabe
Hyogo Dental Hygienists' Association, Kobe, Japan

N. Takimoto
Ishihara Dental Clinic, Nagoya, Japan

M. Kawaguchi
Department of Hematology, Kobe University Hospital, Kobe, Japan

T. Sasano et al. (eds.), *Interface Oral Health Science 2009*,
DOI 10.1007/978-4-431-99644-6_102, © Springer 2010

1 Introduction

To promote oral health, it is important that the elderly should be aware of their symptoms and continue monitoring them, to practice proper health behaviors such as oral cleaning and visiting a dentist. This study aimed to clarify the hypothesis that symptom awareness can be a clue to find oral diseases and the promotion of good oral health behaviors among the elderly.

2 Symptom Awareness

Subjects of the study consisted of 459 individuals 60 years and over, who were asked about subjective symptoms and oral health behavior, and given an oral health examination [1]. Even though most subjects (345 individuals = 75.2%) had subjective symptoms, 55.7% of them did not think of them as health problems. The level of those who had complaints or subjective symptoms went down with age as described previously [2], and the level of those who sensed dry mouth or bad breath were more frequent among females than among males.

3 Relationship Between Symptoms and Oral Diseases

Logistic regression analysis revealed that there were no relationships between oral complaints and oral diseases, however, those who had subjective symptoms were at higher risk to have decayed teeth, periodontitis, and missing teeth ($p < 0.01$–0.05).

4 Relationship Between those Symptoms and Oral Health Behaviors

The elderly who had oral complaints used an interdental brush or a dental floss and cleaned their tongue much more often than those who did not ($p < 0.05$), but they showed negative responses regarding visiting a dentist ($p < 0.05$).

Likewise, the elderly who had subjective symptoms used an interdental brush or dental floss and cleaned their tongue much more often than those who did not ($p < 0.05$). No relationship was observed between having subjective symptoms and visiting a dentist.

5 Conclusions

Based on these findings, it may be possible to manage oral health using subjective symptoms as a start. Symptom awareness also can be a clue to promote effective oral health behaviors for oral hygiene but cannot be used as a reliable criterion for dental checkups. Still, more than half of the elderly did not think that the symptoms were indicative of health problems, and the level of those who were aware of the symptom went down with age.

References

1. Sakashita R, Otsuka K, Watanebe K et al (2009) Study on relationships among oral complaint, subjective symptom and oral disease to promote oral health behavior of the elderly. UH CNAS, RINPC Bulletin 16:1–12
2. Nishiura S, Nakashima Y, Mori K et al (2008) Life span study of exploratory eye movements in healthy subjects. Kurume Med J 54:65–72

The study of mandibular position applied to oral appliance for treatment of obstructive sleep apnea syndrome

Toshimi Ito, Toru Ogawa, Tasuku Suzuki, Michikazu Matsuda, and Keiichi Sasaki

Abstract. Oral appliances are generally designed to displace one's mandible forward and downward, thus increasing the airway patency of the patient. Although it has been reported that the activity of the muscle genioglossus (GG) is increased by displacing the mandible protrusively, there is no study evaluating the relation between GG activity and the vertical displacement of mandible.

Key words. OAs, mandibular position, obstructive sleep apnea syndrome

1 Introduction

Oral Appliances (OAs) are generally designed to displace one's mandible forward and downward so as to increase the airway patency of the patient anatomically. It is also suggested that the genioglossus (GG) muscle may play an important role on the airway patency. Although it has been reported that the activity of GG is increased by displacing the mandible protrusively, there is no study evaluating the relation between the GG activity and the vertical displacement of mandible. Therefore, the aim of this study was to examine the relationship between GG activity and mandibular displacement considering both protrusive and vertical movements.

2 Recordings of GG Activity at Predetermined Mandibular Position

Nine healthy male adults (age: 27.5 ± 1.30) were employed for the study. The individualized maxillary and mandibular OAs were manufactured with 2 mm-thick thermoplastic resin plates (ERKODUR®, ERKODENT) for each subject, involving

T. Ito (✉), T. Ogawa, T. Suzuki, M. Matsuda, and K. Sasaki
Division of Advanced Prosthetic Dentistry, Tohoku University Graduate School of Dentistry, 4-1 Seiryo-machi, Aoba-ku, Sendai 980-3375, Japan
e-mail: toshimi-thk@umin.ac.jp

T. Sasano et al. (eds.), *Interface Oral Health Science 2009*,
DOI 10.1007/978-4-431-99644-6_103, © Springer 2010

all the existing teeth in the arc. The muscular activity of the left GG was recorded using the custom-made bipolar silver ball electrodes. The EMG activity was identified as GG activity by tonic burst activity during tongue protrusion and maximum mouth opening and rhythmic phasic activity coinciding with inspiration in upright body position. Breathing activity was recorded by a disposable air flow sensor (Dymedix Corporation). The GG activity was recorded in six mandibular positions (2, 4, 8, 12 mm open position and 50% protrusive position, maximum protrusive position) and two body positions (upright and supine positions) by using the OAs and silicone bite blocks.

3 Relationship Between GG Activity and Mandibular Position

In supine body position, the EMG activities of GG, during inspiration at 12 mm open position were significantly larger than those at 4 mm open position. At 12 mm open position, the GG activities during inspiration in supine position were significantly larger than those in upright position. Furthermore, the GG activities at 12 mm open position can be recognized as comparable to those recorded at 50% protrusion.

On the other hand, the air flow activities were not affected significantly by the mandibular position and body position.

These results indicate that the GG activity was influenced not only by protrusive mandibular positioning, but also by vertical mandibular positioning, i.e., jaw opening. Furthermore, the GG activity might represent a compensatory system as an attempt to overcome the obstructive effect of a vertical mouth opening, thus maintaining the pattern of respiratory function. From the clinical point of view, the present study suggested that OAs for obstructive sleep apnea syndrome patients may be effective once they provide an appropriate mandibular positioning, considering both protrusive and vertical mandibular positioning.

Prediction of future number of remaining teeth of Japanese elderly, based on data from the national survey of dental diseases in Japan

Katsuhiko Taura, Yudai Yamada, Jun Suzuki, Emi Ito, and Takeyoshi Koseki

Abstract. "8020" Movement (retain 20 natural teeth at the age of 80) has proposed from 1989 for prolonging the life span of teeth. The purpose of this study is to predict the attainment time "8020". We analyze the data of the nine national surveys of dental diseases in Japan conducted every 6 years from 1957 through 2005 to determine the actual trend of retention of natural teeth. The average numbers of remaining teeth of ten 6-year stratified age groups are determined by calculating the rates of tooth loss in 6 years of corresponding age groups for the period 1957 through 2005. Future rates of tooth loss are then estimated using regression analysis on the assumption that the present rate of improvements in dental care, preventive measures, and public awareness for oral health would continue. Our estimation suggests that the "8020" goal would not be attained until after 2029 or 2035.

Key words. "8020", prediction of future trend, survey of dental diseases in Japan, number of remaining teeth

1 Introduction

"8020" Movement has developed from 1989 for prolonging the life span of teeth. We reported that the "8020" goal would not be attained until after 2065 or 2071, based on the national surveys of dental disease in 1995. The purpose of this study is to predict the attainment time "8020" using recent national surveys of dental disease in Japan.

K. Taura (✉) and Y. Yamada
Division of Preventive Dentistry, Department of Oral Health Enhancement, Dental Care Center, Tohoku University Hospital, Sendai, Japan
e-mail: ktaura@mail.tains.tohoku.ac.jp

J. Suzuki, E. Itoh, and T. Koseki
Division of Preventive Dentistry, Department of Oral Health and Development Science, Tohoku University Graduate School of Dentistry, Sendai 980-8575, Japan

T. Sasano et al. (eds.), *Interface Oral Health Science 2009*,
DOI 10.1007/978-4-431-99644-6_104, © Springer 2010

2 Materials and Methods

We used the data of the nine national surveys of dental diseases in Japan conducted every 6 years from 1957 through 2005 to determine the actual trend of retention of natural teeth. The average numbers of remaining teeth of ten 6-year stratified age groups were determined by calculating the rates of tooth loss in 6 years of corresponding age groups for the period 1957 through 2005. Future rates of tooth loss were then estimated using regression analysis of the rates of each age group.

3 Results and Discussion

Mean number of teeth for the group aged 74–79 in the 1987 survey remained 6.1 teeth, and increased to 7.4 and 11.4 in the 1993 and 2005 surveys, respectively. In three age groups, 56–61, 62–67, and 68–73, mean remaining teeth increased at 6.1, 6.8, and 6.9, respectively, between the 1987 and 2005 surveys. This increase in mean from 1987 through 2005 was about six times compared to the 1.1 teeth increase during the period between 1957 and 1987 reported in a survey. Future rates of tooth loss were estimated using regression analysis of the rates of each age group on the assumption that the present rate of improvements in dental care, preventive measures, and public awareness for oral health would continue after 2005. This projection of future remaining teeth in the elderly is dependent upon numerous assumptions. According to the results of those projections for the year 2011, 2017, 2023, 2029, and 2035, and mean remaining teeth for age-group 74–79 estimated 22.3 teeth, we estimated that the attainment time of 8020 will be in 2030s.

The projection of remaining teeth in the present study exceeds 0.5–2.5 teeth compared to our previous study. These improvements will be expected in the near future.

The goal of Healthy Japan 21 (Health promotion activity in twenty-first century Japan) is that more than 20% of the elderly population, persons more than 80-years old, retain at least 20 natural teeth in the national survey of dental diseases. We attained this goal in 2005 national survey.

These values of oral health have improved in a couple of recent surveys. It depends upon oral health consciousness and secondary prevention of tooth loss under 8020 campaign these days. However, because of aging, a few number of intact teeth is several and severe periodontal disease, these factors might be influenced on the assumption of tooth loss rate after age of 60. The projections of attainment time “8020” includes some problems. With an effort related to dentistry, it may be possible that tooth retention rate will continue to improve and attain 8020 in near future. In this calculation, we suggest that the“8020” goal would not be attained until after 2029 or 2035.

National survey on the school-based fluoride mouth rinsing program in Japan: proposition regarding final assessment of Healthy Japan 21 in 2010, and in 2020

Katsuhiko Taura, Kazunari Kimoto, Satoru Haresaku, Osamu Sakai, and Takeyoshi Koseki

Abstract. The purpose of this study is to clarify the spread conditions of the school-based fluoride mouth rinsing (s-FMR) program since 1983 and to estimate the total number of schools and children who would be participating in the s-FMR for proposition regarding final assessment of Healthy Japan 21 in 2010, and to calculate the estimated number in 2020. In 2008, the total number of schools and children who participated in the program were 6,433 and 674,141, respectively. It shows that the number of children participating in the program corresponds to 5.1% of the same-aged population. The estimated number of schools and children participating in the program would approximately be 7,262 and 7,386,231 in 2010, and 14,770, and 1,523,632 in 2020, respectively, calculated on basis of data from 1983 to 2008. We need the cooperation of dental organizations, dental schools, and municipal corporations to achieve the goals in the s-FMR.

Key words. fluoride mouth rinsing, school-based, questionnaire, public dental care

---

K. Taura (✉)
Division of Preventive Dentistry, Department of Oral Health Enhancement, Dental Care Center, Tohoku University Hospital, Sendai, Japan
e-mail: ktaura@mail.tains.tohoku.ac.jp

K. Kimoto
Division of Oral Health, Department of Health Science, Kanagawa Dental College, Yokusuka, Japan

S. Haresaku
Department of Preventive and Public Health Dentistry, Fukuoka Dental College, Fukuoka, Japan

O. Sakai
Non-profit Japanese Conference on the Promotion of the Use of Fluoride in Caries Prevention (NPO-JPUF), Yokosuka, Japan

T. Koseki
Division of Preventive Dentistry, Department of Oral Health and Development Science, Tohoku University Graduate School of Dentistry, Sendai, Japan

T. Sasano et al. (eds.), *Interface Oral Health Science 2009*,
DOI 10.1007/978-4-431-99644-6_105,

1 Introduction

School-based fluoride mouth rinsing (s-FMR) is a safe, effective, easy, and an inexpensive caries-preventive public health method to be learnt and practiced by preschoolers and school going children. The purpose of this study is to clarify the spread conditions of the s-FMR program since 1983 and to estimate the total number of schools and children participating in the s-FMR for proposition regarding final assessment of Healthy Japan 21 in 2010. We calculate the estimated number in 2020 and set up the next s-FMR goal in 2020 using these results to promote the s-FMR throughout Japan.

2 Methods

Data were collected by questionnaire surveys regarding the schools and children participating in the s-FMR program. The other contents of the questionnaire for the s-FMR were source of financial support frequency of rinsing per week, fluoride concentration in the rinsing solution, and used agent for mouth rinsing. Questionnaires were sent to the key persons and dentists of NPO-JPUF by mail or e-mail nearly every 2 years. The latest survey was conducted in 2008, cooperating with NPO-JPUF, 8020 Promotion Foundation, and WHO Collaborating Center for Translation of Oral Health Science. Additionally, the estimated number in 2010 and 2020 were determined using polynominal regression analysis based on data from 1983 to 2008, by Stat View 5.0.

3 Results and Discussion

All of 47 prefectures in Japan adopted in s-FMR after 2005. It shows the increasing number of schools and children in s-FMR nearly every 2 years (1983–2008). In 2008, the total number of schools and children participating in the program were 6,433 and 674,141, respectively. It shows that the s-FMR rate by the total schools in the nation consists of 11.1% of nursery school and kindergartens, 9.1% of primary schools, 2.7% of secondary schools, and 3.1% of special schools, respectively (9.0% of schools). Furthermore, the number of children participating in the s-FMR program corresponds to about 5.1% of the same-aged population. The estimated numbers of schools and children who will participate in s-FMR program are 7,262 and 7,386,231 in 2010, and 14,770 and 1,523,632 in 2020, respectively, calculated on basis of data from 1983 to 2008. These results suggest that cooperation among dental organizations, dental schools, and municipal corporations can play an important role in order to spread s-FMR in Japan. We propose that a goal, the item with regarding to caries prevention, for s-FMR should be adopted in Healthy Japan 21 at the final assessment in 2010 in order to attain the goal, which caries experience (DMFT) of permanent teeth is below 1.0 at 12 years old. Furthermore, some public health measures based on health promotion should be provided, in order to make a big jump for s-FMR in the future.

A new intra-oral pressure monitor for screening swallowing dysfunction

Tatsuo Aoba, Jun Suzuki, Naoko Tanda, Kyoko Ikawa, Katsuhiko Taura, Emi Ito, and Takeyoshi Koseki

Abstract. Swallowing, which is important for eating and drinking for survival, is one of the basic functions of the mouth. The purpose of this study was to develop a safe and easily-handleable measuring device to evaluate swallowing processes in younger generations. An intra-oral pressure-measuring device estimated the intra-oral positive and negative pressures produced by sucking or blowing. The results indicated that the maximal positive pressure or minimal negative pressure increased or decreased depending upon age. This suggests that the function to produce intra-oral negative pressure, which was made with lip and faucial isthmus closure for sucking, and positive pressure, which was made with lip closure for blowing, was developed with age or physical strength. This simple measuring system of oral pressure was useful in fieldworks for screening oral dysfunction among all generations, not depending upon the condition of tooth or dentition.

Key words. intra-oral pressure, swallowing dysfunction, screening, sucking, blowing

1 Introduction

Swallowing is one of the basic functions of the mouth to eat or drink for their living. The impairment of any part of the swallowing process, dysphagia, increases the risk of aspiration pneumonia in all generations. Intra-oral pressure made with lip and tongue muscles are important to stabilize the intra-oral structure, including the form of dentition. Simple and easily-handled tool to measure intra-oral pressure is expected, especially for using in the field. The purpose of this study was to develop a safe and easily-handleable measuring device to evaluate the swallowing processes in younger generations.

T. Aoba, J. Suzuki, N. Tanda, K. Ikawa, K. Taura, E. Ito, and T. Koseki (✉)
Division of Preventive Dentistry, Department of Oral Health and Development Sciences, Tohoku University Graduate School of Dentistry, 4-1 Seiryo-machi, Aoba-ku, Sendai 980-8575, Japan
e-mail: tkoseki@mail.tains.tohoku.ac.jp

T. Sasano et al. (eds.), *Interface Oral Health Science 2009*,
DOI 10.1007/978-4-431-99644-6_106, © Springer 2010

2 Relation Between Positive and Negative Intra-oral Pressure

Intra-oral pressures were measured by THU meter on healthy children and youth. The results indicated that the maximal positive pressure (blowing) or minimal negative pressure (sucking) increased or decreased depending upon their age. Positive oral pressure (blowing) was almost under 5 kPa for 2–5 year olds and increased to over 10 kPa during teenage. Negative oral pressure (sucking) was about −5 kPa at a young age and decreased to −15 kPa during teenage. Therefore, a child who could make higher positive pressure was also making lower negative pressure.

3 Monitoring the Developing Process of Swallowing

At newborn, the primitive reflection induced by breast-feeding progresses into sucking and swallowing till about six months after birth. Positioning of dentition is determined by intra-oral muscle, tongue, and extra-oral muscle, lip/cheek, which are trained by breast-feeding. Bottle-feeding is not enough to exercise these intra/extra muscles because the baby needs only a leaking force for sucking. The function to produce intra-oral negative pressure, which was made with lip and facial isthmus closure for sucking, and positive pressure, which was made with lip closure for blowing, is developed with age or physical strength, after the establishment of the swallowing process in infants. We need further studies about the influence of breast/bottle-feeding to study the function of swallowing in younger generation.

4 Need for Screening Device in General Health Check-up

The most definitive diagnosis system for the disorder of swallowing is considered to be Videofluorography (VF). This contrast radiography is a useful system where patients swallow contrast medium such as barium, and then muscle movements during swallowing can be observed. But it is not available in field study or in general health check-up. One of the most common tests for swallowing is drinking-water test; however, it is used just as a screening test to find the cases of disabilities. Because oral function is strongly related to oral muscle, and also to intra-oral pressure, the intra-oral pressure-measuring device will be one of the easiest and useful tools to estimate the oral function of swallowing. Because almost all methods, except the THU meter, utilize the parameter of tooth or dentition, there are few methods to estimate oral function throughout life, from a newborn baby to an elderly person who has no teeth.

5 Conclusion

This simple measuring system of oral pressure will be useful in fieldworks for screening oral dysfunction for all generations and assess the value not depending upon the condition of tooth or dentition.

A numerical simulation method for dental occlusion with forces applied to the tooth in mandible

Tokumasa Akashi, Yoshihiro Takao, Masahiko Terazima, Wen-Xue Wang, and Akihiko Nakashima

Abstract. This chapter presents the way to simulate dental occlusion by using a three-dimensional finite element method. A series of conventional linear elastic analysis is proposed. Thus, two issues are discussed: how to define occlusal loads and how to realize the delicately distributed loads at the top surface of tooth by a small number of elements there.

Key words. finite element method, occlusion, mandible

1 Introduction

Recently, the number of patients with mandibular defect is increasing and the orthognathic surgery has become a common treatment. If the occlusal system balance is disturbed after surgery, some negative effects may take place. Thus, it is necessary to simulate the dental occlusion realized after and before operations for each patient by using both preoperative and postoperative mandible finite element method (FEM) models by using a quick and easy construction of models.

No previous literature performs the precise simulation of occlusion process. Thus, the simulation is a new and challenging subject and the contact analysis is expected.

T. Akashi
Graduate School of Engineering, Kyushu University, 744 Motooka, Nishi-ku, Fukuoka 819-0395, Japan

Y. Takao (✉) and W.-X. Wang
Research Institute for Applied Mechanics, Kyushu University, 6-1 Kasuga-koen, Kasuga, Fukuoka 816-8580, Japan
e-mail: ytakao@riam.kyushu-u.ac.jp

M. Terazima and A. Nakashima
Department of Orthodontics, Faculty of Dentistry, Kyushu University, 3-11 Maidashi, Higashi-ku, Fukuoka 812-8582, Japan

T. Sasano et al. (eds.), *Interface Oral Health Science 2009*,
DOI 10.1007/978-4-431-99644-6_107,

However, it is a Himalayan task due to a complicated contact surface topology, which needs tremendous resources and time. An alternative method is proposed successfully, which is reviewed here.

2 Numerical Method and Results

A three dimensional FEM mandible model composed of seven different materials with around 10,800 elements and 13,000 nodes, including periodontal ligament, was obtained quickly and easily by a mapping formula through around 120 representative points. Most of the elements are hexagonal with eight nodes. A model of the second molar was easily picked up from the mandible model that was made manually to prepare for the future local modification. Though the analysis was performed on the second molar, the procedure could be applied to other teeth. The numbers of node and element for the second molar are 456 and 358, respectively.

A series of conventional linear elastic analysis may be acceptable if the occlusal load is obtained as a function of time and applied to the top surface of tooth. There are several items to consider [1]. One is how to obtain the occlusal load, which is assumed to be related to the gap between teeth of the mandible and maxilla. The force F_1 and half of F_1 are assigned inside the contour of 0 mm gap and at the zone between 0 and 0.6 mm gap, respectively. Next is how to apply the delicately distributed occlusal load to the top surface divided into only 12 surfaces here. The number of elements at the top surface is increased by 2^2, 3^2, and 6^2 times, where the displacement of a new node is connected to one of old nodes (tied), appropriately. It was obtained that there is no practical difference in the equivalent stress at PDL between tied and conventionally refined mesh cases. Then, the recursion of the tied model to the original one is performed successfully by checking equivalent stresses in PDL. This yields a force conversion scheme to each node on the top surface. Another recursion with concentrated force and moment is performed too, which yields one force vector and moment vector to the model. That is, three methods are proposed, such as tied model, original model with a force conversion scheme, and original model with a concentrated force and moment.

The sliding process is realized by introducing the shear force normal to contact load and adversely parallel to the sliding direction, where the shear force is parallel to the top surface of each top element and is the product of normal force and the coefficient of friction.

Reference

1. Akashi T, Takao Y, Terajima M et al (2009) A numerical simulation method for dental occlusion with forces applied to mandible with tooth. Proceedings of biomechanical engineering conference, Sapporo, Japan, pp 327–328 (Japanese)

Tohoku-Forsyth Symposium

Osteopontin and CSF-1 in bone resorption

Susan R. Rittling

Abstract. Osteopontin (OPN) is a secreted phosphoprotein that is found in high levels in bone tissues. Expression of OPN is required for optimal bone resorption under pathological conditions, with a notable exception being in the case of bone metastases, where OPN deficiency does not affect tumor-associated bone loss. It is suggested that tumor cells are the primary source of CSF-1 in tumors metastatic to the bone, and that CSF-1 and integrin signaling pathways crosstalk in osteoclasts in the vicinity of tumors expressing CSF-1. Thus, high-level expression of CSF-1 in tumor cells may override the osteoclast defect seen in OPN-deficient mice.

Key words. osteopontin, bone metastasis, CSF-1, integrin, signaling

1 Introduction

Osteopontin (OPN) is a major noncollagenous protein of bone as its name implies [1]: it is a component of the bone matrix, accumulating at highest levels in cement lines and laminae limitantes [2, 3]. OPN is secreted by the three major bone cell types: osteoblasts, where it is a frequently assessed as a marker of osteoblast differentiation [4]; osteocytes, where its expression is regulated by mechanical stress [5, 6]; and osteoclasts, where its synthesis and deposition onto recently resorbed surfaces may facilitate further bone resorption [7]. The protein has a high affinity for hydroxyapatite, due to its acidic nature and poly-Asp sequence; it is likely that OPN secreted by any cells in the presence of exposed bone mineral will be quickly adsorbed to the mineral surface.

S.R. Rittling
The Forsyth Institute, 140 The Fenway, Boston, MA 02115, USA
e-mail: srittling@forsyth.org

T. Sasano et al. (eds.), *Interface Oral Health Science 2009*,
DOI 10.1007/978-4-431-99644-6_108, © Springer 2010

2 Osteopontin and Bone Resorption

Studies in the OPN knockout mice have shown in a series of different models that OPN is required for optimal bone resorption. The bone phenotype of young healthy mice is not affected structurally by a lack of OPN [8, 9], although there are differences in the crystal structure [10]. However, in pathologic circumstances, bone resorption is impaired in the absence of OPN. In ovariectomized mice, the extensive bone loss seen in wild-type (WT) mice does not occur in OPN-deficient mice [11]. Similarly, after hindlimb immobilization, relevant to patients restricted to bed rest and astronauts, OPN deficiency protects against bone loss in the trabecular bone of the distal femur and proximal tibia [12]. In an in vitro model of bone resorption, PTH-stimulated calcium release is suppressed in bones that lack OPN [13]. During fracture healing, there is reduced resorption of the fracture callus that is attributed to a defect in osteoclast function [14]. And finally, age-associated bone loss in female mice was not observed in OPN-deficient animals [9]. A notable exception to the effect of OPN on pathologic bone resorption was identified in the case of bone metastases. In these lesions, tumor growth within the bone marrow cavity, typically at the distal femur and proximal tibia, results in massive pathologic bone resorption over a short time period. This bone resorption is similar in WT and OPN-deficient mice [15], suggesting that tumor cells are able to override the osteoclast defect.

The resistance to bone resorption seen in the OPN-deficient mice is explained by in vitro experiments with osteoclasts. OPN-deficient osteoclasts have both differentiation and functional defects [7, 16]. In vivo, the extent of the ruffled border is reduced in OPN-deficient mice, and the clear zone appears reduced [9]. Osteoclasts differentiated from OPN–/– bone marrow precursors in vitro in the presence of RANKL and M-CSF are smaller and have fewer nuclei than their WT counterparts [16]. Bone resorption by OPN-deficient osteoclasts is impaired in vitro as well: pits formed in the absence of OPN are shallower and less complex than pits made by WT cells. This defect is partially due to a migratory defect, which can be corrected by exogenous OPN [7, 16], but the depth of the pits is not restored by exogenous protein: maximal pit depth may require OPN secretion by the osteoclasts themselves onto the resorption surface [7]. These observations are consistent with earlier work describing signaling pathways in osteoclasts that are activated by OPN [17, 18] and clearly indicate the importance of the protein for maximal osteoclast function.

3 Osteopontin Receptors

OPN binds to several different integrins, most notably those of the α_v class ($\alpha_v\beta_3$, $\alpha_v\beta_5$, and $\alpha_v\beta_1$) [19], but the protein also binds to a series of β_1-containing integrins, including $\alpha_9\beta_1$, $\alpha_5\beta_1$, $\alpha_4\beta_1$, as well as $\alpha_4\beta_7$ [20], reviewed in [21]. The binding of OPN to the latter four integrins is mediated by an SVVYGLR sequence in the NH_2-half of human OPN: this sequence is cryptic in the intact molecule [22, 23]. In addition, OPN has been shown to interact with CD44 [24], and the presence of the v6 and/or the v7 exons of CD44 appears to be required for this interaction [24, 25]. The nature of the interaction of OPN with CD44 is still not well understood: it has been suggested that OPN may not interact directly with CD44, but rather that the appropriate form(s) of CD44 stimulate interaction between OPN and β_1-containing integrins, but this point remains controversial [25–27]. In addition, a sequence in the N-terminal half of human OPN has been identified as important in lymphocyte migration [28]. Osteoclasts interact primarily with OPN through the $\alpha_v\beta_3$ integrin, with additional involvement of CD44 [29]. Thus, defects in osteoclast function in OPN-deficient osteoclasts may partially reflect a deficiency in $\alpha_v\beta_3$ integrin signaling.

4 CSF-1 in Tumor Cells

The ability of tumor cells to override the osteoclast defect in OPN-deficient mice may be related to overexpression of osteoclast-stimulating factors secreted by tumor cells. There are two molecules that are secreted by these tumor cells that are likely candidates: CSF-1 and VEGF: we have focused on the potential role of tumor cell-derived CSF-1, also known as M-CSF. CSF-1 is an obligate growth factor for cells of the monocyte lineage, which includes osteoclasts. Thus, young CSF-1-deficient mice are devoid of osteoclasts, although some osteoclast development occurs late in life in these mice [30]. CSF-1 is expressed as both a cell surface and a secreted protein [31, 32], and numerous tumor cell lines express the secreted form [33]. In vitro experiments demonstrated that the signaling pathways initiated by the $\alpha_v\beta_3$ integrin and by the CSF-1 receptor, c-FMS, interact, with PLCgamma2 being a site of interaction of the two pathways [34]; indeed, the two receptors are associated with each other on the osteoclast membrane [35]. These observations suggest that overstimulation of the c-FMS signaling pathway might substitute for deficient signaling through the $\alpha_v\beta_3$ integrin. Such an interaction has been observed in the differentiation in vitro of β_3-deficient osteoclasts, which is supported by high concentrations (100 ng/ml) CSF-1, but not in by lower levels (10 ng/ml)of the cytokine that are adequate for differentiation of control cells [36].

CSF-1 is found as both a secreted cytokine, and as a cell-surface-associated form: both these forms are required for the complete reconstitution of CSF-1-deficient mice [37]. We demonstrated that the bone metastatic cell line r3T expresses both cell-surface-associated and secreted CSF-1 [38], and that the cell-surface-associated form can support osteoclast differentiation in vitro (Fig. 1). In addition, high concentrations of CSF can protect osteoclast precursors from cell death associated with TGF-β treatment. These results, together with data demonstrating elevated CSF-1 expression in bone metastases and in serum of tumor-bearing mice, strongly suggest that tumors metastatic to the bone are the predominant source of CSF-1 in these lesions, and that overexpression of this cytokine by tumor cells contributes to the development of bone metastasis and to the override of the osteoclast defect in OPN-deficient mice.

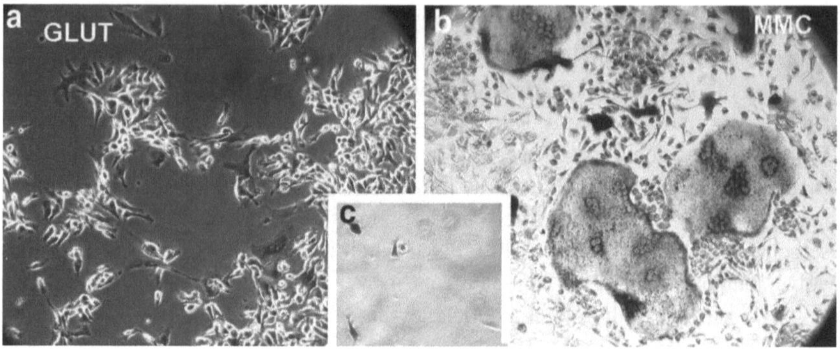

Fig. 1 r3T metastatic tumor cells support osteoclastogenesis in the presence of RANKL. r3T cells treated with glutaraldehyde (**a**) express only the cell-surface form of CSF-1, while cells treated with mitomycin C (**b**) express both the cell surface and the secreted form. Treated cells were cocultured with osteoclast precursors (OCPs) for 3 days, and osteoclasts identified by TRAP staining. Few TRAP-positive cells were seen when OCPs were incubated with RANKL in the absence of added cells (**c**)

5 Conclusion

Based on these observations, we suggest that tumor-specific expression of CSF-1 may contribute to the development of bone metastases. Therefore, CSF-1 or its receptor may be a therapeutic target in bone metastasis.

References

1. Oldberg A, Franzen A, Heinegard D (1986) Cloning and sequence analysis of rat bone sialoprotein (osteopontin) cDNA reveals an Arg-Gly-Asp cell-binding sequence. Proc Natl Acad Sci USA 83:8819–8823
2. McKee MD, Farach-Carson MC, Butler WT et al (1993) Ultrastructural immunolocalization of noncollagenous (osteopontin and osteocalcin)and plasma (albumin and α_2HS-glycoprotein) in rat bone. J Bone Miner Res 8:485–496
3. McKee MD, Nanci A (1996) Osteopontin at mineralized tissue interfaces in bone, teeth, and osseointegrated implants: ultrastructural distribution and implications for mineralized tissue formation, turnover, and repair. Microsc Res Tech 33:141–164
4. Lian JB, Stein GS (1992) Concepts of osteoblast growth and differentiation: basis for modulation of bone cell development and tissue formation. Crit Rev Oral Biol Med 3:269–305
5. Klein-Nulend J, Roelofsen J, Semeins CM et al (1997) Mechanical stimulation of osteopontin mRNA expression and synthesis in bone cell cultures. J Cell Physiol 170:174–181
6. Terai K, Takano-Yamamoto T, Ohba Y et al (1999) Role of osteopontin in bone remodeling caused by mechanical stress. J Bone Miner Res 14:839–849
7. Chellaiah MA, Kizer N, Biswas R et al (2003) Osteopontin deficiency produces osteoclast dysfunction due to reduced CD44 surface expression. Mol Biol Cell 14:173–189
8. Rittling SR, Matsumoto HN, McKee MD et al (1998) Mice lacking osteopontin show normal development and bone structure but display altered osteoclast formation in vitro. J Bone Miner Res 13:1101–1111

9. Franzen A, Hultenby K, Reinholt FP et al (2008) Altered osteoclast development and function in osteopontin deficient mice. J Orthop Res 26:721–728
10. Boskey AL, Spevak L, Paschalis E et al (2002) Osteopontin deficiency increases mineral content and mineral crystallinity in mouse bone. Calcif Tissue Int 71:145–154
11. Yoshitake H, Rittling SR, Denhardt DT, Noda M (1999) Osteopontin-deficient mice are resistant to ovariectomy-induced bone resorption. Proc Natl Acad Sci USA 96:8156–8160 [published erratum appears in Proc Natl Acad Sci U S A 1999 Sep 14;96(19):10944]
12. Ishijima M, Tsuji K, Rittling SR et al (2002) Resistance to unloading-induced three-dimensional bone loss in osteopontin-deficient mice. J Bone Miner Res 17:661–667
13. Ihara H, Denhardt DT, Furuya K et al (2001) Parathyroid hormone-induced bone resorption does not occur in the absence of osteopontin. J Biol Chem 276:13065–13071
14. Duvall CL, Taylor WR, Weiss D et al (2007) Impaired angiogenesis, early callus formation, and late stage remodeling in fracture healing of osteopontin-deficient mice. J Bone Miner Res 22:286–297
15. Natasha T, Kuhn M, Kelly O, Rittling SR (2006) Override of the osteoclast defect in osteopontin-deficient mice by metastatic tumor growth in the bone. Am J Pathol 168:551–561
16. Suzuki K, Zhu B, Rittling SR et al (2002) Colocalization of intracellular osteopontin with CD44 is associated with migration, cell fusion, and resorption in osteoclasts. J Bone Miner Res 17:1486–1497
17. Chellaiah M, Kizer N, Silva M et al (2000) Gelsolin deficiency blocks podosome assembly and produces increased bone mass and strength. J Cell Biol 148:665–678
18. Chellaiah MA, Soga N, Swanson S et al (2000) Rho-A is critical for osteoclast podosome organization, motility, and bone resorption. J Biol Chem 275:11993–12002
19. Denhardt DT, Noda M, O'Regan AW et al (2001) Osteopontin as a means to cope with environmental insults: regulation of inflammation, tissue remodeling, and cell survival. J Clin Invest 107:1055–1061
20. Ito K, Kon S, Nakayama Y et al (2009) The differential amino acid requirement within osteopontin in [alpha]4 and [alpha]9 integrin-mediated cell binding and migration. Matrix Biol 28:11–19
21. Ramaiah SK, Rittling S (2007) Pathophysiological role of osteopontin in hepatic inflammation, toxicity and cancer. Toxicol Sci 103:4–13
22. Yokosaki Y, Matsuura N, Sasaki T et al (1999) The integrin alpha(9)beta(1) binds to a novel recognition sequence (SVVYGLR) in the thrombin-cleaved amino-terminal fragment of osteopontin. J Biol Chem 274:36328–36334
23. Barry ST, Ludbrook SB, Murrison E, Horgan CM (2000) A regulated interaction between alpha5beta1 integrin and osteopontin. Biochem Biophys Res Commun 267:764–769
24. Weber GF, Ashkar S, Glimcher MJ, Cantor H (1996) Receptor-ligand interaction between CD44 and osteopontin (ETA-1). Science 271:509–512
25. Katagiri YU, Sleeman J, Fujii H et al (1999) CD44 variants but not CD44s cooperate with beta1-containing integrins to permit cells to bind to osteopontin independently of arginine-glycine-aspartic acid, thereby stimulating cell motility and chemotaxis. Cancer Res 59:219–226
26. Smith LL, Greenfield BW, Aruffo A, Giachelli CM (1999) CD44 is not an adhesive receptor for osteopontin. J Cell Biochem 73:20–30
27. Lin YH, Yang-Yen HF (2001) The osteopontin-CD44 survival signal involves activation of the phosphatidylinositol-3-kinase/Akt signaling pathway. J Biol Chem 276:46024–46030
28. Cao Z, Dai J, Fan K et al (2008) A novel functional motif of osteopontin for human lymphocyte migration and survival. Mol Immunol 45:3683–3692
29. Chellaiah MA, Hruska KA (2003) The integrin alpha(v)beta(3) and CD44 regulate the actions of osteopontin on osteoclast motility. Calcif Tissue Int 72:197–205
30. Yoshida H, Hayashi S, Kunisada T et al (1990) The murine mutation osteopetrosis is in the coding region of the macrophage colony stimulating factor gene. Nature 345:442–444
31. Rubin J, Biskobing DM, Jadhav L et al (1998) Dexamethasone promotes expression of membrane-bound macrophage colony-stimulating factor in murine osteoblast-like cells. Endocrinology 139:1006–1012

32. Halasy-Nagy J, Hofstetter W (1998) Expression of colony-stimulating factor-1 in vivo during the formation of osteoclasts. J Bone Miner Res 13:1267–1274
33. Schwaninger R, Rentsch CA, Wetterwald A et al (2007) Lack of noggin expression by cancer cells is a determinant of the osteoblast response in bone metastases. Am J Pathol 170:160–175
34. Nakamura I, Lipfert L, Rodan GA, Le TD (2001) Convergence of {alpha}}v{beta}3 integrin- and macrophage colony stimulating factor-mediated signals on phospholipase C{{gamma} in prefusion osteoclasts. J Cell Biol 152:361–374
35. Elsegood CL, Zhuo Y, Wesolowski GA et al (2006) M-CSF induces the stable interaction of cFms with alphaVbeta3 integrin in osteoclasts. Int J Biochem Cell Biol 38:1518–1529
36. Faccio R, Takeshita S, Zallone A et al (2003) c-Fms and the alphavbeta3 integrin collaborate during osteoclast differentiation. J Clin Invest 111:749–758
37. Dai XM, Zong XH, Sylvestre V, Stanley ER (2004) Incomplete restoration of colony-stimulating factor 1 (CSF-1) function in CSF-1-deficient Csf1op/Csf1op mice by transgenic expression of cell surface CSF-1. Blood 103:1114–1123
38. Yagiz K, Rittling SR (2009) Both cell-surface and secreted CSF-1 expressed by tumor cells metastatic to bone can contribute to osteoclast activation. Exp Cell Res 315:2442–2452

Role of amelogenin self-assembly in protein-mediated dental enamel formation

Henry C. Margolis, Felicitas B. Wiedemann-Bidlack, Barbara Aichmayer, Peter Fratzl, Seo-Young Kwak, Elia Beniash, Yasuo Yamakoshi, and James P. Simmer

Abstract. Extracellular matrix molecules play a critical role in regulating biological mineralization. This brief review describes recent advances made in our laboratory on elucidating the role of amelogenin, the predominant enamel matrix protein, in regulating the formation of the highly organized enamel tissue. Based on results obtained using dynamic light scattering, transmission electron microscopy, and small-angle X-ray scattering, we have found that the self-assembly of amelogenins is primarily regulated by pH and that the full-length amelogenin can form higher-order chain-like structures. These latter structures are comprised of uniformly sized aggregates of amelogenin called nanospheres. Based on in vitro mineralization studies, full-length amelogenin was also found to uniquely regulate the formation of parallel arrays of hydroxyapatite crystals. The unique ability of full-length amelogenin to form chain-like structures and induce organized mineral formation was dependent upon the presence of its hydrophilic C terminus.

Key words. amelogenesis, amelogenin, dental enamel, mineralization, calcium phosphates

---

H.C. Margolis (✉), F.B. Wiedemann-Bidlack, and S.-Y. Kwak
Department of Biomineralization, The Forsyth Institute, 140 The Fenway, Boston, MA 02115, USA
e-mail: hmargolis@forsyth.org

B. Aichmayer and P. Fratzl
Department of Biomaterials, Max Planck Institute of Colloids and Interfaces, 14424 Potsdam, Germany

E. Beniash
Department of Oral Biology, University of Pittsburgh, Pittsburgh, PA 48109, USA

Y. Yamakoshi and J.P. Simmer
Dental Research Laboratory, University of Michigan, Ann Arbor, MI 15261, USA

T. Sasano et al. (eds.), *Interface Oral Health Science 2009*,
DOI 10.1007/978-4-431-99644-6_109, © Springer 2010

1 Introduction

Biological mineralization is primarily regulated by extracellular matrix molecules that control crystal size, shape, and organization during mineralized tissue formation. Such regulation is of particular importance in the formation of the highly ordered dental enamel [1] that is unique in a number of ways in comparison to other vertebrate mineralized tissues. Mature enamel differs distinctly from bone and dentin in that it contains little to no residual matrix protein (1–2% by weight) and is essentially comprised of carbonated hydroxyapatite crystals (>95% by weight). Unlike the other mineralized tissues, enamel consists of extremely long, thin, and aligned apatitic crystals packed in parallel arrays (called enamel rods), which form intricate interwoven patterns. Importantly, the proper formation of dental enamel has been shown [2] to be critically dependent upon the secretion of the predominant enamel matrix protein, amelogenin, which makes up >90% of the enamel matrix. Although the mechanism of enamel formation is still not completely understood, considerable progress has been made toward elucidating the mechanism by which enamel matrix proteins regulate enamel mineralization. This brief review will highlight recent advances in this area that have been made in our laboratories using native and recombinant enamel matrix proteins in vitro.

2 Primary Sequence and Properties of Amelogenin

Full-length amelogenins from pig (P173) and mouse (M180) are highly homologous proteins with 173 and 180 amino acids, respectively [1]. Amelogenins are characterized by three prominent amino acid domains: a hydrophobic 45-amino acid N-terminal domain rich in tyrosine; a large central predominantly hydrophobic proline-rich domain; and an 11-amino acid C-terminal domain that is charged and hydrophilic. The primary structure of the N- and C-terminal regions of amelogenin is almost completely conserved across mammalian species, suggesting that these segments play an important role in amelogenesis, while variations occur in the central portion of the protein [3]. The hydrophilic C-terminal domain of amelogenin is lost first during proteolytic processing, for example, resulting in P148 (loss of the 25 amino acid C terminus), the predominant amelogenin degradation product found in secretory pig enamel. The native amelogenins also contain a single phosphate group on serine-16. Notably, recombinant (r) amelogenins studied here are not phosphorylated and lack the N-terminal methionine found in native amelogenin. Hence, full-length recombinant mouse (rM179) and pig (rP172) amelogenins are comprised of 179 and 172 amino acids, respectively.

3 Self-Assembly of Amelogenin

Earlier studies [4–7] showed that the predominant enamel matrix protein amelogenin can self-assemble to form nanometer-sized aggregates called nanospheres. These findings led investigators to suggest that amelogenin nanospheres could further aggregate into higher-order structures to guide crystal growth and organization [8–13]. These suggestions were also based on transmission electron microscopy (TEM) observations suggesting that chains of nanometer-sized spheres are associated with developing dental enamel [5, 14]. Initial observations of amelogenin aggregation by TEM and atomic force microscopy [4] were subsequently confirmed in solution using dynamic light scattering (DLS) under a variety of experimental conditions with variations in pH and temperature [1]. In general, using DLS, which provides indirect measures of particle sizes expressed as hydrodynamic radii, nanospheres under basic conditions (pH$\geq$8) were significantly larger than those observed under acidic conditions (pH$\leq$5.9). In addition, it was found under specified basic conditions using DLS [6, 11] that the size of full-length amelogenin (rM179) particles increased significantly (from around 15 to 60 nm) with an increase in temperature. DLS alone, however, provided little insight into the structure of the amelogenin aggregates. Recently, combined approaches of DLS and small-angle X-ray scattering (SAXS) provided new insight into the structure of amelogenin nanospheres [11]. As illustrated in Fig. 1a, the observed behavior of the full-length recombinant protein was best explained by a core-shell model for the nanospheres, where hydrophilic and negatively charged side chains (the shell) prevent the agglomeration of hydrophobic cores of the protein nanospheres at lower temperatures, while clusters consisting of several nanospheres start to form at elevated temperatures. Conclusions based on modeling studies of the SAXS data were also found to be consistent with the formation of chain-like assemblies of the nanospheres at elevated temperatures (Fig. 1b). In contrast, while capable of forming nanospheres, rM166 (that lacks the hydrophilic C terminus) showed very different aggregation behavior resulting in the formation of larger precipitates just above room temperature. These results, together with recent observations that rM179, unlike rM166, can regulate mineral organization

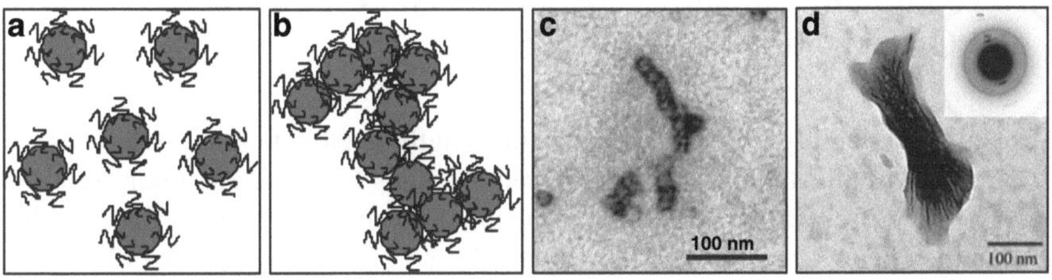

Fig. 1. Amelogenin self-assembly and regulation of mineralization. (**a**) Schematic showing a core-shell model for full-length mouse amelogenin (rM179) nanospheres at lower temperature deduced from combined dynamic light scattering (DLS) and small-angle X-ray scattering (SAXS) data. Hydrophilic and negatively charged side chains (the shell) prevent the agglomeration of hydrophobic cores of the protein nanospheres [11]. (**b**) Schematic of chain-like structures comprising amelogenin nanospheres formed at higher temperature based on modeling of DLS and SAXS data [11]. (**c**) Transmission electron microscopy (TEM) image of full-length pig amelogenin (rP172; 2 mg/mL) at pH 7.2 and 37°C showing tightly connected and elongated (*chain-like*) assemblies of amelogenin nanospheres [15], consistent with the DLS/SAXS findings (**b**). (**d**) TEM image showing high magnification of a parallel array of apatitic crystals [confirmed by selected area electron diffraction (*inset*)] formed in the presence of rM179 when mineralization and amelogenin assembly are carried out simultaneously [12]

in vitro (see below), suggest that the unique aggregation of nanospheres of the full-length amelogenin rM179 into higher-order structures may play an essential role in regulating the growth and organization of enamel crystals during enamel formation. Subsequent studies in our laboratory were then carried out to further examine the effect of temperature and pH on amelogenin self-assembly and structure under physiological pH conditions in vitro, using DLS, turbidity measurements, and TEM [15]. Full-length recombinant amelogenins from mouse (rM179) and pig (rP172) were investigated, along with proteolytic cleavage products (rM166 and native P148) lacking the hydrophilic C terminus of parent molecules. Results indicated that the self-assembly of full-length amelogenin nanospheres to larger aggregate structures is primarily triggered by the same pH (pH 7.2) in the wide temperature range from 13 to 37°C, and not by temperature. Furthermore, it was found that larger assemblies of all proteins studied formed through the rearrangement of nanospherical particles, although the onset of such further assembly took place at different pH values: pH 7.7 (P148), pH 7.5 (rM166), pH 7.2 (rP172), and pH 7.2 (rM179). In general, differences in pH values at which higher-order assembly occurred correlated with differences in the isoelectric points of the proteins studied. Importantly, structural differences were also observed. The full-length molecules formed tightly connected elongated, high-aspect ratio (chain-like) assemblies comprised of small spheres (Fig. 1c), while the amelogenin cleavage products appeared as loosely associated spherical particles [15], suggesting that the hydrophilic C terminus plays an essential role in higher-order amelogenin assembly, as was previously concluded based on combined SAX/DLS data [11]. Hence, tight pH control during secretory amelogenesis at near-neutral pH values [16–19] may serve to regulate the functions of both full-length and cleaved amelogenins.

4 Regulation of Calcium Phosphate Crystal Growth and Organization by Amelogenins

The unique assembly properties of the full-length amelogenin are also reflected in its ability to affect mineralization. Recent findings from our laboratory (and by others) on the regulation of calcium phosphate formation by a surfactant [20] and on the effect of amelogenins on silica [21] and calcium phosphate [12] mineralization in vitro have provided new evidence in support of the hypothesis that hierarchical structures in nature result from cooperative interactions between organic assembly and crystal growth. In particular, we have found [12] that calcium phosphate (hydroxyapatite) crystallites organized in parallel arrays (Fig. 1d), similar to those found in developing enamel [1], are formed in vitro when the assembly of full-length mouse amelogenin (rM179) and the spontaneous mineral formation are carried out simultaneously. These findings suggest that nascent enamel structures emerge as a result of cooperative interactions between forming crystals and assembling proteins. These findings and hypothesis are consistent with a model developed by others [22] based on the demonstration that the interplay between self-assembly of organic molecules and mineralization, and the subsequent cooperative reorganization of hybrid inorganic–organic nanostructures, can give rise to the formation of hierarchical structures. In addition, it was found [12] that rM166 that lacks the hydrophilic C terminus did not have the same effect on mineralization in vitro, as that observed with the full-length amelogenin rM179. These results demonstrated that the C-terminal domain in amelogenin is also essential for the alignment of crystals into parallel arrays. Recently, we have found [23] that the full-length recombinant porcine amelogenin (rP172) can similarly regulate the formation of calcium phosphate crystallites organized in parallel arrays. Importantly, as described in greater detail elsewhere, (see the chapter by SY Kwak, this volume), the noted formation of parallel arrays of calcium phosphate crystallites in the presence of rP172 proceeded through the formation and brief stabilization of amorphous calcium phosphate [23].

5 Conclusions

In conclusion, we have found that the self-assembly of both full-length and cleaved amelogenins is tightly regulated by pH. In addition, under physiological pH conditions, full-length amelogenins uniquely form higher-order chain-like structures comprised of uniformly sized amelogenin aggregates. The full-length amelogenin was also shown to uniquely regulate the formation of parallel arrays of hydroxyapatite crystals. The unique assembly and functional capabilities exhibited by the full-length amelogenins were found to be dependent on the presence of its hydrophilic C terminus.

Acknowledgments This work was supported by grant DE-016376 (HCM) from the National Institute of Dental and Craniofacial Research. FBW-B was partially supported by grant T32 DE-007327.

References

1. Margolis HC, Beniash E, Fowler CE (2006) Role of macromolecular assembly of enamel matrix proteins in enamel formation. J Dent Res 85:775–793
2. Gibson CW, Yuan ZA, Hall B et al (2001) Amelogenin-deficient mice display an amelogenesis imperfecta phenotype. J Biol Chem 276:31871–31875
3. Fincham AG, Moradian-Oldak J, Simmer JP (1999) The structural biology of the developing dental enamel matrix. J Struct Biol 126:270–299
4. Fincham AG, Moradian-Oldak J, Simmer JP et al (1994) Self-assembly of a recombinant amelogenin protein generates supramolecular structures. J Struct Biol 112:103–109
5. Fincham AG, Moradian-Oldak J, Diekwisch TG et al (1995) Evidence for amelogenin "nanospheres" as functional components of secretory-stage enamel matrix. J Struct Biol 115:50–59

6. Moradian-Oldak J, Leung W, Fincham AG (1998) Temperature and pH-dependent supramolecular self-assembly of amelogenin molecules: a dynamic light-scattering analysis. J Struct Biol 122:320–327
7. Moradian-Oldak J, Paine ML, Lei YP et al (2000) Self-assembly properties of recombinant engineered amelogenin proteins analyzed by dynamic light scattering and atomic force microscopy. J Struct Biol 131:27–37
8. Moradian-Oldak J, Simmer JP, Lau EC et al (1994) Detection of monodisperse aggregates of a recombinant amelogenin by dynamic light-scattering. Biopolymers 34:1339–1347
9. Moradian-Oldak J, Simmer JP, Lau EC et al (1995) A review of the aggregation properties of a recombinant amelogenin. Connect Tissue Res 32:125–130
10. Fincham AG, Luo W, Moradian-Oldak J et al (2000) Enamel biomineralization: the assembly and disassembly of the protein extracellular organic matrix. In: Smith MM, Teaford MF, Ferguson MWJ (eds) Development, function, and evolution of teeth. Cambridge University Press, Cambridge, pp 65–81
11. Aichmayer B, Margolis HC, Sigel R et al (2005) The onset of amelogenin nanosphere aggregation studied by small-angle X-ray scattering and dynamic light scattering. J Struct Biol 151:239–249
12. Beniash E, Simmer JP, Margolis HC (2005) The effect of recombinant mouse amelogenins on the formation and organization of hydroxyapatite crystals *in vitro*. J Struct 149:182–190
13. Du C, Falini G, Fermani S et al (2005) Supramolecular assembly of amelogenin nanospheres into birefringent microribbons. Science 307:1450–1454 [published erratum in Science 309(5744):2166]
14. Robinson C, Fuchs P, Weatherell JA (1981) The appearance of developing rat incisor enamel using a freeze fracturing technique. J Cryst Growth 53:160–165
15. Wiedemann-Bidlack FB, Beniash E, Yamakoshi Y et al (2007) pH triggered selfassembly of native and recombinant amelogenins under physiological pH and temperature *in vitro*. J Struct Biol 160:57–69
16. Aoba T, Moreno EC (1987) The enamel fluid in the early secretory stage of porcine amelogenesis: chemical composition and saturation with respect to enamel mineral. Calcif Tissue Int 41:86–94
17. Smith CE (1998) Cellular and chemical events during enamel maturation. Crit Rev Oral Biol Med 9:128–161
18. Smith CE, Issid M, Margolis HC et al (1996) Developmental changes in the pH of enamel fluid and its effects on matrix resident proteinases. Adv Dent Res 10:159–169
19. Smith CE, Chong DL, Bartlett JD et al (2005) Mineral acquisition rates in developing enamel on maxillary and mandibular incisors of rats and mice: implications to extracellular acid loading as apatite crystals mature. J Bone Miner Res 20:240–249
20. Fowler CE, Li M, Mann S et al (2005) Influence of surfactant assembly on the formation of calcium phosphate materials-a model for dental enamel formation. J Mater Chem 15:3317–3325
21. Fowler CE, Beniash E, Yamakoshi Y et al (2006) Co-operative mineralization and protein self-assembly in amelogenesis: silica mineralization and assembly of recombinant amelogenins *in vitro*. Eur J Oral Sci 114(Suppl 1):297–303
22. Colfen H, Mann S (2003) Higher-order organization by mesoscale selfassembly and transformation of hybrid nanostructures. Angew Chem Int Ed Engl 42:2350–2365
23. Kwak SY, Wiedemann-Bidlack FB, Beniash E et al (2009) Role of 20 kDa amelogenin (P148) phosphorylation in calcium phosphate formation *in vitro*. J Biol Chem 284:18972–18979

The human genetics of amelogenesis imperfecta

John D. Bartlett

Abstract. Amelogenesis imperfecta (AI) results from the mutation of gene(s) that affect enamel development. The resulting enamel may have alterations of enamel quantity, composition, and/or structure and typically appears abnormally thin, soft, rough, and/or stained. AI is defined as enamel defects occurring in the absence of any other generalized or systemic disease. Mutations in five different genes (*AMEL*, *ENAM*, *MMP20*, *KLK4*, and *FAM83H*) have been demonstrated to cause AI. However, mutations of these genes alone are not responsible for all known cases of AI. Therefore, AI causative genes remain to be discovered. This chapter summarizes the mutations that cause AI (genotype) and the effect of each mutation on the appearance of the dental enamel (phenotype). The goal of this chapter is to provide clinicians with a concise guide that will suggest possible genotypes by careful examination of an individual's AI phenotype and pattern of malformed enamel inheritance.

Key words. enamel, amelogenesis imperfecta, mutation

1 Introduction

Inherited enamel defects that occur in the absence of a generalized syndrome are collectively designated as AI. The classification of AI into fourteen distinct subtypes, based on clinical phenotype and mode of inheritance [1], can be broadly distilled down to three main types of AI. These are hypoplastic, hypomaturation, and hypocalcified AI. Dental enamel starts as a protein-rich matrix and ends as a virtually protein-free mineral. Although proteins are not part of the final product, they play an essential role in enamel development. Hypoplastic enamel is thin and

J.D. Bartlett
Department of Cytokine Biology, Forsyth Institute, 140 The Fenway, Boston, MA 02115, USA
e-mail: jbartlett@forsyth.org

T. Sasano et al. (eds.), *Interface Oral Health Science 2009*,
DOI 10.1007/978-4-431-99644-6_110, © Springer 2010

is associated with defective matrix synthesis that occurs as the enamel increases in thickness. Hypomaturation enamel is soft and typically stained, but it is of normal thickness and is associated with a failure to remove enamel matrix proteins. It is necessary to remove these proteins so that the enamel crystals can grow to fill the spaces between the crystallites to allow the enamel to achieve its final hardened form. Hypocalcified enamel is the most severe and appears to represent a more fundamental disturbance that affects both the early and late stages of enamel development. Hypocalcified enamel is soft, rough, and is rapidly lost by attrition. Hypoplastic, hypocalcified, and hypomaturation enamel phenotypes are each caused by a different set of gene mutations or are caused by mutations located in different locations within a gene.

AI can be inherited by autosomal dominant (ADAI), autosomal recessive (ARAI), and X-linked modes of transmission (reviewed in [2]). This chapter focuses on the genetic mutations that cause AI and on how these mutations affect an individual's teeth.

2 The Genetics of Amelogenesis Imperfecta

2.1 *AMELX (MIM 300391)*

Amelogenin is the most abundant protein present within the developing enamel matrix. It represents approximately 90% of total matrix protein. The functional amelogenin gene is located on the X-chromosome (*AMELX*, Xp22.3–p22.1). Therefore, all the daughters and none of the sons of affected males will inherit the mutant gene. A second amelogenin gene is expressed on the Y-chromosome (*AMELY*), but this gene is expressed at low levels [3], does not appear necessary for proper dental enamel formation, and does not contribute to the etiology of AI. X-linked AI accounts for about 5% of all AI cases [4, 5], and fifteen different mutations in the *AMELX* gene are currently known to cause AI. Of these 15 mutations, eight are point mutations. Thus, the substitution of just one amino acid for another at key locations within the amelogenin protein will cause AI. In general, *AMELX* mutations are associated with a smooth hypoplastic phenotype, but can also be associated with a hypomineralization/hypomaturation AI phenotype with discolored enamel [6].

Potentially more diagnostic than the pattern of inheritance (which may not be obvious in small families with few affected members) is the distinctive vertical banding pattern on the enamel of affected heterozygous females. The vertical bands are deposited because alternating bands of ameloblasts secrete normal or defective amelogenin after randomly inactivating either the defective or normal X-chromosome [7, 8]. Amelogenin deficient mouse teeth display a chalky-white discoloration and have disorganized hypoplastic enamel with a thickness less than 10% of normal [9].

2.2 *ENAM (MIM 606585)*

Enamelin is present within the developing enamel matrix and represents approximately 1–5% of total matrix protein. Mutations in a single copy of the enamelin gene (*ENAM*, 4q21) cause hypoplastic enamel. Currently, nine enamelin mutations are known to cause ADAI. The enamelin mutations result in enamel that is hypoplastic. However, those heterozygous for the p.422fsX488 mutation have a mild phenotype consisting of localized pitting defects [10] and those with the p.N197fsX227 mutation have a moderate phenotype consisting of grooves or shallow pits formed in parallel horizontal lines [11]. When both alleles are defective, there is virtually no enamel layer. Mice heterozygous for functional *Enam* have nearly normal maxillary incisors, but their mandibular incisors have chalky-white enamel that becomes severely abraded. All incisors of *Enam* deficient mice were affected. There was no true enamel, and the teeth had a white opaque appearance with severe dentin abrasion [12].

2.3 *AMBN (MIM 601259)*

Ameloblastin is present within the developing enamel matrix and represents approximately 5–10% of total matrix protein. While no AI-causing mutations have yet been identified in *AMBN*, it is likely only a matter of time before one is found, as a recessive AI phenotype consisting of severe enamel hypoplasia is manifested by *Ambn* null mice [13]. Recently, it was discovered that the *Ambn* null mice are not truly null for *Ambn*. Only exons five and six were deleted from the transcript [14]. It will be interesting to determine if complete deletion of *Ambn* results in an altered phenotype compared to the exon five and six deletion.

2.4 *MMP20 (MIM 604629)*

Matrix metalloproteinase-20 is expressed during early enamel development and cleaves the secreted enamel matrix proteins including amelogenin, enamelin, and ameloblastin. Relative to the matrix proteins, MMP20 is present in trace amounts within the developing enamel matrix. *MMP20* localizes to chromosome 11q22.3–q23, and three different MMP20 mutations are known to cause ARAI. Two of these mutations caused pigmented hypomaturation AI [15, 16], and one resulted in hypoplastic-hypomaturation AI [17]. In all three cases, the teeth were normal in size, but the enamel layer did not contrast well with dentin on radiographs, and the enamel tended to chip away from the underlying dentin. The hypoplastic-hypomaturation enamel also had surface roughness and a yellowish-brown pigmentation that was present during tooth eruption, suggesting the stain was intrinsic and not acquired [17].

The *Mmp20* null mouse does not process amelogenin properly, possesses an altered enamel protein and associated rod pattern, has hypoplastic enamel, has enamel that delaminates from the dentin, and has a deteriorating tooth morphology as enamel development progresses [18].

2.5 *KLK4 (MIM 603767)*

Kallikrein-4 is expressed during mid-late stage enamel development when the enamel has reached its full thickness and when proteins are actively removed from the hardening matrix. KLK4 degrades proteins to facilitate their removal from the enamel matrix. The Kallikrein-4 gene is on chromosome 19q13.3–q13.4 and just one KLK4 mutation is known to cause ARAI. Like the *MMP20* mutations, the proband with the *KLK4* mutation had normal sized teeth that were discolored yellow-brown; the enamel had only a slight increase in opacity over that of dentin as observed on radiographs, and the enamel fractured off the occlusal surfaces of the molars. The affected enamel appeared to be of full thickness so the proband with the homozygous *KLK4* mutation was designated as having an autosomal recessive hypomaturation phenotype [19].

The *Klk4* null mouse has enamel with a high protein content that rapidly abrades following weaning. Although the enamel was of normal thickness, the individual enamel crystallites failed to grow together, interlock, and function as a unit [20].

2.6 *FAM83H (MIM 611927)*

Family with sequence similarity 83 member H (*FAM83H*) appears to be ubiquitously expressed based on expressed sequence tag analysis. Interestingly, *FAM83H* mutations only affect dental enamel [21–24]. *FAM83H* is not secreted, but associates with intracellular vesicles and is speculated to play a role in intracellular trafficking or cytoskeletal organization [25]. *FAM83H* localizes to chromosome 8q24.3, and *Fam83H* mutations cause autosomal-dominant hypocalcified AI (ADHCAI), which is the most common form of AI in the United States [8]. Individuals with *FAM83H* mutations may have enamel that is cheesy soft in consistency, light yellow in shade, and of nearly normal thickness. However, the abraded teeth lose contour and often become tapered toward the incisal edge or occlusal surface. The abraded surfaces are rough in texture, take up stain rapidly, and are sensitive to thermal changes [23].

The first reported *FAM83H* mutation occurred in 2008, and now, just over a year later, fourteen total *FAM83H* mutations have been reported. Therefore, *FAM83H* mutations are responsible for a significant percentage of ADHCAI cases. Interestingly, all the *FAM83H* mutations occur in the last exon, and all are predicted to express a truncated protein. This type of mutation pattern suggests strongly that the truncated *FAM83H* protein acts in a dominant-negative fashion. It remains unclear why just the dental enamel is affected.

Although they do not strictly adhere to the definition of AI, due to their association with syndromes, it is worth noting that two other mutated genes cause enamel malformations. The genes are *DLX3* and *CNNM4*.

2.7 *DLX3 (MIM 600525)*

Distal-less homeobox 3 localizes to chromosome *(DLX3)*17q21.3–q22, and *DLX3* mutations cause autosomal dominant tricho-dento-osseous syndrome (TDO). TDO is characterized by enamel hypoplasia with root taurodontism, kinky curly hair at birth, and increased thickness and density of cranial bones [26–28]. Enamel hypoplasia and taurodontism are consistently found, but the nondental features may be minimal [29]. This has lead to some confusion as to whether specific *DLX3* mutations may be strictly classified as causing AI. However, it appears that detailed phenotyping of individuals with minimal nondental features reveals the TDO phenotype [30]. Therefore, *DLX3* mutations are now considered to cause TDO.

2.8 *CNNM4 (MIM 607805)*

Cyclin M4 maps to chromosome 2p12–p11.2 and *CNNM4* mutations cause autosomal recessive cone-rod dystrophy and amelogenesis imperfecta (Jalili syndrome). Affected individuals have hypomineralized enamel that is similar to AI resulting from *MMP20* and *KLK4* mutations. The enamel is soft, has a high protein content, and is stained, and the teeth demonstrate significant occlusal attrition. Unlike the *MMP20* and *KLK4* AI phenotypes, *CNNM4* tooth roots are taurodont, and, significantly, the patients suffer from an eye pathology termed cone-rod dystrophy. Considerable visual impairment occurs in infancy or early childhood, with progressive loss of vision with advancing age. The function of CNNM4 is not currently known, but based on its similarity (68.0%) and homology (56.5%) to CNNM2, which is a magnesium transporter, it is thought that CNNM4 is also a metal transporter. Two articles were published in early 2009 that together identify ten different mutations that cause autosomal recessive cone-rod dystrophy and amelogenesis imperfecta [31, 32].

3 Summary

In conclusion, mutations in five different genes (*AMEL*, *ENAM*, *MMP20*, *KLK4*, and *FAM83H*) have been demonstrated to cause AI. Also, two additional genes (*DLX3* and *CNNM4*) cause syndromic dental malformations. Although some gene mutations result in AI phenotypes that are very similar to each other, such as the

MMP20 and *KLK4* mutations, each gene mutation does result in a different overall phenotype that, based on the mode of inheritance, can be suggestive of an AI genotype. Interestingly, it is estimated that the genetic mutations known to cause AI are responsible for approximately 50% of all known AI kindreds [33]. Clearly, more genes remain to be discovered that when mutated will cause AI. The genetics of AI will be a fruitful research field for years to come.

Acknowledgment This investigation was supported by research grant DE016276 (JDB) from the National Institute of Dental and Craniofacial Research.

References

1. Witkop CJ Jr (1988) Amelogenesis imperfecta, dentinogenesis imperfecta and dentin dysplasia revisited: problems in classification. J Oral Pathol 17:547–553
2. Becerik S, Cogulu D, Emingil G et al (2009) Exclusion of candidate genes in seven Turkish families with autosomal recessive amelogenesis imperfecta. Am J Med Genet A 149A: 1392–1398
3. Salido EC, Yen PH, Koprivnikar K et al (1992) The human enamel protein gene amelogenin is expressed from both the X and the Y chromosomes. Am J Hum Genet 50:303–316
4. Backman B, Holmgren G (1988) Amelogenesis imperfecta: a genetic study. Hum Hered 38:189–206
5. Backman B (1988) Amelogenesis imperfecta – clinical manifestations in 51 families in a northern Swedish county. Scand J Dent Res 96:505–516
6. Wright JT, Hart PS, Aldred MJ et al (2003) Relationship of phenotype and genotype in X-linked amelogenesis imperfecta. Connect Tissue Res 44(Suppl 1):72–78
7. Berkman MD, Singer A (1971) Demonstration of the lyon hypothesis in X-linked dominant hypoplastic amelogenesis imperfecta. Birth Defects Orig Artic Ser 7:204–209
8. Witkop CJ, Sauk JJ (1976) Heritable defects of enamel. In: Stewart RE, Prescott GH (eds) Oral facial genetics. C.V. Mosby Co., St. Louis, pp 151–226
9. Gibson CW, Yuan ZA, Hall B et al (2001) Amelogenin-deficient mice display an amelogenesis imperfecta phenotype. J Biol Chem 276:31871–31875
10. Hart TC, Hart PS, Gorry MC et al (2003) Novel ENAM mutation responsible for autosomal recessive amelogenesis imperfecta and localised enamel defects. J Med Genet 40:900–906
11. Kida M, Ariga T, Shirakawa T et al (2002) Autosomal-dominant hypoplastic form of amelogenesis imperfecta caused by an enamelin gene mutation at the exon-intron boundary. J Dent Res 81:738–742
12. Hu JC, Hu Y, Smith CE et al (2008) Enamel defects and ameloblast-specific expression in Enam knock-out/lacz knock-in mice. J Biol Chem 283:10858–10871
13. Fukumoto S, Kiba T, Hall B et al (2004) Ameloblastin is a cell adhesion molecule required for maintaining the differentiation state of ameloblasts. J Cell Biol 167:973–983
14. Wazen RM, Moffatt P, Zalzal SF et al (2009) A mouse model expressing a truncated form of ameloblastin exhibits dental and junctional epithelium defects. Matrix Biol 28:292–303
15. Kim JW, Simmer JP, Hart TC et al (2005) MMP-20 mutation in autosomal recessive pigmented hypomaturation amelogenesis imperfecta. J Med Genet 42:271–275
16. Ozdemir D, Hart PS, Ryu OH et al (2005) MMP20 active-site mutation in hypomaturation amelogenesis imperfecta. J Dent Res 84:1031–1035
17. Papagerakis P, Lin HK, Lee KY et al (2008) Premature stop codon in MMP20 causing amelogenesis imperfecta. J Dent Res 87:56–59
18. Caterina JJ, Skobe Z, Shi J et al (2002) Enamelysin (matrix metalloproteinase 20)-deficient mice display an amelogenesis imperfecta phenotype. J Biol Chem 277:49598–49604

19. Hart PS, Hart TC, Michalec MD et al (2004) Mutation in kallikrein 4 causes autosomal recessive hypomaturation amelogenesis imperfecta. J Med Genet 41:545–549
20. Simmer JP, Hu Y, Lertlam R et al (2009) Hypomaturation enamel defects in Klk4 knockout/LacZ knockin mice. J Biol Chem 284:19110–19121
21. Hart PS, Becerik S, Cogulu D et al (2009) Novel FAM83H mutations in Turkish families with autosomal dominant hypocalcified amelogenesis imperfecta. Clin Genet 75:401–404
22. Kim JW, Lee SK, Lee ZH et al (2008) FAM83H mutations in families with autosomal-dominant hypocalcified amelogenesis imperfecta. Am J Hum Genet 82:489–494
23. Lee SK, Hu JC, Bartlett JD et al (2008) Mutational spectrum of FAM83H: the C-terminal portion is required for tooth enamel calcification. Hum Mutat 29:E95–E99
24. Wright JT, Frazier-Bowers S, Simmons D et al (2009) Phenotypic variation in FAM83H-associated amelogenesis imperfecta. J Dent Res 88:356–360
25. Ding Y, Estrella MRP, Hu YY et al (2009) Fam83h is associated with intracellular vessicles and ADHCAI. J Dent Res 88:991–6
26. Haldeman RJ, Cooper LF, Hart TC et al (2004) Increased bone density associated with DLX3 mutation in the tricho-dento-osseous syndrome. Bone 35:988–997
27. Price JA, Bowden DW, Wright JT et al (1998) Identification of a mutation in DLX3 associated with tricho-dento-osseous (TDO) syndrome. Hum Mol Genet 7:563–569
28. Price JA, Wright JT, Kula K et al (1998) A common DLX3 gene mutation is responsible for tricho-dento-osseous syndrome in Virginia and North Carolina families. J Med Genet 35:825–828
29. Price JA, Wright JT, Walker SJ et al (1999) Tricho-dento-osseous syndrome and amelogenesis imperfecta with taurodontism are genetically distinct conditions. Clin Genet 56:35–40
30. Wright JT, Hong SP, Simmons D et al (2008) DLX3 c.561_562delCT mutation causes attenuated phenotype of tricho-dento-osseous syndrome. Am J Med Genet A146:343–349
31. Parry DA, Mighell AJ, El-Sayed W et al (2009) Mutations in CNNM4 cause Jalili syndrome, consisting of autosomal-recessive cone-rod dystrophy and amelogenesis imperfecta. Am J Hum Genet 84:266–273
32. Polok B, Escher P, Ambresin A et al (2009) Mutations in CNNM4 cause recessive cone-rod dystrophy with amelogenesis imperfecta. Am J Hum Genet 84:259–265
33. Kim JW, Simmer JP, Lin BP et al (2006) Mutational analysis of candidate genes in 24 amelogenesis imperfecta families. Eur J Oral Sci 114(Suppl 1):3–12

Porphyromonas gingivalis: surface polysaccharides as virulence determinants

Annette Arndt and Mary Ellen Davey

Abstract. The proliferation of anaerobic bacteria in the subgingival crevice of the oral cavity is central to the progression of periodontal disease, with *Porphyromonas gingivalis* being implicated as one of the key pathogens. Research has clearly shown that surface polysaccharides are key virulence determinants for this organism. Here, we review what is known about the different surface polysaccharides synthesized by *P. gingivalis* as well as their role in the pathogenicity of this oral anaerobe.

Key words. *Porphyromonas gingivalis*, exopolysaccharides, lipopolysaccharides, K-antigen capsule

1 Introduction

Porphyromonas gingivalis is a Gram-negative obligate anaerobe that persists as a natural member of the oral microbiome, yet outgrowth of this commensal is associated with severe periodontal disease, resulting in destruction of the tissues supporting the gums, and ultimately, exfoliation of the teeth [1–5]. Research has clearly shown the destructive capabilities of *P. gingivalis*. This oral anaerobe can invade and multiply within gingival epithelial cells as well as penetrate into deeper epithelial cell layers, potentially releasing the whole organism and/or virulence factors into the blood stream (reviewed in [6]). Furthermore, in addition to its ability to cause disease in the mouth, there is data indicating its potential role in systemic disease, including its ability to invade endothelial cells [7–9] and to cause aggregation of platelets [10]. Recently, it was discovered that *P. gingivalis* can be detected in human atheromas [11], and that outer membrane components can induce macrophage transformation into foam cells [12]. Hence, although periodontal disease is best described as a polymicrobial disease, this oral anaerobe clearly has the ability to be highly destructive. For many pathogenic bacteria, surface polysaccharides play a key role

A. Arndt and M.E. Davey (✉)
Department of Molecular Genetics, The Forsyth Institute, Boston, MA 02115, USA
e-mail: mdavey@forsyth.org

T. Sasano et al. (eds.), *Interface Oral Health Science 2009*,
DOI 10.1007/978-4-431-99644-6_111, © Springer 2010

in immune modulation and evasion [13]. Since surface polysaccharides form the outermost layer of the cell envelope, they are often involved in direct interactions with the host and are therefore typically highly antigenic. Several aspects define their antigenicity, including monosaccharide composition, glycosidic linkages, configuration of the anomeric center of the sugars and/or conformation of the polymer (reviewed in [13]). Bacterial polysaccharides can be divided into three distinct types: exopolysaccharides (EPS), lipopolysaccharides (LPS), and capsular polysaccharides (CPS). Here, we discuss the role of these different surface polysaccharides in the virulence of this organism.

2 Exopolysaccharide and Lipopolysaccharides

EPS are large molecules (>500 kD) that are loosely associated with the bacterial cell envelope and often lead to a mucoid, slimy colony phenotype; however, a tight association of certain EPS polymers to the cell surface without the existence of membrane anchoring has been reported [14, 15]. In most biofilm communities, EPS is the major constituent of the extracellular matrix. Although the composition of the matrix produced by *P. gingivalis* during biofilm formation has not been determined, the composition of an EPS produced by *P. gingivalis* ATCC strain 53978 (W50) was determined. The main constituent of this EPS is mannose (Man), with low levels of rhamnose (Rha), glucose (Glc), galactose (Gal), and 2-acetamido-2-deoxy-D-glucose [16]. While the biochemical composition of EPS has been elucidated, the genes required for synthesis and transport of this EPS have not been identified. In contrast to EPS, LPS molecules are tightly associated with the outer membrane. LPS molecules are covalently bound to lipid A, a component of the cell envelope (reviewed in [17]). Studies by Farquharson et al. [16] determined that the O-LPS of *P. gingivalis* can be split into an O-antigen polysaccharide and a core oligosaccharide whose monosaccharide composition are quite similar (Rha, Man, Gal, Glc, and GlcNAc), yet the proportions of these sugars are distinct. Just recently, it was discovered that *P. gingivalis* synthesizes an additional surface polysaccharide, which can clearly be distinguished from O-LPS and K-antigen capsule (discussed below) [18, 19]. Because this newly discovered surface polysaccharide is anchored to the surface by lipid A, it has been categorized as an LPS molecule. The anionic polysaccharide part (APS) of this LPS molecule (A-LPS) is made of a phosphorylated branched mannan. Interestingly, studies have shown that A-LPS is immunologically related to the surface-associated proteases, Arg-gingipains, since a monoclonal antibody (MAb 1B5) raised to RgpA, cross-reacts with A-LPS [18, 20]. As shown by studies of Slaney et al. [21], deletion mutants in *porR* (PG1138) and *wbpB* (PG2119) lack APS, indicating an involvement of these genes in APS biosynthesis; however, further studies are required to determine which genes are required for synthesis of this novel LPS. Interestingly, the A-LPS mutants are more susceptible to killing by the host complement system, indicating an important role of APS in resistance to the host [21, 22]. Surprisingly, the loss of K-antigen capsule does not affect resistance to killing by the host complement system [22]. Comparisons of the lipid-A components of total lipid A of *P. gingivalis* ATCC 53978 wild-type and a *porR* (PG1138) mutant (which shows no cross-reactivity to MAb 1B5 [23]) was used to evaluate lipid-A forms specific to A-LPS. This analysis determined that O-LPS and A-LPS are linked to lipid A core molecules that have different fatty acid compositions and different degrees of phosphorylation. However, surprisingly, the same O-antigen ligase (PG1051) is responsible for linking these two distinct LPS molecules to their respective lipid A cores [19]. In addition, although both of these LPS structures induce host immune response, the response to lipid A of total LPS is significantly stronger than that of A-LPS alone, which is likely because of the chemical and structural differences of the two LPS molecules [19].

3 K-antigen Capsule

Similar to LPS, capsules are also covalently attached to either phospholipid or lipid-A in the outer membrane [14, 24, 25]. Furthermore, both types of polysaccharides are accountable for the serotype-specificity. To date, three different O-antigen serotypes and at least six K-antigen serotypes have been identified in *P. gingivalis* [26, 27]. The K-antigen capsule of *P. gingivalis* is considered a key virulence factor. The majority of isolates from patients with chronic periodontal disease are encapsulated [28], and while nonencapsulated strains typically result in localized abscesses, encapsulated *P. gingivalis* strains cause a spreading type of infection in mice with recovery from blood, spleen, and kidneys, and therefore, these strains are often described as more virulent [29, 30]. In general, capsules are made up of a hydrated dense matrix of polysaccharide composed of repeating single sugar molecules joined by glycosidic linkages, which can contain both organic and inorganic modifications. CPS can protect the bacterial cell from desiccation and can either facilitate or prevent the adherence to abiotic and biotic surfaces. Furthermore, together with LPS, CPS typically confers resistance to nonspecific and specific host immunity. Interestingly, divergence in *P. gingivalis* ability to adhere seems to be serotype specific. Rosen and Sela [31] showed that CPS, and to a lesser extent LPS, of the encapsulated *P. gingivalis* strain PK1924 act as receptors mediating the coaggregation between this member of the K5 serotype family and *Fusobacterium nucleatum* PK 1594. Free hydroxyl groups present on D-galactose (a component of the K5 capsule) are responsible for the binding of *P. gingivalis* PK 1924 to a fusobacterial lectin. The coaggregation between these two oral organisms is likely an important factor in the development of the oral biofilm in vivo. In contrast, neither the encapsulated K1-antigen strains (ATCC 53978 and W83) nor the nonencapsulated strains seem to form coaggregates with *F. nucleatum* [31]. Although at least six different *P. gingivalis K*-antigen serotypes exist, only K1 has been studied in detail. As shown by Farquharson et al. [16], K1 capsule of *P. gingivalis* ATCC 53978 consists of primarily uronic acids, including mannuronic acid, glucuronic acid, and galacturonic acid. Interestingly, the mechanism by which K-antigen capsule is anchored to the surface has not been determined. Biosynthesis and assembly of this complex polymer requires a variety of enzymes. Typically, genes involved in polysaccharide biosynthesis are clustered in large operons encoding genes involved in the synthesis of precursor molecules, glycosyltransferases, as well as proteins for export and assembly. Sequence analysis of the *P. gingivalis* strain W83 genome identified multiple potential polysaccharide biosynthesis loci [32]. Recently, it was confirmed that the region of PG0106–PG0120 encode the enzymes involved in K1-capsule synthesis, since deletion of regions within this locus [33] resulted in K1 capsule negative phenotypes. The region encodes several glycosyltransferases, a serine acetyltransferase, a GlcNAc epimerase, an acetyltransferase, a UDP-Glc (GDP-Man) dehydrogenase, and a flippase as well as multiple conserved hypothetical proteins. This locus encodes one large transcript (~17 kb), but transcriptional analysis determined that some of the genes are also expressed as separate transcriptional units [34]. Davey and Duncan [34] also showed that a PG0106 mutation in strain W83 results in a nonencapsulated mutant that gained the ability to form a biofilm; hence, K-antigen capsule expression blocks biofilm formation. As expected, the K1 antigen locus in different serotypes is not conserved, and variations in the region are likely responsible for the loss of K1 antigen capsule [33].

Annotation of the W83 genome identified at least three additional loci encoding genes involved in polysaccharide synthesis. Unlike the K-antigen capsule locus (PG0106–PG0120) described above, the coding sequence of the polysaccharide synthesis locus (PG1135–PG1142) is highly conserved in various strains [33]. Two additional loci in the chromosome that show sequence similarity to polysaccharide biosynthesis regions of other bacteria were also identified, including PG0435–PG0437, a small locus comprised of three genes encoding a putative polysaccharide biosynthesis protein (PG0435), a putative polysaccharide transport protein (PG0436), and a polysaccharide export protein (PG0437), and PG1560–PG1565, which also consists of genes encoding proteins with sequence similarity to genes involved in the synthesis of glycans. However, a direct involvement of those two loci in EPS, APS/LPS, or CPS synthesis has not yet been reported. Further investigation is necessary to reveal the possible participation of these loci in the synthesis of surface polysaccharide structures.

4 Regulation of Surface Polysaccharide Expression

Since attachment is critical for bacteria that persist in the oral cavity and capsule production can modulate cell–cell and cell–surface interactions, it follows that expression of surface polysaccharides is tightly regulated. Recently, Ltp1, a tyrosine phosphatase encoded by PG1641, was shown to play a role in the development of biofilm communities [35, 36], and moreover, it was shown to regulate expression of all of the loci encoding genes involved in the synthesis of surface polysaccharides. Ltp1 is the first regulatory protein for which a direct involvement in the regulation of surface polysaccharides in *P. gingivalis* has been determined.

Since regulatory proteins, such as Ltp1, have the potential to modulate both biofilm development and expression of virulence determinants, such proteins are prime targets for drug therapy. A future objective in periodontal research is to identify the regulatory proteins controlling expression of surface polysaccharides.

References

1. Choi JI, Nakagawa T, Yamada S et al (1990) Clinical, microbiological and immunological studies on recurrent periodontal disease. J Clin Periodontol 17:426–434
2. Dzink JL, Socransky SS, Haffajee AD (1988) The predominant cultivable microbiota of active and inactive lesions of destructive periodontal diseases. J Clin Periodontol 15:316–323
3. Grossi SG, Zambon JJ, Ho AW et al (1994) Assessment of risk for periodontal disease. I. Risk indicators for attachment loss. J Periodontol 65:260–267
4. Lamont RJ, Jenkinson HF (2000) Subgingival colonization by *Porphyromonas gingivalis*. Oral Microbiol Immunol 15:341–349
5. Moore WE, Moore LH, Ranney RR et al (1991) The microflora of periodontal sites showing active destructive progression. J Clin Periodontol 18:729–739
6. Yilmaz O (2008) The chronicles of *Porphyromonas gingivalis*: the microbium, the human oral epithelium and their interplay. Microbiology 154:2897–2903
7. Dorn BR, Burks JN, Seifert KN et al (2000) Invasion of endothelial and epithelial cells by strains of *Porphyromonas gingivalis*. FEMS Microbiol Lett 187:139–144
8. Dorn BR, Harris LJ, Wujick CT et al (2002) Invasion of vascular cells in vitro by *Porphyromonas endodontalis*. Int Endod J 35:366–371
9. Jandik KA, Belanger M, Low SL et al (2008) Invasive differences among *Porphyromonas gingivalis* strains from healthy and diseased periodontal sites. J Periodontal Res 43:524–530
10. Pham K, Feik D, Hammond BF et al (2002) Aggregation of human platelets by gingipain-R from *Porphyromonas gingivalis* cells and membrane vesicles. Platelets 13:21–30
11. Kozarov EV, Dorn BR, Shelburne CE et al (2005) Human atherosclerotic plaque contains viable invasive *Actinobacillus actinomycetemcomitans* and *Porphyromonas gingivalis*. Arterioscler Thromb Vasc Biol 25:e17–e18
12. Qi M, Miyakawa H, Kuramitsu HK (2003) *Porphyromonas gingivalis* induces murine macrophage foam cell formation. Microb Pathog 35:259–267
13. Comstock LE, Kasper DL (2006) Bacterial glycans: key mediators of diverse host immune responses. Cell 126:847–850
14. Roberts IS (1996) The biochemistry and genetics of capsular polysaccharide production in bacteria. Annu Rev Microbiol 50:285–315
15. Troy FA, Frerman FE, Heath EC (1971) The biosynthesis of capsular polysaccharide in *Aerobacter aerogenes*. J Biol Chem 246:118–133
16. Farquharson SI, Germaine GR, Gray GR (2000) Isolation and characterization of the cell-surface polysaccharides of *Porphyromonas gingivalis* ATCC 53978. Oral Microbiol Immunol 15:151–157
17. Ogawa T, Asai Y, Makimura Y et al (2007) Chemical structure and immunobiological activity of *Porphyromonas gingivalis* lipid A. Front Biosci 12:3795–3812
18. Paramonov N, Rangarajan M, Hashim A et al (2005) Structural analysis of a novel anionic polysaccharide from *Porphyromonas gingivalis* strain W50 related to Arg-gingipain glycans. Mol Microbiol 58:847–863
19. Rangarajan M, Aduse-Opoku J, Paramonov N et al (2008) Identification of a second lipopolysaccharide in *Porphyromonas gingivalis* W50. J Bacteriol 190:2920–2932
20. Curtis MA, Thickett A, Slaney JM et al (1999) Variable carbohydrate modifications to the catalytic chains of the RgpA and RgpB proteases of *Porphyromonas gingivalis* W50. Infect Immun 67:3816–3823

21. Slaney JM, Gallagher A, Aduse-Opoku J et al (2006) Mechanisms of resistance of *Porphyromonas gingivalis* to killing by serum complement. Infect Immun 74:5352–5361
22. Slaney JM, Curtis MA (2008) Mechanisms of evasion of complement by *Porphyromonas gingivalis*. Front Biosci 13:188–196
23. Shoji M, Ratnayake DB, Shi Y et al (2002) Construction and characterization of a nonpigmented mutant of *Porphyromonas gingivalis*: cell surface polysaccharide as an anchorage for gingipains. Microbiology 148:1183–1191
24. Whitfield C (2006) Biosynthesis and assembly of capsular polysaccharides in *Escherichia coli*. Annu Rev Biochem 75:39–68
25. Whitfield C, Valvano MA (1993) Biosynthesis and expression of cell-surface polysaccharides in gram-negative bacteria. Adv Microb Physiol 35:135–246
26. Sims TJ, Schifferle RE, Ali RW et al (2001) Immunoglobulin G response of periodontitis patients to *Porphyromonas gingivalis* capsular carbohydrate and lipopolysaccharide antigens. Oral Microbiol Immunol 16:193–201
27. van Winkelhoff AJ, Appelmelk BJ, Kippuw N et al (1993) K-antigens in *Porphyromonas gingivalis* are associated with virulence. Oral Microbiol Immunol 8:259–265
28. Laine ML, Appelmelk BJ, van Winkelhoff AJ (1997) Prevalence and distribution of six capsular serotypes of *Porphyromonas gingivalis* in periodontitis patients. J Dent Res 76:1840–1844
29. Dierickx K, Pauwels M, Laine ML et al (2003) Adhesion of *Porphyromonas gingivalis* serotypes to pocket epithelium. J Periodontol 74:844–848
30. Laine ML, van Winkelhoff AJ (1998) Virulence of six capsular serotypes of *Porphyromonas gingivalis* in a mouse model. Oral Microbiol Immunol 13:322–325
31. Rosen G, Sela MN (2006) Coaggregation of *Porphyromonas gingivalis* and *Fusobacterium nucleatum* PK 1594 is mediated by capsular polysaccharide and lipopolysaccharide. FEMS Microbiol Lett 256:304–310
32. Nelson KE, Fleischmann RD, DeBoy RT et al (2003) Complete genome sequence of the oral pathogenic bacterium *Porphyromonas gingivalis* strain W83. J Bacteriol 185:5591–5601
33. Aduse-Opoku J, Slaney JM, Hashim A et al (2006) Identification and characterization of the capsular polysaccharide (K-antigen) locus of *Porphyromonas gingivalis*. Infect Immun 74:449–460
34. Davey ME, Duncan MJ (2006) Enhanced biofilm formation and loss of capsule synthesis: deletion of a putative glycosyltransferase in *Porphyromonas gingivalis*. J Bacteriol 188:5510–5523
35. Maeda K, Tribble GD, Tucker CM et al (2008) A *Porphyromonas gingivalis* tyrosine phosphatase is a multifunctional regulator of virulence attributes. Mol Microbiol 69:1153–1164
36. Simionato MR, Tucker CM, Kuboniwa M et al (2006) *Porphyromonas gingivalis* genes involved in community development with *Streptococcus gordonii*. Infect Immun 74:6419–6428

Building the genomic base-layer of the oral "omic" world

The Forsyth Metagenomic Support Consortium* and Jacques Izard

Abstract. With the shift of molecular technologies directed toward the understanding of greater biological complexity of the oral cavity, a knowledge gap was created by the lack of genomic data from the diverse oral microorganisms. To facilitate and enable the interpretation of metagenomic, transcriptomic, and proteomic data generated or soon to be generated from oral biofilms, we are providing reference genomic information from phylogenetically diverse oral bacterial isolates. This work, initiated by the National Institute of Dental and Craniofacial Research as an isolated effort, is now part of the Human Microbiome Project. The goal of this effort is the public release of genomic data in support of functional and phylogenetic analyses of the complex oral microbiome. The genomic information acquired will be a key component in understanding the interaction of the oral biofilms with the human host and in developing novel healthcare strategies to prevent and treat oral diseases.

Key words. oral microbiome, oral biofilm, bacterial phylogeny, metagenome, bacterial diversity

1 Introduction

The human oral biofilms are readily accessible complex bacterial communities. Although it is one of the best-characterized microbiomes at the phylogenetic level, its internal dynamics and relationship to the host remain a mystery. This abundant self-renewable biofilm is responsible for oral health as well as diseases of both

J. Izard (✉)
Department of Molecular Genetics, The Forsyth Institute, Boston, MA 02135, USA
e-mail: JIzard@forsyth.org

*The Forsyth Metagenomic Support Consortium included Oxana V. Baranova, Derek Spencer, Tsute Chen, Wenhan Yu, Alvin R. Plummer, William G. Wade, and Floyd E. Dewhirst at The Forsyth Institute, Boston, MA, USA; Emmanuel F. Mongodin, Derrick E. Fouts, and Karen Nelson at the J. Craig Venter Institute, Rockville, MD, USA.

T. Sasano et al. (eds.), *Interface Oral Health Science 2009*,
DOI 10.1007/978-4-431-99644-6_112, © Springer 2010

hard and soft tissues of the oral cavity. In addition, evidence is accumulating to indicate the microbiome's significant influence on overall health, through a direct interactive exchange between the host response and the members of the oral microbiome [1–3]. Bacterial complexes are involved in those processes, and no individual pathogen can be singled out [4–7].

Bacterial diversity is a key characteristic of dental sub and supragingival plaque as well as the biofilms on other oral surfaces including the gingiva, tongue, hard palate, and cheeks. To date, over 600 bacterial species and a single *archaea* species have been identified as members of the oral community [5, 8, 9]. This number is increasing as more studies are performed all over the world. The added diversity may be related to disease status, regional environmental factors, or diet. For example, one might expect that the oral flora of a subject with a diet rich in raw fish with high fatty acid content will differ from the flora of a vegan subject. Such population changes have been well demonstrated in subjects with periodontitis compared with healthy subjects [4, 6, 10]. This underscores the interplay between the bacterial members of the oral biofilms, the host's physiological reactions and the fact that the mouth is an open system to the environment. Foods and fluids that we ingest as well as air that we breathe are significant sources of new bacterial challenges on a daily basis.

The extent of the morphological diversity of the bacteria in the oral cavity is astonishing. The highly motile spiral of the treponemes might coexist with the gliding multicellular-filamentous species of the genus *Simonsiella*, the corncob arrangements of *Corynebacterium matruchotii* or *Fusobacterium nucleatum* with *Streptococcus sanguis*, as well as cocci, short and long rods [11–14]. The phylogenetic diversity encompasses at least 12 phyla and over 170 genera [8]. The genera include cultivable named species (47%), cultivable yet-to-be-named species (18%), and yet-to-be-cultivated phylotypes, which often wear the inaccurate label of uncultivable in the literature (35%). These numbers are always fluctuating as the naming of organisms is an ongoing process [14, 15], and novel culture methods are being developed.

The variety of organisms, mentioned above, results in a diverse genetic potential that is mostly untapped and unknown. How those bacteria produce pathogenic factors, resist antibiotics, evade host immune response, communicate with each others, or simply use nutrients for their own energy production is mostly a mystery. The genetic characteristics of these oral bacteria are unknown because most of them are considered commensal and have not been the focus of many studies. Other factors detrimental to their study are their fastidious growth characteristics compared to *Escherichia coli* or *Bacillus subtilis* and their lack of established genetic systems.

While metagenomics, transcriptomics, and proteomics are part of the next key steps of understanding complex systems, they all rely on availability of the genomic sequence to decipher the content of the dataset via similarities to better known systems. These similarities provide the original clues toward function and phylogenetic attribution. With the absence of such data for most of the oral microbiome members, the scientific community was likely at risk to miss some of the benefits from the ongoing technological advances in the omic world. Those advances touch all aspects of fields looking at a large-scale analysis of a genome, a proteome, or a metabolome, just to name a few. Prior to joining the Human Microbiome project, we provided genome surveys to the community (Table 1). Genome surveying is a

Table 1 Genome survey data produced and released in Genbank

Phylum	Genus	Species	Strain	Taxon ID[a]	SEQF ID[b]	No. of contigs and singlets	Combined length (Kbp)	Genbank accession number
Bacteroidetes	*Prevotella*	*sp. oral taxon 302*	F0020	302	1,020	404	362	FI090687-FI091098
Firmicutes	*Bulleidia*	*extructa*	W1219	603	1,088	350	385	ET629122-ET629476
Firmicutes	*Eubacterium*	*infirmum*	ATCC 700433	105	1,108	413	409	ET630587-ET631005
Firmicutes	*Solobacterium*	*moorei*	W5408	678	1,152	338	354	ET631874-ET632213
Firmicutes	*Veillonella*	*parvula*	ATCC 17745	161	1,058	353	372	ET632214-ET632569
Fusobacteria	*Leptotrichia*	*buccalis*	ATCC 14201	563	1,028	345	392	ET631526-ET631873
Proteobacteria	*Campylobacter*	*gracilis*	ATCC 33236	623	1,005	384	402	ET629477-ET629866
Proteobacteria	*Campylobacter*	*rectus*	ATCC 33238	748	1,089	356	353	ET629867-ET630228
Proteobacteria	*Campylobacter*	*showae*	ATCC 51146	763	1,091	352	340	ET630229-ET630586
Spirochaetes	*Treponema*	*lecithinolyticum*	OMZ 684T	653	1,060	1,377	1,469	ET632570-ET633946
Synergistetes	*Jonquetella*	*anthropi*	E3_33 E1	777	1,476	509	618	ET631006-ET631525
Synergistetes	*Pyramidobacter*	*piscolens*	W5455	357	1,541	852	615	DU723013-DU723395
						Total	6,033	

[a]Taxon ID refers to a phylotype designation as part of the investigation of phylogenetic diversity of the oral microbiome (http://www.homd.org)[b]SEQF ID is a unique identifier of bacterial genomes sequenced from a unique isolate from a specific laboratory (http://www.homd.org)

low cost approach to provide anchoring genetic data for metagenomic analysis of obscure branches of the phylogenetic tree. Genetic libraries were created, and 12 organisms from 6 phyla were sequenced (Table 1). A minimum of 350,000 bases were recovered and annotated per genome (Table 1). A novel annotation pipeline was created to accommodate the short contigs generated by low sequencing coverage of random genomic libraries (Dewhirst et al. unpublished data) [8, 15]. This took advantage of the lengths of the singlets (single sequencing read) and the contigs (assembly of two or more sequencing reads) that may not cover the full length of the open reading frame but are sufficient to ascertain similarities and to provide matching sequences to pyrosequencing generated short reads. The dynamic annotation, updated periodically, for each genome survey is available online at the Human Oral Microbiome database (http://www.homd.org) [8]. The manual annotation of *Pyramidobacter piscolens* genome survey data was published when the organism was named [16].

With the announcement of the Human Microbiome Project, another era of bacterial genome sequencing began [17]. Becoming a member of this scientific endeavor allowed the transition from the production of genome surveys to the sequencing of full genomes by pyrosequencing. This is being done in partnership with four genomic centers: the Broad Institute of MIT and Harvard, the J. Craig Venter Institute, the Genome Sequencing Center at Washington University, and the Human Genome Sequencing Center at the Baylor College of Medicine. Now, with the first 50 bacterial genomes provided to the centers for sequencing, we look forward to closing the gap in describing genetic and phylogenetic diversity. The progress done can be monitored at the Human Microbiome Project website (http://www.hmpdacc.org) and at the Human Oral Microbiome database. In addition, the bacterial strains are provided to the Biodefense and Emerging Infections Research Resources Repository at ATCC (http://www.beiresources.org) for availability to the scientific community at large. For phylogenetic studies, the essentially full sequences of the 16S ribosomal RNA gene are also provided to GenBank repository.

How will this benefit the patient? With the acceptance that oral bacteria are part of the equation for good oral health, it becomes important to ensure that day-to-day treatments (toothpastes, oral rinses, etc.) are not damaging the host immune response or suppressing of the most critical group of bacteria involved in oral health. Moving toward prevention requires an unprecedented effort to understand what health is. It is also important to understand what the organisms' proportions are, which bacterial complexes are prevalent, which metabolic pathways are shared, which part of the genomic potential is expressed (transcriptomics), what the composition of a bacteria in the biofilm (proteomics) is, and consequently which abundant potential peptide targets are available for drug design and which metabolites are released (metabolomics) and could be targets for inhibitors. Genetic studies of both the normal and the pathogenic flora will be, as they have been in the past, a key element to develop drug therapies [18, 19]. Discovering the components contributing to health at the microbiome level can lead to clinical strategies to maintain a healthy balance between the microbiome and the host response.

The next steps are for the scientific community to expand the work of isolating new strains and providing genomic sequence information. This process will improve our ability to understand how the oral microbiome self-organizes and interacts with the host. This host interaction is a key component in the dynamic interplay, leading to the maintenance of oral health as well as the population shifts resulting in chronic diseases such as caries and periodontitis. Increased understanding of the behavior of the oral biofilms through transcriptomics, proteomics, interactomics, functomics, and many other "omic" disciplines relies on the available genomic data, and will open new avenues for treatment strategies.

Acknowledgment This work was supported by the grants DE017106 (JI) and DE16937 (Floyd E. Dewhirst) from the National Institute of Dental and Craniofacial Research, National Institutes of Health Bethesda, MD. We would like to thank the collaborators that are now actively involved in producing the pyrosequencing data at the Broad Institute of MIT and Harvard, the J. Craig Venter Institute, the Genome Sequencing Center at Washington University, and the Human Genome Sequencing Center at the Baylor College of Medicine.

References

1. Mealey BL, Oates TW (2006) Diabetes mellitus and periodontal diseases. J Periodontol 77(8):1289–1303
2. Cavrini F, Sambri V, Moter A et al (2005) Molecular detection of *Treponema denticola* and *Porphyromonas gingivalis* in carotid and aortic atheromatous plaques by FISH: report of two cases. J Med Microbiol 54(1):93–96
3. Davenport ES, Williams CE, Sterne JA et al (1998) The East London Study of Maternal Chronic Periodontal Disease and Preterm Low Birth Weight Infants: study design and prevalence data. Ann Periodontol 3(1):213–221
4. Socransky SS, Haffajee AD, Cugini MA et al (1998) Microbial complexes in subgingival plaque. J Clin Periodontol 25(2):134–144
5. Paster BJ, Boches SK, Galvin JL et al (2001) Bacterial diversity in human subgingival plaque. J Bacteriol 183(12):3770–3783
6. Marsh PD (1994) Microbial ecology of dental plaque and its significance in health and disease. Adv Dent Res 8(2):263–271
7. Sakamoto M, Huang Y, Ohnishi M et al (2004) Changes in oral microbial profiles after periodontal treatment as determined by molecular analysis of 16S rRNA genes. J Med Microbiol 53(Pt 6):563–571
8. Dewhirst FE, Izard J, Paster BJ et al (2008) The Human Oral Microbiome Database. http://www.homd.org
9. Lepp PW, Brinig MM, Ouverney CC et al (2004) *Methanogenic archaea* and human periodontal disease. Proc Natl Acad Sci USA 101(16):6176–6181
10. Aas JA, Paster BJ, Stokes LN et al (2005) Defining the normal bacterial flora of the oral cavity. J Clin Microbiol 43(11):5721–5732
11. Izard J, Hsieh CE, Limberger RJ et al (2008) Native cellular architecture of Treponema denticola revealed by cryo-electron tomography. J Struct Biol 163:10–17
12. Saglie R, Newman MG, Carranza FA Jr et al (1982) Bacterial invasion of gingiva in advanced periodontitis in humans. J Periodontol 53(4):217–222
13. Lancy P Jr, Dirienzo JM, Appelbaum B et al (1983) Corncob formation between *Fusobacterium nucleatum* and *Streptococcus sanguis*. Infect Immun 40(1):303–309
14. Lancy P Jr, Appelbaum B, Holt SC et al (1980) Quantitative in vitro assay for "corncob" formation. Infect Immun 29(2):663–670

15. Chen T, Abbey K, Deng W et al (2005) The bioinformatics resource for oral pathogens. Nucleic Acids Res 33:W734–W740
16. Downes J, Vartoukian SR, Dewhirst FE et al (2009) *Pyramidobacter piscolens gen. nov., sp. nov., a member of the phylum 'Synergistetes' isolated from the human oral cavity*. Int J Syst Evol Microbiol 59:972–980
17. Turnbaugh PJ, Ley RE, Hamady M et al (2007) The human microbiome project. Nature 449(7164):804–810
18. Stokes NR, Sievers J, Barker S et al (2005) Novel inhibitors of bacterial cytokinesis identified by a cell-based antibiotic screening assay. J Biol Chem 280(48):39709–39715
19. Hung DT, Shakhnovich EA, Pierson E et al (2005) Small-molecule inhibitor of *Vibrio cholerae* virulence and intestinal colonization. Science 310(5748):670–674

Cariogenic microflora and the immune response

Daniel J. Smith and Martin A. Taubman

Abstract. *Streptococcus mutans* has been associated with dental caries, especially in children. These cariogenic microorganisms are one of many species of oral streptococci, some of which begin to colonize the oral cavity and induce mucosal immune responses shortly after birth. Mucosal SIgA responses to early colonizing oral streptococcal epitopes, which are shared with *S. mutans* virulence antigens, may influence the rate of mutans streptococcal colonization. Especially important may be responses to adhesin, glucosyltransferase and glucan-binding protein epitopes. Active and passive immune strategies have been pursued in animals and in human clinical trials to explore the ability of orally available antibody to interfere with the colonization and accumulation of cariogenic streptococci.

Key words. *Streptococcus mutans*, dental caries, mucosal immunity, SIgA

1 Introduction

Streptococci comprise the bulk of the microorganisms that initially colonize the human oral cavity [1]. The interplay between these, mostly commensal bacteria, and the host is a mystery currently being explored. Species within the oral biofilm become increasingly diverse during the first 2 years of life, including colonization by acidogenic mutans streptococci, which can cause dental caries if given the opportunity to accumulate on dental surfaces. At the same time, the mucosal immune system is actively responding to the emerging biofilm microbiota, in part, through the synthesis and secretion of secretory IgA antibody. Since both systemic and mucosal microbial infections have been managed by immunization, investigators have attempted to adapt these strategies to the control of dental caries caused by acidogenic bacteria. The most frequent target has been *Streptococcus mutans*,

D.J. Smith (✉) and M.A. Taubman
Department of Immunology, The Forsyth Institute, 140 The Fenway, Boston, MA 02115, USA
e-mail: dsmith@forsyth.org

DOI 10.1007/978-4-431-99644-6_113, © Springer 2010

using proteins involved either in the adhesion of the bacteria to salivary proteins or proteins which participate in the glucan-mediated accumulation process as antigens. This manuscript will provide a brief review of the natural SIgA antibody response characteristics to oral streptococci together with active and passive immune approaches to the control of dental caries caused by *S. mutans* infection.

2 Natural History of Mutans Streptococcal Infection

Mutans streptococci (primarily *S. mutans* and *Streptococcus sobrinus*) initially colonize the oral cavity during the first years of life [2, 3]. Infection generally occurs by vertical transmission from the primary caregiver although horizontal transmission seems responsible in limited instances [4]. Permanent colonization normally transpires in the second or third year of life and is influenced by dental anatomy, bacterial dose, amount and frequency of fermentable carbohydrate exposure, and expression of several *S. mutans* virulence factors. Once infection occurs, these cariogenic streptococci then become permanent members of the oral biofilm. Interestingly, the primary teeth of some children remain free of (or are colonized at very low levels with) mutans streptococci despite high maternal levels of these organisms [2, 3].

Although the pathogenicity of mutans streptococci results from metabolic (lactic) acid production, bacterial adhesins, glucosyltransferases (Gtf) and glucan-binding proteins (Gbp) facilitate mutans streptococcal colonization and accumulation [5]. Their respective genes display significant diversity whose expression can lead to appreciable differences in the ability to form experimental biofilms [6, 7]. Recent evidence also suggests that the expression of some of these virulence factors is under the control of two-component response regulators [8].

3 Ontogeny of Mucosal Immunity on the Oral Cavity

Mucosal immunity in the oral cavity is manifest, in part, as secretory IgA antibody secretion in major and minor salivary gland fluids. Initially, following birth, the oral cavity is colonized primarily by oral streptococci, such as *Streptococcus mitis* and *Streptococcus salivarius* [1, 9]. The oral flora becomes much more complex as teeth erupt, diets change, and immune systems mature. Adaptive secretory immune responses occur soon after colonization. How these responses help to influence or control colonization and growth of commensal or pathogenic oral microflora during emergence and maturation of the oral biofilm is currently being explored.

Several early salivary antibody response characteristics have been described [10]. Most principal early colonizing commensals induce an immune response, initially IgM but soon thereafter SIgA. Most infants secrete both IgA1 and IgA2 subclasses early in life, although IgA2 secretion may be delayed for several months. Antigenic load appears to influence salivary IgA concentrations. As a result, by the end of the first year of life, salivary IgA concentrations, although not at adult levels, have increased substantially from those of the first few months of life. Mucosal immune responses at this time are directed primarily to protein epitopes; responses to T cell-independent polysaccharide antigens are likely to be poor as they are for systemic responses early in life.

Salivary IgA antibody responses to commensal microorganisms vary in amount and, often, in kind among children. Even siblings, who are presumably exposed to the same bacterial "collection" from their mother, can display substantially different sets of salivary IgA antibody specificities to the same species [3]. Differences in bacterial genotype, host biofilm, salivary protein expression, and innate and adaptive immune characteristics likely play a role in the ultimate salivary IgA antibody specificity profile.

Mucosal immune responses to mutans streptococci following their permanent incorporation into the oral biofilm appear to be directed mainly to the aforementioned components associated with initial adhesion or subsequent accumulation (adhesins, Gtfs, and Gbps). Responses to a few other bacterial antigens are consistently observed but remain to be identified. As discussed below, the rate and specificity of response may be modulated by preexisting flora.

Naturally induced salivary immune responses may influence *S. mutans* colonization. Oral streptococci often have similar metabolic and colonization mechanisms. Thus, components associated with pioneer streptococci may share antigenic epitopes with similar functioning proteins in later colonizing oral flora [11]. For example, the principal adhesin of *S. mutans* (antigen I/II) shares at least 17 similar MHC class II epitopes with earlier colonizing *Streptococcus gordonii* or *Streptococcus sanguinis*. Likewise, *S. mutans* Gtf shares 13 similar or identical MHC class II binding peptides with Gtf from *S. sanguinis* or *S. salivarius*. *S. mutans* GbpB and proteins involved in cell wall synthesis of pioneer *S. mitis* and *S. sanguinis* also share epitopes. This suggests the possibility that antibody induced by "pioneer proteins" could react with counterparts of later-colonizing *S. mutans*, possibly modifying their colonization potential.

A recent study by Nogueira and colleagues supports this hypothesis [12]. They compared the salivary IgA antibody profiles of 21 pairs of 5–13 month old children whose pairing was based on similar ages, salivary IgA concentrations, tooth eruption patterns, and racial background, but were different for the presence of *S. mutans*. Over 75% of children in whom no *S. mutans* could be detected culturally displayed salivary SIgA antibody to *S. mutans* Gbp, whereas antibody of this specificity could be detected in only 38% of the respective infected child of the pair. Although most of the children went on to be colonized with *S. mutans*, the data suggest that an ability to respond to *S. mutans* GbpB, perhaps as a "recall" response to an earlier exposure to a similar commensal bacterial protein, may modulate initial *S. mutans* infection.

The salivas of young children who were colonized by *S. mutans* were analyzed for IgA antibody specificities to Gtf epitopes putatively associated with enzymatic function and GbpB epitopes corresponding to MHC class II peptides, using synthetic peptides in multiplex-binding assays [13]. Children often recognized and responded to glucan-binding epitopes of Gtf and several MHC class II peptides of both virulence proteins. Interestingly, some of these salivary IgA antibody specificities were reactive with epitopes that seem to be shared by these *S. mutans* components and corresponding proteins of pioneer streptococci, providing more evidence for a "primed" response to *S. mutans* antigens, such as Gtf and GbpB.

Experimental models would be useful to help unravel the interplay between mucosal immune responses and acquisition and maintenance of oral biofilms. Since rats can be colonized with human strains of mutans streptococci, salivas taken before and after infection with *S. mutans* were analyzed for IgA antibody to Gtf and GbpB epitopes [14]. After an approximately 2 month infection period, most measured rodent antibody specificities to *S. mutans* Gtf and GbpB epitopes paralleled the pattern seen in young children, suggesting use of this model to explore the role of natural mucosal immunity in the oral cavity.

4 Active and Passive Vaccines Against Dental Caries

Over the past several decades, a variety of approaches have been used to induce adaptive host responses which could interfere with the establishment of cariogenic streptococci or reduce their cariogenic effects [15, 16]. Each of the virulence antigens discussed above has been used as intact proteins, as recombinant or synthetic peptides, or as the respective DNA to induce protective host responses in experimental systems [15, 17, 18]. Combining targets in one vaccine entity has been reported to enhance protection. Local delivery of antigen or DNA to regional mucosal sites in nonhost reactive vehicles, such as liposomes or PLGA, has proven useful to mount detectable and effective immune responses. However, few clinical trials have been performed to assess the effectiveness of the active immune approach in humans. In phase I trials in young adults oral or topical application of *S. sobrinus* Gtf, was shown to be safe and induced a short-term salivary IgA anti-Gtf immune response, which could be associated with delayed recolonization of biofilms with indigenous mutans streptococci [19]. Application of virulence antigen entities to nasal or tonsilar tissues of older children was only modestly effective in IgA response induction [20].

Passive immune applications of nonhost-derived antibody have also been evaluated for modulation of *S. mutans* colonization or their cariogenic effects [21]. Benefits of this approach are (1) potentially effective antibody can be delivered regardless of the immune status of the host, (2) antibody can be directed only to putatively important epitopes, and (3) the route to human use may be less encumbered by regulation than adaptive approaches. Disadvantages are that (1) significant amounts of antibody are needed, which require regular (probably daily) application, and (2) effective delivery techniques need to be developed, especially for young children experiencing initial mutans streptococcal challenge. Possively applied antibody to mutans streptococcal adhesins, Gtfs and Gbps, or epitopes associated with these components, have been effective using this approach. Mouse monoclonal (IgG) [22], chicken egg yolk

(IgY) antibodies [23, 24] or transgenically manufactured secretory IgA/G hybrid monoclonal antibody in tobacco plants [25] have been used for passive application. In the rodent model, antibody has been delivered in the diet and drinking water during the initial mutans streptococcal infection period or throughout the full course of infection. Most experimental approaches have reported a significant reduction in dental caries and, occasionally, reduced biofilm accumulation of cariogenic streptococci. Recently, single-chain antibody Fv, which has specificity for an *S. mutans* adhesin epitope, has been expressed in lactobacilli and, after oral lactobacillus administration, has been associated with lower *S. mutans* levels and reduced dental caries following infection of rats with these cariogenic streptococci [26].

In human pilot studies using the passive approach, young adults were provided with mouse monoclonal or transgenic immune reagents in applicator trays after a thorough, several week, treatment of dental surfaces with chlorohexidine [25]. These passive treatments were reported to block the reacquisition of indigenous *S. mutans*; no new dental caries were reported in the follow-up period. Although a similar trial using the same transgenic reagent did not support these results [27], differences in patient compliance may have mitigated a positive result. Novel methods of human monoclonal antibody selection to epitopes of *S. mutans* virulence components may provide reagents with significant passive effect [28].

Acknowledgments Grant support for the authors' research has come from the U.S. Public Health Service (DE-06133, DE-04733, DE/AI-12434, and TW-06324).

References

1. Cole MF, Bryan S, Evans MK et al (1999) Humoral immunity to commensal oral bacteria in human infants: salivary secretory immunoglobulin A antibodies reactive with *Streptococcus mitis* biovar 1, *Streptococcus oralis*, *Streptococcus mutans*, and *Enterococcus faecalis* during the first two years of life. Infect Immun 67:1878–1886
2. Caufield PW, Cutter GR, Dasanayake AP (1993) Initial acquisition of mutans streptococci by infants: evidence for a discrete window of infectivity. J Dent Res 72:37–45
3. Smith DJ, King WF, Akita H et al (1998) Association of salivary IgA antibody and initial mutans streptococcal infection. Oral Microbiol Immunol 13:278–285
4. Mattos-Graner RO, Li Y, Caufield P et al (2001) Genotypic diversity of mutans streptococci in Brazilian nursery children suggests horizontal transmission. J Clin Microbiol 39:2313–2316
5. Hamada S, Slade HD (1980) Biology, immunology and cariogenicity of *Streptococcus mutans*. Microbiol Rev 44:331–384
6. Mattos-Graner RO, Jin S, King WF et al (2001) Cloning of the *Streptococcus mutans* gene encoding glucan binding protein B and analysis of genetic diversity and protein production in clinical isolates. Infect Immun 69:6931–6941
7. Mattos-Graner RO, Napimoga MH, Fukushima K et al (2004) Comparative analysis of Gtf isozyme production and diversity in isolates of *Streptococcus mutans* with different abilities of growth in biofilms. J Clin Microbiol 42:4586–4592
8. Stipp RN, Gonçalves RB, Höfling JF et al (2008) Transcriptional analysis of gtfB, gtfC, and gbpB and their putative response regulators in several isolates of *Streptococcus mutans*. Oral Microbiol Immunol 23:466–473

9. Smith DJ, Taubman MA (1993) Emergence of immune mechanisms in saliva. Crit Rev Oral Biol Med 4:335–341
10. Cripps AW, Gleeson M (2005) Ontogeny of mucosal immunity and aging. In: Mucosal immunology, vol 1. pp 303–321
11. King WF, Nogueira RD, Mattos-Graner RO et al (2008) Epitopes shared among pioneer oral flora and *Streptococcus mutans* GbpB. AADR 87th general session, Dallas, TX, 1–4 April 2008
12. Nogueira RD, Alves AC, Napimoga MH et al (2005) Characterization of salivary IgA responses in children heavily exposed to the oral bacteria *Streptococcus mutans*: influence of specific antigen recognition in infection. Infect Immun 73:5675–5684
13. Nogueira RD, Alves AC, King WF et al (2007) Age-specific salivary IgA response to *Streptococcus mutans* GbpB and reactivity with GbpB-derived synthetic peptides. Clin Vaccine Immunol 14:804–807
14. Nogueira RD, King WF, Gunda G et al (2008) Mutans streptococcal infection induces salivary antibody to virulence proteins and associated functional domains. Infect Immun 76:3606–3613
15. Smith DJ (2002) Dental caries vaccines: prospects and concerns. Crit Rev Oral Biol Med 13:335–349
16. Taubman MA, Nash D (2006) The scientific and public-health imperative for a vaccine against dental caries. Nat Rev Immunol 6:555–563
17. Wu HY, Russell MW (1993) Induction of mucosal immunity by intranasal application of a streptococcal surface protein antigen with the cholera toxin B subunit. Infect Immun 61:314–322
18. Xu QA, Yu F, Fan MW et al (2007) Protective efficacy of a targeted anti-caries DNA plasmid against cariogenic bacteria infections. Vaccine 25:1191–1195
19. Smith DJ, Taubman MA (1987) Oral immunization of humans with *Streptococcus sobrinus* glucosyltransferase. Infect Immun 55:2562–2569
20. Childers NK, Tong G, Mitchell S et al (1999) A controlled clinical study of the effect of nasal immunization with a *Streptococcus mutans* antigen alone or incorporated into liposomes on induction of immune responses. Infect Immun 67:618–623
21. Smith DJ, Godiska R (2006) Passive approaches for dental caries prevention. In: Sim JS, Sunwoo HH (eds) The amazing egg. University of Alberta Press, Canada, pp 341–354
22. Lehner T, Russell MW, Challacombe SJ et al (1978) Passive immunisation with serum and immunoglobulins against dental caries in rhesus monkeys. Lancet 1:693–695
23. Otake S, Nishihara Y, Makimura M et al (1991) Protection of rats against dental caries by passive immunization with hen-egg-yolk antibody (IgY). J Dent Res 70:162–166
24. Smith DJ, King WF, Godiska R (2001) Passive transfer of immunoglobulin Y antibody to *Streptococcus mutans* glucan binding protein B can confer protection against experimental dental caries. Infect Immun 69:3135–3142
25. Ma JK, Hikmat BY, Wycoff K et al (1998) Characterization of a recombinant plant monoclonal secretory antibody and preventive immunotherapy in humans. Nat Med 4:601–606
26. Kruger C, Hu Y, Pan Q et al (2002) In situ delivery of passive immunity by lactobacilli producing single-chain antibodies. Nat Biotechnol 20:702–706
27. Weintraub JA, Hilton JF, White JM et al (2005) Clinical trial of a plant-derived antibody on recolonization of mutans streptococci. Caries Res 39:241–250
28. Sui J, King WF, Smith DJ et al (2008) Isolation of high affinity human monoclonal antibodies that inhibit mutans streptococcal glucosyltransferase. AADR 87th general session, Dallas, TX, 1–4 April 2008

Porphyromonas gingivalis infection elicits immune-mediated RANKL-dependent periodontal bone loss in rats

Xiaozhe Han, Xiaoping Lin, Toshihisa Kawai, Karen B. LaRosa, and Martin A. Taubman

Abstract. This study sought to determine the role of receptor activator of nuclear factor-kappaB ligand (RANKL) in the *Porphyromonas gingivalis* infection-associated periodontal bone loss. Rowett rats were infected orally on Day 0–3 with 10^8 live *Pg* (ATCC 33277). Separate groups received anti-RANKL antibody, osteoprotegerin (OPG)-Fc, or control L6-Fc injection into gingival papilla in addition to *P. gingivalis* infection. After 4 weeks, robust immune responses to *P. gingivalis* were detected in infected rats. The majority of RANKL-expressing cells in rat gingival tissues are T and B cells. Soluble RANKL in gingival homogenates and periodontal bone loss were significantly elevated in *P. gingivalis* infected rats compared to uninfected rats ($p<0.05$). Injection of anti-RANKL antibody or OPG-Fc into rat gingival papillae after *P. gingivalis* infection significantly reduced periodontal bone loss ($p<0.05$). This study suggests that *P. gingivalis* infection-associated bone loss is mediated by immune cells in a RANKL-dependent manner.

Key words. receptor activator of nuclear factor-kappaB ligand, *Porphyromonas gingivalis*, periodontal bone loss, periodontitis, immune cells

1 Introduction

T and B lymphocytes appear to be prominent in chronic periodontal lesions in humans and rodents. Receptor activator of nuclear factor-kappaB ligand (RANKL) is a major factor in the regulation of osteoclast differentiation. RANKL, its receptor RANK, and a decoy receptor osteoprotegerin (OPG) are three key molecules that

X. Han(✉), X. Lin, T. Kawai, K.B. LaRosa, and M.A. Taubman
Department of Immunology, The Forsyth Institute, 140 The Fenway, Boston, MA 02115, USA
e-mail: xhan@forsyth.org

X. Lin
Department of Stomatology, Shengjing Hospital of China Medical University, Shenyang, 110004, Liaoning, China

T. Sasano et al. (eds.), *Interface Oral Health Science 2009*,
DOI 10.1007/978-4-431-99644-6_114, © Springer 2010

regulate osteoclast recruitment and function [1]. We have found that T and B lymphocytes in human periodontal lesions bear abundant surface RANKL [2]. Furthermore, our previous studies demonstrated that antigen-specific RANKL-expressing T and B lymphocytes can directly mediate bone resorption in the rat adoptive transfer/gingival antigen injection model [3, 4]. *Porphyromonas gingivalis* (*P. gingivalis*) is one of the constellations of oral organisms associated with immune-mediated human chronic periodontitis. The purpose of this study was to establish a *P. gingivalis* model of periodontal infection in rats and to characterize the role of immune cell RANKL expression in *P. gingivalis* infection-associated periodontal bone loss.

2 Results

Robust serum IgG and salivary IgA antibody response ($p<0.01$) and T cell proliferation ($p<0.01$) to *P. gingivalis* bacteria were detected in rats 4 weeks after infection, but were not detectable in noninfected rats. After 4 weeks, the concentration of sRANKL in rat gingival homogenates ($p<0.01$) was significantly elevated in the *P. gingivalis* infected group compared to the uninfected group. Periodontal bone loss ($p<0.05$) in rats infected with *P. gingivalis* was significantly increased compared to those in uninfected rats. Four weeks after infection, the majority of RANKL-expressing cells in rat gingival tissues are T and B cells, not NK cells or macrophages, as determined by flow cytometry. Rat gingival homogenate-induced TRAP+ cell formation was inhibited by anti-RANKL IgG in vitro. Injection of anti-RANKL antibody ($p<0.05$) or OPG-Fc ($p<0.01$), but not a control protein L6-Fc, into gingival papillae significantly reduced periodontal bone loss of *P. gingivalis* infected rats.

3 Conclusion

We have established a rat periodontal infection model that elicits potent host immune responses, immune-mediated RANKL expression, and subsequent periodontal bone resorption. This study suggests that *P. gingivalis* infection-associated periodontal bone resorption is mediated by immune cells in a RANKL-dependent manner. Therefore, intervention with RANKL-dependent bone resorption by anti-RANKL measures could be a potential therapeutic regimen used in the treatment of periodontal disease.

Reference

1. Suda T, Takahashi N, Udagawa N et al (1999) Modulation of osteoclast differentiation and function by the new members of the tumor necrosis factor receptor and ligand families. Endocr Rev 20:345–357

2. Kawai T, Matsuyama T, Hosokawa Y et al (2006) B and T lymphocytes are the primary sources of RANKL in the bone resorptive lesion of periodontal disease. Am J Pathol 169:987–998
3. Valverde P, Kawai T, Taubman MA (2004) Selective blockade of voltage-gated potassium channels reduces inflammatory bone resorption in experimental periodontal disease. J Bone Miner Res 19:155–164
4. Han X, Kawai T, Eastcott JW et al (2006) Bacterial-responsive B lymphocytes induce periodontal bone resorption. J Immunol 176:625–631

Is RANKL shedding involved in immune cell-mediated osteoclastogenesis?

Hiroyuki Kanzaki, Xiaozhe Han, Xiaoping Lin, Toshihisa Kawai, and Martin A. Taubman

Abstract. Host immune responses play a key role in promoting bone resorption in periodontitis. RANKL-positive T- and B-lymphocytes have been described in periodontal lesions, along with induction of osteoclastogenesis. RANKL initially exists as a membrane-bound form, and cell-to-cell contact is required for inducing osteoclastogenesis. However, membrane-bound RANKL can be cleaved by enzymes, such as TNF-alpha-converting enzyme (TACE), and its product soluble RANKL (sRANKL) also has osteoclastogenic activity. We hypothesized that TACE might function as a main RANKL-shedding enzyme from activated immune cells, and that sRANKL cleaved by TACE might be important in osteoclastogenesis. To clarify our hypothesis, in vitro experiments using human peripheral blood mononuclear cells were performed. Consistent with a previous report (Lum et al. J Biol Chem 274:13613–13618, 1999), our data suggested that TACE might be the only functionally active sheddase for RANKL expressed by immune cells. sRANKL cleaved from activated immune cells by TACE could play a pivotal role in osteoclastogenesis.

Key words. osteoclastogenesis, TACE, sRANKL, periodontitis, immune cell

1 Introduction

Our recent studies demonstrated that host immune responses play a prominent role in promoting bone resorption in periodontitis [1]. T- and B-lymphocytes found in periodontal lesions express abundant RANKL, and stimulation of RANKL expression by these lymphocytes can induce osteoclastogenesis [2]. RANKL initially exists as a membrane-bound form, and cell-to-cell contact is required for induction of osteoclastogenesis [3]. Osteoclast precursors and osteoclasts express RANK, the

H. Kanzaki (✉), X. Han, X. Lin, T. Kawai, and M.A. Taubman
Department of Immunology, The Forsyth Institute, 140 The Fenway, Boston, MA, USA
e-mail: hkanzaki@forsyth.org

T. Sasano et al. (eds.), *Interface Oral Health Science 2009*,
DOI 10.1007/978-4-431-99644-6_115, © Springer 2010

receptor for RANKL [4], and RANK–RANKL signaling is blocked by osteoprotegerin (OPG) [5]. A balance between RANKL and OPG contributes to regulation of bone resorption [6].

2 RANKL Shedding

There is some evidence that membrane-bound RANKL can be cleaved by enzymes, such as TNF-alpha-converting enzyme (TACE) while maintaining biological activity [7, 8]. Soluble RANKL (sRANKL) can play a role in bone resorption as a soluble remote factor in the remodeling of bone metabolism [9]. Clinical reports have shown that production of sRANKL in the gingival crevicular fluid from periodontitis patients correlated with the severity of periodontitis [10, 11]. In addition, the RANKL:OPG ratio inclined to the RANKL side with the progress of periodontitis both at the mRNA level [12–14] and at the protein level [15]. The concentration of OPG was also reported to be decreased with periodontitis [16], indicating that sRANKL is not neutralized by OPG in the inflamed periodontium, and that sRANKL can induce osteoclastogenesis from a distant site. We hypothesized that TACE might play a prominent role in RANKL shedding by activated immune cells, and that sRANKL cleaved by TACE might be functionally significant in osteoclastogenesis.

In addition to TACE, other enzymes such as MMP-7, 14, and ADAM-10 have also been reported as potential RANKL-sheddases [17, 18]. To investigate which enzyme(s) is responsible for RANKL cleavage, expression of potential RANKL-shedding enzymes by human peripheral blood mononuclear cells (PBMCs) was observed by mRNA and protein level in vitro. PBMC expressed all of the potential RANKL-sheddases including TACE, ADAM-10, MMP-7, and MMP-14. However, specific antibody neutralizing experiments for each sheddase demonstrated that TACE was the only functionally active sheddase for RANKL expressed by PBMC. Only specific anti-TACE antibody, but not other antienzyme antibodies tested, resulted in significant reduction of sRANKL. These results suggest that TACE is the most potent RANKL-sheddase and that sRANKL cleaved by TACE might play an important role in immune cell-mediated osteoclastogenesis.

3 Conclusion

In pathological conditions such as periodontitis, sRANKL, cleaved from the plasma membrane of activated PBMC (including T and B cells) and other immune cells by TACE, could play a functional role in osteoclastogenesis.

References

1. Taubman MA, Valverde P, Han X et al (2005) Immune response: the key to bone resorption in periodontal disease. J Periodontol 76:2033–2041
2. Kawai T, Matsuyama T, Hosokawa Y et al (2006) B and T lymphocytes are the primary sources of RANKL in the bone resorptive lesion of periodontal disease. Am J Pathol 169:987–998
3. Yasuda H, Shima N, Nakagawa N et al (1998) Osteoclast differentiation factor is a ligand for osteoprotegerin/osteoclastogenesis-inhibitory factor and is identical to TRANCE/RANKL. Proc Natl Acad Sci USA 95(7):3597–3602
4. Nakagawa N, Kinosaki M, Yamaguchi K et al (1998) RANK is the essential signaling receptor for osteoclast differentiation factor in osteoclastogenesis. Biochem Biophys Res Commun 253:395–400
5. Simonet WS, Lacey DL, Dunstan CR et al (1997) Osteoprotegerin: a novel secreted protein involved in the regulation of bone density. Cell 89(2):309–319
6. Hofbauer LC, Khosla S, Dunstan CR et al (2000) The roles of osteoprotegerin and osteoprotegerin ligand in the paracrine regulation of bone resorption. J Bone Miner Res 15:2–12
7. Patel I, Attur MG, Patel RN et al (1998) TNF-alpha convertase enzyme from human arthritis-affected cartilage: isolation of cDNA by differential display, expression of the active enzyme, and regulation of TNF-alpha. J Immunol 160:4570–4579
8. Lum L, Wong BR, Josien R et al (1999) Evidence for a role of a tumor necrosis factor-alpha (TNF-alpha)-converting enzyme-like protease in shedding of TRANCE, a TNF family member involved in osteoclastogenesis and dendritic cell survival. J Biol Chem 274:13613–13618
9. Mizuno A, Kanno T, Hoshi M et al (2002) Transgenic mice overexpressing soluble osteoclast differentiation factor (sODF) exhibit severe osteoporosis. J Bone Miner Metab 20(6):337–344
10. Kawai T, Matsuyama T, Hosokawa Y et al (2006) B and T lymphocytes are the primary sources of RANKL in the bone resorptive lesion of periodontal disease. Am J Pathol 169:987–998
11. Vernal R, Dutzan N, Hernández M et al (2006) High expression levels of receptor activator of nuclear factor-kappa B ligand associated with human chronic periodontitis are mainly secreted by CD4+ T lymphocytes. J Periodontol 77(10):1772–1780
12. Liu D, Xu JK, Figliomeni L et al (2003) Expression of RANKL and OPG mRNA in periodontal disease: possible involvement in bone destruction. Int J Mol Med 11(1):17–21
13. Bostanci N, Ilgenli T, Emingil G et al (2007) Differential expression of receptor activator of nuclear factor-kappaB ligand and osteoprotegerin mRNA in periodontal diseases. J Periodontal Res 42(4):287–293
14. Wara-aswapati N, Surarit R, Chayasadom A et al (2007) RANKL upregulation associated with periodontitis and *Porphyromonas gingivalis*. J Periodontol 78(6):1062–1069
15. Bostanci N, Ilgenli T, Emingil G et al (2007) Gingival crevicular fluid levels of RANKL and OPG in periodontal diseases: implications of their relative ratio. J Clin Periodontol 34(5):370–376
16. Mogi M, Otogoto J, Ota N et al (2004) Differential expression of RANKL and osteoprotegerin in gingival crevicular fluid of patients with periodontitis. J Dent Res 83(2):166–169
17. Lynch CC, Hikosaka A, Acuff HB et al (2005) MMP-7 promotes prostate cancer-induced osteolysis via the solubilization of RANKL. Cancer Cell 7(5):485–496
18. Hikita A, Yana I, Wakeyama H et al (2006) Negative regulation of osteoclastogenesis by ectodomain shedding of receptor activator of NF-kappaB ligand. J Biol Chem 281(48):36846–36855

Possible IgG transportation mechanism mediated by neonatal Fc receptor expressed in gingival epithelial cells

Kazuhisa Ouhara, Mikihito Kajiya, Philip Stashenko, Martin A. Taubman, and Toshihisa Kawai

Abstract. The neonatal Fc receptor (FcRn) for immunoglobulin G (IgG) expressed in intestinal epithelium has been shown to be responsible for IgG transport and to be involved in IgG catabolism (Simister, Vaccine 21:3365–3369, 2003; Ghetie and Ward, Immunol Res 25:97–113, 2002). In this study, we first examined the expression of FcRn in normal gingival epithelial cells (GECs) and then tested the possibility that IgG transportation is mediated by such FcRn expressed in GEC. To make these determinations, experiments were carried out to: (1) evaluate the expression pattern of FcRn using immunohistochemical staining of FcRn in human gingival tissue, (2) measure the expression of FcRn-mRNA in the cultured GEC, and (3) examine the transport of IgG via FcRn expressed in the GEC cultured in cell culture insert. Immunohistochemistry detected the expression of FcRn in human GEC. Primary culture of both GEC and human GEC line (OBA9) demonstrated the expression of FcRn-mRNA. Since RNAi-based inhibition of FcRn expression resulted in the diminished transportation of IgG, it was determined that FcRn expressed in GEC was responsible for the transport of IgG. Strikingly, the transport of IgG mediated by epithelial FcRn was pH-dependent. These results suggested that FcRn expressed in GEC may contribute to the recycling of IgG in the oral cavity.

Key words. immunoglobulin, IgG, neonatal Fc receptor, gingival epithelial cells, pH-dependent transportation

1 Introduction

Neonatal Fc receptor (FcRn) plays a pivotal role in maternal immunoglobulin G (IgG) transportation across the placenta during pregnancy [1]. The FcRn expressed in gut epithelium binds maternal IgG present in ingested milk in the gut and delivers

K. Ouhara, M. Kajiya, P. Stashenko, M.A. Taubman, and T. Kawai (✉)
The Forsyth Institute, Boston, MA, USA
e-mail: tkawai@forsyth.org

T. Sasano et al. (eds.), *Interface Oral Health Science 2009*,
DOI 10.1007/978-4-431-99644-6_116, © Springer 2010

it to the bloodstream of the newborn [2]. The key phenomenon of this IgG transportation mediated by FcRn is pH dependency [3]. More specifically, low pH (pH 6.0–6.5) in the intestinal epithelium is required for the FcRn-mediated transport of IgG into laminar propria. Furthermore, this intestinal FcRn has been suggested to play a role in IgG recycling, as well as protecting IgG from degradation [4]. For example, FcRn-KO mice have a significantly diminished level of serum IgG compared to wild-type mice [5]. However, it is unclear if FcRn is also expressed in gingival epithelium. Since IgG is present in both saliva, as well as gingival crevice fluid, it is speculated that such IgG present in the fluid of oral cavity may also be transported by FcRn to lamina propria. Therefore, this study aimed to confirm (1) the expression of FcRn in the gingival epithelium and (2) the possible enrolment of FcRn expression in GEC as a mediator of IgG transportation.

2 Approaches

The expression of FcRn protein in rat and human gingival tissue was evaluated by immunohistochemical (IHC) staining using anti-FcRn antibody (Santa Cruz). IgG was isolated from both rat and human sera using protein-A/G column purification kit (Pierce). Primary culture of rat GEC and human GEC cell line (OBA9) was cultured in keratinocyte-serum-free medium (K-SFM, Invitrogen), and their expression of FcRn-mRNA was measured by RT-PCR. To examine in vitro if GEC can transport IgG, OBA9 cells were grown in cell culture insert (0.4-μm pore size, Falcon) and placed in the wells of a 24-well plate filled with K-SFM at different pH levels (pH 6.5 or 7.4). The IgG in K-SFM (pH 6.5 or 7.4) was applied to the culture insert containing confluent OBA9. The amount of IgG transported to the bottom well was monitored by ELISA. In order to silence FcRn-mRNA, siRNA specific to human FcRn (Santa Cruz) was mixed with DharmaFECT transfection reagent (Thermo Scientific), and the solution was applied to the OBA9 culture. As a control, scrambled irrelevant sequence of siRNA with transfection reagent was applied. The effects of siRNA for FcRn were confirmed by both RT-PCR and Western blot to determine the expression of FcRn-mRNA and FcRn protein, respectively.

3 Results

According to IHC staining, the expression of FcRn protein was found in the epithelium of normal healthy gingival tissue of both human and rats. Expression of FcRn-mRNA was detected in both human OBA9 cells and rat primary culture of GEC. More importantly, the FcRn-mediated IgG transportation through GEC was pH dependent. Specifically, when cultured on cell culture insert, FcRn-mediated IgG transportation occurs from the medium at lower pH (pH 6.5), where the IgG is applied to the OBA9 cells in culture insert, to the medium of the bottom compartment where the medium is adjusted at neutral pH (pH 7.4). However, IgG placed to OBA9 cells in culture insert with the medium at higher pH (pH 7.4) was not transported to the bottom compartment at lower pH (pH 6.5). Such IgG transportation across OBA9 cells was significantly suppressed by the pretreatment of OBA9 cells with siRNA for FcRn, suggesting that FcRn is responsible for the IgG transportation across GEC.

4 Conclusion

FcRn protein expression was found in the epithelium of normal healthy gingival tissue. FcRn expressed in GEC can transport IgG in a pH-dependent manner. Therefore, since it is reported that pH in human saliva is within the pH range of 6.0–6.8, while the pH in gingival laminar propria is around 7.3, it is very plausible that FcRn-mediated IgG transportation may readily occur between gingival surface and laminar propria.

Acknowledgment This study was supported by IADR/GlaxoSmithKline Innovation in Oral Care Award grant.

References

1. Simister NE (2003) Placental transport of immunoglobulin G. Vaccine 21:3365–3369
2. Israel EJ, Patel VK, Taylor SF, Marshak-Rothstein A, Simister NE (1995) Requirement for a beta 2-microglobulin-associated Fc receptor for acquisition of maternal IgG by fetal and neonatal mice. J Immunol 154:6246–6251
3. Raghavan M, Gastinel LN, Bjorkman PJ (1993) The class I major histocompatibility complex related Fc receptor shows pH-dependent stability differences correlating with immunoglobulin binding and release. Biochemistry 32:8654–8660
4. Ghetie V, Ward ES (2002) Transcytosis and catabolism of antibody. Immunol Res 25:97–113
5. Chaudhury C, Mehnaz S, Robinson JM et al (2003) The major histocompatibility complex-related Fc receptor for IgG (FcRn) binds albumin and prolongs its lifespan. J Exp Med 197:315–322

Effects of extracellular adenosine on sRANKL production from activated T cells

Marcelo José Silva, Harrison E. Mackler, Kazuhisa Ouhara, Cristina Ribeiro Cardoso, Martin A. Taubman, and Toshihisa Kawai

Abstract. This study investigated the effects of extracellular adenosine and four adenosine receptor agonists on the proliferation of activated T cells isolated from mice, as well as the production of soluble receptor activator of NF-kappa B ligand (sRANKL) derived from the activated T cells. All five compounds were applied to the T cells stimulated in vitro by TCR/CD28 engagement. The culture supernatants from the in-vitro-activated T cells were collected for the measurement of sRANKL production using ELISA, and proliferation of activated T cells was analyzed using the ^{3}H thymidine incorporation assay. The results obtained from this study indicated that extracellular adenosine and adenosine receptor agonists can suppress both proliferation of activated T cells and sRANKL production by the activated T cells, suggesting that adenosine receptor agonists may potentially lead to the development of new therapies for the prevention and treatment of inflammatory diseases affecting bone remodeling.

Key words. T cells, T cell receptor (TCR), CD28, receptor activator of NF-kappa B ligand (RANKL), adenosine

1 Introduction

Understanding the regulatory mechanisms that underlie bone formation and resorption is vital to the discovery of treatment modalities for bone destructive diseases, including periodontal disease. It is accepted that overactivation of T lymphocytes plays a part in the development of inflammatory bone resorption in the periodon-

M.J. Silva, H.E. Mackler, K. Ouhara, M.A. Taubman, and T. Kawai (✉)
The Forsyth Institute, Boston, MA, USA
e-mail: tkawai@forsyth.org

C.R. Cardoso
Universidade Federal do Triângulo Mineiro, Uberaba, Minas Gerais, Brazil

T. Sasano et al. (eds.), *Interface Oral Health Science 2009*,
DOI 10.1007/978-4-431-99644-6_117, © Springer 2010

tally diseased lesion by the increased production of soluble receptor activator of NF-kB ligand (sRANKL), a cytokine that induces osteoclastogenesis [1, 2]. Current drug treatments for periodontal disease consist of antibiotics and nonspecific anti-inflammatory drugs, none of which are capable of suppressing the bone resorption that occurs in periodontal disease.

While it is well documented that intracellular adenosine plays an important role in biochemical processes, such as energy transfer in the forms of adenosine triphosphate (ATP) and adenosine diphosphate (ADP), extracellular adenosine has also been found to have diverse effects on physiological processes, including immune and inflammatory functions [3]. T cells, for example, express two adenosine receptor isoforms, A2a and A2b, among a total four isoforms of cell-surface receptors for adenosine [4]. Specifically, it is reported that activation of A2a and A2b confers inhibitory effects on immune and inflammatory responses [4]. Therefore, we have hypothesized that adenosine or adenosine receptor agonists can suppress sRANKL produced by activated T lymphocytes in proinflammatory states. The aim of this study was to evaluate the effects of extracellular adenosine on sRANKL production and proliferation of T cells stimulated by TCR/CD28 engagement.

2 Methods

2.1 Reagents

The adenosine receptor agonists chosen for this study were 2-chloro-N6-cyclopentyladenosine (CCPA), 2-*p*-(2-carboxyethyl)phenethylamino-5′-*N*-ethylcarboxamidoadenosine (CGS-21680), 5′-*N*-Ethylcarboxamidoadenosine (NECA), and 1-Deoxy-1-[6-[[(3-iodophenyl)methyl]amino]-9H-purin-9-yl]-*N*-methyl-b-D-ribofuranuronamide (IB-MECA) (Sigma, St. Louis, MO) as the high affinity agonists for A1, A2a, A2b, and A3 receptors, respectively.

2.2 T Lymphocyte Stimulation Assay

Lymphocytes were isolated from the lymph nodes of C57BL/6 mice ($n=2$) by gradient centrifugation using Histopaque™ (Sigma, St. Louis, MO). The protocol to obtain the mouse lymph nodes was approved by Forsyth IACUC. The 96-well tissue culture plate was coated with anti-CD28 MAb (2 μg/mL) and anti-TCR MAb (5 μg/mL) (BD Biosciences, San Jose, CA) by incubation at 37°C for 1 h. Single-cell suspensions of lymphocytes were incubated in 96-well plates coated with or without anti-TCR/CD28 MAbs in RPMI 1640 medium (Invitrogen, Carlsbad, CA) supplemented with 10% FBS and antibiotics (2×10^5 cells/well) for 6 days. At day 0, cells were treated with adenosine and the adenosine receptor agonists at different concentrations, respectively. On day 4, the cells were harvested for reverse transcriptase polymerase chain reaction (RT-PCR). On day 5, culture supernatant was collected for measurement of sRANKL using an ELISA kit (PeproTech). To determine the proliferation of lymphocytes, ^{3}H thymidine (0.5 μCi) was added to the culture during the last 12 h of a total 6-day culture, and the radioactivity incorporated in the growing cells was measured using a radioscintillation counter.

3 Results

The extracellularly applied adenosine (1 and 10 μM) inhibited the production of sRANKL from activated T cells by TCR/CD28 engagement. We also sought to determine if adenosine receptor agonists could inhibit sRANKL production by the activated T cells. The results demonstrated that the A2b agonist had the highest inhibition of sRANKL production, when compared to adenosine and the other agonists tested, reaching 80% of inhibition compared to the no-drug control. The agonists for A1 and A2a had a level of inhibition similar to that of adenosine, whereas the A3 agonist appeared to be less efficient in the inhibition of sRANKL produced by activated T cells compared to the other drugs tested.

In addition, the inhibitory effects of extracellular adenosine were also examined on the proliferation of activated T cells. All tested agonists for A1, A2a, A2b, A3, and, to a lesser extent, extracellular adenosine showed comparable suppression effects on the proliferation of activated T cells by TCR/CD28 engagement. All reagents tested showed statistically significant suppression of T cell proliferation compared to the no-drug control.

4 Conclusion

In conclusion, our data suggested that extracellular adenosine can suppress sRANKL production by TCR/CD28-activated T cells, along with the suppression of the growth of such activated lymphocytes. The expression of sRANKL and proliferation of T cells showed a similar trend in response to the adenosine and adenosine receptor agonists, whereas there were subtle differences in the efficiency of these reagents tested. Since extracellular adenosine is known to be degraded by adenosine-deaminase and since the intracellular signaling mediated by each of four adenosine receptors is different, the pharmacokinetics of these reagents in the in vivo context would be different from the in vitro culture system. Nonetheless, these results indicate that extracellular adenosine and adenosine receptor agonists can be utilized to develop a therapeutic approach for the pathological bone resorption induced by the activation of lymphocytes in the context of periodontal disease.

Acknowledgment This study was supported by NIH grant DE-18499.

References

1. Kawai T, Matsuyama T, Hosokawa Y, Makihira S, Seki M, Karimbux NY, Goncalves RB, Valverde P, Dibart S, Li YP, Miranda LA, Ernst CW, Izumi Y, Taubman MA (2006) B and T lymphocytes are the primary sources of RANKL in the bone resorptive lesion of periodontal disease. Am J Pathol 169:987–998
2. Kawai T, Paster BJ, Komatsuzawa H, Ernst CW, Goncalves RB, Sasaki H, Ouhara K, Stashenko PP, Sugai M, Taubman MA (2007) Cross-reactive adaptive immune response to oral commensal bacteria results in an induction of receptor activator of nuclear factor-kappaB ligand (RANKL)-dependent periodontal bone resorption in a mouse model. Oral Microbiol Immunol 22:208–215
3. Linden J (2001) Molecular approach to adenosine receptors: receptor-mediated mechanisms of tissue protection. Annu Rev Pharmacol Toxicol 41:775–787
4. Brown RA, Spina D, Page CP (2008) Adenosine receptors and asthma. Br J Pharmacol 153(Suppl 1):S446–456

Activation of the critical enamel protease kallikrein-4

Coralee E. Tye and John D. Bartlett

Abstract. Kallikrein-4 (KLK4) is a serine protease expressed during enamel maturation, which is critical for proper enamel formation. KLK4 is secreted as an inactive zymogen and identification of its activator remains elusive. Herein we discuss what is currently known about the activation of pro-KLK4.

Key words. enamel, kallikrein-4, matrix metalloproteinase-20, dipeptidyl peptidase I

1 Introduction

Kallikrein-4 (KLK4) is a serine protease expressed during the transition and maturation stages of enamel development. Proteolytic processing of the enamel matrix by KLK4 is critical for proper enamel formation, and homozygous mutation of KLK4 causes *Amelogenesis imperfecta*, where the enamel is hypomineralized and is protein-rich [1]. KLK4 null mice have a hypomaturation phenotype, and enamel is rapidly abraded in these animals [2].

2 Activation of Kallikreins

KLK4, like other kallikreins, is secreted as an inactive zymogen that is proteolytically activated by specific release of its amino-terminal propeptide. Activation of the 15-member kallikrein family has been examined [3]; however, identification of the activator of KLK4 remains unclear.

C.E. Tye and J.D. Bartlett (✉)
Department of Cytokine Biology, Forsyth Institute, 140 Fenway, Boston, MA 02115, USA
Department of Developmental Biology, Harvard School of Dental Medicine, Boston, MA 02115, USA
e-mail: jbartlett@forsyth.org

T. Sasano et al. (eds.), *Interface Oral Health Science 2009*,
DOI 10.1007/978-4-431-99644-6_118, © Springer 2010

Many kallikreins are activated as part of a proteolytic cascade. KLK4 is unique because it does not appear to be activated by other kallikreins. However, KLK4 does activate numerous kallikreins, including KLKs 1–3, 5, 6, 9, and 11–15. Pro-KLK4 is cleaved by KLK11; however, residues beyond the propeptide are also removed, making KLK11 an unlikely in vivo activator [3]. Initiators of proteolytic cascades are commonly autocatalytic and several kallikreins self-activate; however, KLK4 does not appear to be autocatalytic [4].

3 Activation of KLK4 by matrix metalloproteinase-20

The tooth-specific protease matrix metalloproteinase-20 (MMP20) has been shown to activate KLK4 in vitro [4]; however, KLK4 and MMP20 expression overlap only briefly during enamel development and KLK4 is not autocatalytic, which suggests that another protease activates KLK4 in vivo. KLK4 expression is also detected in prostate, ovary, breast, endometrium, liver, muscle, skin, thyroid, salivary gland, and is upregulated in numerous cancers [5]. Normal MMP20 expression within these tissues is unlikely, indicating the strong possibility of additional activator(s) of KLK4.

4 Activation of KLK4 by dipeptidyl peptidase I

We recently identified the expression of dipeptidyl peptidase I (DPPI) in mouse enamel organ and demonstrated in vitro that DPPI activates KLK4 [6, 7]. Microhardness testing of mature enamel from DPPI null mice revealed that loss of DPPI expression significantly reduced enamel hardness. DPPI (cathepsin C) is a ubiquitously expressed cysteine aminopeptidase, which sequentially removes two N-terminal amino acid residues from proteins. In addition to lysosomal activity, DPPI is secreted and activates proenzymes of chymotrypsin-like serine proteases, including granzymes A and B, cathepsin G, elastase, and chymases [8].

5 Conclusions

Identification of the in vivo activator of KLK4 is important for furthering not only the understanding of enamel formation but also its exciting potential as a therapeutic target for numerous cancers. Current studies are examining the importance of MMP20 and DPPI on the in vivo activation of KLK4.

Acknowledgments Research presented in this review was supported in part by research grant DE016276 from the National Institute of Dental and Craniofacial Research, National Institutes of Health.

References

1. Hart PS, Hart TC, Michalec MD et al (2004) Mutation in kallikrein 4 causes autosomal recessive hypomaturation amelogenesis imperfecta. J Med Genet 41:545–549
2. Simmer JP, Hu Y, Lertlam R et al (2009) Hypomaturation enamel defects in Klk4 knockout/LacZ knockin mice. J Biol Chem 284:19110–19121
3. Yoon H, Laxmikanthan G, Lee J et al (2007) Activation profiles and regulatory cascades of the human kallikrein-related peptidases. J Biol Chem 282:31852–31864
4. Ryu O, Hu JC, Yamakoshi Y et al (2002) Porcine kallikrein-4 activation, glycosylation, activity, and expression in prokaryotic and eukaryotic hosts. Eur J Oral Sci 110:358–365
5. Obiezu CV, Shan SJ, Soosaipillai A et al (2005) Human kallikrein 4: quantitative study in tissues and evidence for its secretion into biological fluids. Clin Chem 51:1432–1442
6. Tye CE, Lorenz RL, Bartlett LD (2009) Lysosomal protease expression in mature enamel. Cells Tissues Organs 189:111–114
7. Tye CE, Pham CT, Simmer JP et al (2009) DPPI may activate KLK4 during enamel formation. J Dent Res 88:323–327
8. Turk B, Turk D, Dolenc I et al (2004) Dipeptidyl peptidase I. In: Barrett AJ, Rawlings ND, Woessner JF (eds) Handbook of proteolytic enzymes. Elsevier Academic, San Diego, pp 1192–1196

Epitopes shared among pioneer oral flora and *Streptococcus mutans* GbpB

William F. King, Tsute Chen, Ruchele Nogueira, Renata Mattos-Graner, and Daniel J. Smith

Abstract. The establishment of microorganisms in emerging oral biofilms of humans is likely to be modulated by a constellation of infant, maternal, and microbial factors. We have shown that pioneer microbiota on epithelial surfaces (e.g., *Streptococcus mitis*, *Streptococcus salivarius*) and initially erupting dental surfaces (e.g., *Streptococcus sanguinis*, *Streptococcus gordonii*, and *Streptococcus oralis*) induce mucosal SIgA antibody in saliva. The resulting immune responses may modulate subsequent colonization of bacteria that later join the oral biofilm (e.g., cariogenic *Streptococcus mutans*), a speculation supported by the observation that (a) several *S. mutans* virulence components share MHC Class II binding peptides with pioneer protein homologues, (b) cross-reactive antibody among homologues can be demonstrated and (c) initial *S. mutans* colonization can apparently be delayed when some of these responses are present *a priori*.

Key words. ontogeny, pioneer flora, commensal, *Streptococcus mutans*

1 Narrative

The entry of microorganisms into emerging human oral biofilms is likely to be modulated by a constellation of infant, maternal, and microbial factors. Epithelial receptors, secretory immune responses, maternal flora, breast milk nutrients, cytokines, antibody, bacterial adhesins, and competence/quorum-sensing mechanisms are all likely to play some role in the kind and number of bacteria that eventually colonize the oral cavity.

W.F. King, T. Chen, and D.J. Smith (✉)
The Forsyth Institute, Boston, MA 02115, USA
e-mail: dsmith@forsyth.org

R. Nogueira and R. Mattos-Graner
UNICAMP, Piracicaba, São Paulo, Brazil

T. Sasano et al. (eds.), *Interface Oral Health Science 2009*,
DOI 10.1007/978-4-431-99644-6_119, © Springer 2010

Initially, biofilms form on oral mucosa and the tongue. *Streptococcus mitis* and *Streptococcus salivarius* are often detected at this time. Following eruption of teeth midway through the first year of life, organisms such as *Streptococcus sanguinis* begin to form a community of flora on the tooth surface [1]. In contrast, *Streptococcus mutans*, implicated in the initiation of dental caries [2], do not normally accumulate on tooth surfaces until the second year of life [3].

Colonization of neonatal oral mucosa by pioneer flora provide immune stimuli to the developing mucosal immune system. Salivary IgA antibody activity to *S. mitis* and *S. salivarius* can be observed within several weeks of birth. Even at this young age, infants differ with respect to amount and kind of "response" to *S. mitis* components. SIgA antibody levels to early stimuli continue to increase during the first years of life, in part, as a result of the overall expansion of the mucosal immune capacity [1].

Many of the components of the pioneer flora have significant sequence homology with components of similar function in bacterial members (e.g., *S. sanguinis*) of the initial dental biofilm. Thus, it is possible that secretory immune responses to oral pioneer bacterial components might recognize and/or be enhanced by similar epitopes on components associated with subsequently colonizing microbiota, including potentially cariogenic *S. mutans*. This speculation is strengthened by the observation that (a) several *S. mutans* virulence components share MHC Class II binding peptides with pioneer protein homologues, (b) cross-reactive antibody among homologues can be demonstrated, and (c) initial *S. mutans* colonization can apparently be delayed when some of these responses are present a priori [4].

Acknowledgments Grant support for the authors' research has come from the U.S. Public Health Service (DE-06133, DE-04733, DE/AI-12434, and TW-06324), FAPESP (proc. 02/07156-1, proc. 04/07425-8) and CAPES (ProDoc, proc. 029/03).

References

1. Smith DJ, Taubman MA (1993) Emergence of immune mechanisms in saliva. Crit Rev Oral Biol Med 4:335–341
2. Hamada S, Slade HD (1980) Biology, immunology and cariogenicity of *Streptococcus mutans*. Microbiol Rev 44:331–384
3. Caufield PW, Cutter GR, Dasanayake AP (1993) Initial acquisition of mutans streptococci by infants: evidence for a discrete window of infectivity. J Dent Res 72:37–45
4. Nogueira RD, Alves AC, Napimoga MH et al (2005) Characterization of salivary IgA responses in children heavily exposed to the oral bacteria *Streptococcus mutans*: influence of specific antigen recognition in infection. Infect Immun 73:5675–5684

Inhibitory effect of porcine amelogenins on spontaneous mineralization

Seo-Young Kwak, Felicitas B. Wiedemann-Bidlack, Amy Litman, Elia Beniash, Yasuo Yamakoshi, James P. Simmer, and Henry C. Margolis

Abstract. The potential role of amelogenin phosphorylation in enamel formation has been elucidated through in vitro mineralization studies. Studies focused on the native 20-kDa porcine amelogenin proteolytic cleavage product P148 that is prominent in developing enamel. Under conditions of spontaneous calcium phosphate precipitation, in the presence of nonphosphorylated recombinant full-length porcine amelogenin, rP172, immediately formed nanoparticles of amorphous calcium phosphate (ACP) were found to transform into parallel arrays of needle-like apatitic crystals. In contrast to these findings, P148, with a single phosphate group on serine-16, was found to inhibit calcium phosphate precipitation and stabilize ACP formation for more than 1 day. The present study has provided evidence suggesting that the proteolytic cleavage product P148 may have an important functional role in regulating mineralization during enamel formation.

Key words. enamel, amelogenin, biomineralization, hydroxyapatite

1 Introduction

Prior studies in our group suggest that full-length recombinant mouse amelogenin (rM179) can provide an organized microstructure that regulates the formation of parallel arrays of apatitic crystals under conditions of spontaneous precipitation [1]. The conserved hydrophilic C-terminus is believed to play a key role in these processes. The present study [2] was carried out to determine the potential role of

S.-Y. Kwak, F.B. Wiedemann-Bidlack, A. Litman, E. Beniash, and H.C. Margolis (✉)
Department of Biomineralization, The Forsyth Institute, 140, Fenway, Boston, MA 02115, USA
e-mail: hmargolis@forsyth.org

Y. Yamakoshi and J.P. Simmer
Dental Research Laboratory, University of Michigan, Ann Arbor, MI 15261, USA

E. Beniash
Department of Oral Biology, University of Pittsburgh, Pittsburgh, PA 48109, USA

T. Sasano et al. (eds.), *Interface Oral Health Science 2009*,
DOI 10.1007/978-4-431-99644-6_120, © Springer 2010

amelogenin phosphorylation in enamel formation. In vitro studies under conditions of spontaneous mineralization were carried out in the absence and presence of a prominent native porcine amelogenin cleavage product (that lacks the hydrophilic C-terminus), P148, with a single phosphate group on serine-16, and a recombinant nonphosphorylated form of P148, rP147, that also lacks the N-terminal methionine. For comparison, the full-length nonphosphorylated recombinant porcine amelogenin, rP172, was also studied.

2 Methods

Native P148 and recombinant rP147 and rP172 amelogenins were prepared and purified using methods referenced in the original study [2]. Calcium and phosphate were sequentially added to ice-cold protein solutions (2 mg/ml in water, $pH<4$) to yield final concentrations of 2.5 mM Ca and 1.5 mM P. The solution was brought to 37°C and the pH was quickly adjusted (KOH) to 7.4 ± 0.1. Changes in pH were monitored continuously and the formed minerals were analyzed using transmission electron microscopy (TEM), selected area electron diffraction, and Fourier transform infrared spectroscopy [2].

3 Results and Discussion

In the absence of protein (control), following an initial induction period of ~15 min that corresponded to the formation of amorphous calcium phosphate (ACP), a marked decrease in pH was observed along with the formation of randomly oriented plate-like crystals of hydroxyapatite. A similar change in pH with time was also observed in the presence of the full-length recombinant amelogenin rP172, although the initial induction period was somewhat longer (~45 min.). From TEM results obtained at various times, rP172 was found to transiently stabilize ACP and to control its transformation into well-oriented needle-like apatitic crystals. However, recombinant rP147 that lacks the hydrophilic C-terminal amino acids had little influence on the rate of mineral formation and the final mineral phase formed, relative to the control. In sharp contrast, in the presence of P148, pH values remained nearly constant for 24 h. Results obtained showed that P148 dramatically inhibited spontaneous calcium phosphate formation and stabilized ACP, thus preventing its transformation to apatitic material, as seen in the control. This effect was shown to be related to the single phosphate group found on serine-16, through comparative studies using the nonphosphorylated recombinant analog rP147 [2].

Acknowledgments This work was supported by grant DE-016376 (HCM) from the National Institute of Dental and Craniofacial Research. FBW-B was also partially supported by grant T32 DE-007327.

References

1. Beniash E, Simmer JP, Margolis HC (2005) The effect of recombinant mouse amelogenins on the formation and organization of hydroxyapatite crystals in vitro. J Struct Biol 149:182–190
2. Kwak SY, Wiedemann-Bidlack FB, Beniash E et al (2009) Role of 20 kDa amelogenin (P148) phosphorylation in calcium phosphate formation in vitro. J Biol Chem 284:18972–18979

A stress-based mechanism to explain dental fluorosis

Ramaswamy Sharma and John D. Bartlett

Abstract. Excess fluoride (F^-) causes dental fluorosis (DF). However, the mechanisms underlying DF are not clear. We have previously shown that F^- affects homeostasis of the endoplasmic reticulum (ER), leading to ER stress and the activation of the unfolded protein response (UPR). The UPR is a signaling pathway designed to restore ER homeostasis. Herein, we describe the UPR and summarize the UPR components activated by F^-.

Key words. fluoride, fluorosis, ER stress, unfolded protein response

1 Introduction

Fluoride (F^-) is a potent anticariogenic agent. However, exposure to levels higher than 1 ppm can lead to dental fluorosis (DF), a condition characterized by increased subsurface porosity of enamel. DF is manifested in varying degrees, ranging from white spots to pitted and darkly stained enamel. DF is considered as an adverse health effect since enamel loss and pitting compromise the ability of the tooth enamel to protect the inner dentin and pulp from decay and infection [1]. DF has increased from 23 to 32% over the past decade [2]; however, the molecular mechanisms underlying DF are not clear.

R. Sharma and J.D. Bartlett (✉)
Departments of Cytokine Biology, The Forsyth Institute, 140 The Fenway, Boston, MA 02115, USA
Department of Developmental Biology, Harvard School of Dental Medicine, Boston, MA 02115, USA
e-mail: jbartlett@forsyth.org

T. Sasano et al. (eds.), *Interface Oral Health Science 2009*,
DOI 10.1007/978-4-431-99644-6_121, © Springer 2010

2 ER Stress and the Unfolded Protein Response

The endoplasmic reticulum (ER) is the first organelle in the secretory pathway with multiple functions. It is the site for post-translational modifications, lipid and sterol synthesis, and importantly, protein folding. Proteins have to be "folded" into their proper conformation to be functional. The ER monitors protein folding and prevents secretion of misfolded proteins from the cell [3]. Accumulation of misfolded proteins within the ER will initiate an ER-to-nucleus signaling pathway, termed the unfolded protein response (UPR; for review see [4]).

The UPR serves three major functions: it upregulates molecular chaperones such as BiP that help augment the folding capacity of the ER; it transiently attenuates protein translation via PERK-mediated phosphorylation of the translation initiation factor, eIF2α, thereby allowing cells to cope with the existing protein load; finally, it activates components of the ER-associated degradative pathway to help degrade the accumulated misfolded proteins. If these pathways succeed in alleviating cell stress, the cell survives; if not, the cell undergoes apoptosis via caspase activation.

3 Fluoride activates the UPR

F^- causes distension of the ER and the accumulation of dense bodies within the ER in ameloblasts of rats [5] – these are hallmarks of ER stress. F^- also upregulates several components of the UPR. These include induction of BiP, activation of the UPR sensors, IRE1 and PERK, induction of their downstream targets such as XBP1 and CHOP, and phosphorylation of eIF2α in LS8 ameloblast-like cells [6, 7]. Importantly, IRE1 and eIF2α phosphorylation also occurs in rodents in vivo, suggesting that F^--mediated activation of the UPR can play an important role in the development of fluorosis.

4 Conclusions

F^- disturbs ER homeostasis in ameloblasts, induces ER stress, and activates the UPR, thereby compromising ameloblast cell function that leads to fluorosed enamel.

Acknowledgments This work was supported in part by grant DE018106 (JDB) from the National Institute of Dental and Craniofacial Research.

References

1. National Research Council (2006) Fluoride in drinking water: a scientific review of EPA's standards. The National Academies, Washington, DC

2. CDC (1995) Engineering and administrative recommendations for water fluoridation Centers for Disease Control and Prevention. MMWR Recomm Rep 44(RR-13):1–40
3. Hammond C, Helenius A (1995) Quality control in the secretory pathway. Curr Opin Cell Biol 7(4):523–529
4. Gow A, Sharma R (2003) The unfolded protein response in protein aggregating diseases. Neuromol Med 4(1–2):73–94
5. Ribeiro DA, Hirota L, Cestari TM et al (2006) J Mol Histol 37:361–367
6. Sharma R, Tsuchiya M, Bartlett JD (2008) Fluoride induces endoplasmic reticulum stress and inhibits protein synthesis and secretion. Environ Health Perspect 116(9):1142–1146
7. Kubota K, Lee DH, Tsuchiya M et al (2005) Fluoride induces endoplasmic reticulum stress in ameloblasts responsible for dental enamel formation. J Biol Chem 280(24):23194–23202

Author Index

Keyword Index

MIX
Papier aus verantwortungsvollen Quellen
Paper from responsible sources
FSC® C105338

If you have any concerns about our products,
you can contact us on
ProductSafety@springernature.com

In case Publisher is established outside the EU,
the EU authorized representative is:
Springer Nature Customer Service Center GmbH
Europaplatz 3, 69115 Heidelberg, Germany

Printed by Libri Plureos GmbH
in Hamburg, Germany